Point and Extended Defects in Semiconductors

NATO ASI Series

Advanced Science Institutes Series

A series presenting the results of activities sponsored by the NATO Science Committee, which aims at the dissemination of advanced scientific and technological knowledge, with a view to strengthening links between scientific communities.

The series is published by an international board of publishers in conjunction with the NATO Scientific Affairs Division

A	**Life Sciences**	Plenum Publishing Corporation
B	**Physics**	New York and London
C	**Mathematical**	Kluwer Academic Publishers
	and Physical Sciences	Dordrecht, Boston, and London
D	**Behavioral and Social Sciences**	
E	**Applied Sciences**	
F	**Computer and Systems Sciences**	Springer-Verlag
G	**Ecological Sciences**	Berlin, Heidelberg, New York, London,
H	**Cell Biology**	Paris, and Tokyo

Recent Volumes in this Series

Series B: Physics

Point and Extended Defects in Semiconductors

Edited by

G. Benedek

Università degli Studi di Milano
Milan, Italy

A. Cavallini

Università degli Studi di Bologna
Bologna, Italy

and

W. Schröter

University of Göttingen
Göttingen, Federal Republic of Germany

Plenum Press
New York and London
Published in cooperation with NATO Scientific Affairs Division

Proceedings of the NATO Advanced Research Workshop/
International School of Materials Science and Technology
Second Workshop on
Point, Extended and Surface Defects in Semiconductors,
held November 2–7, 1988,
in Erice, Italy

Library of Congress Cataloging in Publication Data

Workshop on Point, Extended, and Surface Defects in Semiconductors (2nd: 1988: Erice, Italy)
    Point and extended defects in semiconductors / edited by G. Benedek, A. Cavallini, and W. Schröter.
        p.   cm.—(NATO ASI series. Series B, Physics; v. 202)
    "Proceedings of the NATO Advanced Research Workshop/International School of Materials Science and Technology, Second Workshop on Point, Extended, and Surface Defects in Semiconductors, held November 2–7, 1988, in Erice, Italy"—T.p. verso.
    "Published in cooperation with NATO Scientific Affairs Division."
    Includes bibliographical references.
    ISBN 978-1-4684-5711-7   ISBN 978-1-4684-5709-4 (eBook)
    DOI 10.1007/978-1-4684-5709-4
    1. Semiconductors—Defects—Congresses. I. Benedek, G. (Giorgio) II. Cavallini, A. (Anna) III. Schröter, W. (Wolfgang) IV. International School of Materials Science and Technology (1988: Erice, Italy) V. North American Treaty Organization. Scientific Affairs Division. VI. Title. VII. Series.
    QC611.6.D4W67   1988   89-16332
    621.381'52-dc20   CIP

SPECIAL PROGRAM ON CONDENSED SYSTEMS OF LOW DIMENSIONALITY

This book contains the proceedings of a NATO Advanced Research Workshop held within the program of activities of the NATO Special Program on Condensed Systems of Low Dimensionality, running from 1983 to 1988 as part of the activities of the NATO Science Committee.

Other books previously published as a result of the activities of the Special Program are:

SPECIAL PROGRAM ON CONDENSED SYSTEMS OF LOW DIMENSIONALITY

Volume 200   GROWTH AND OPTICAL PROPERTIES OF WIDE-GAP II-VI
             LOW-DIMENSIONAL SEMICONDUCTORS
             edited by T. C. McGill, C. M. Sotomayor Torres,
             and W. Gebhardt

PREFACE

The systematic study of defects in semiconductors began in the early fifties. From that time on many questions about the defect structure and properties have been answered, but many others are still a matter of investigation and discussion. Moreover, during these years new problems arose in connection with the identification and characterization of defects, their role in determining transport and optical properties of semiconductor materials and devices, as well as from the technology of the ever increasing scale of integration.

This book presents to the reader a view into both basic concepts of defect physics and recent developments of high resolution experimental techniques. The book does not aim at an exhaustive presentation of modern defect physics; rather it gathers a number of topics which represent the present-time research in this field.

The volume collects the contributions to the Advanced Research Workshop "Point, Extended and Surface Defects in Semiconductors" held at the Ettore Majorana Centre at Erice (Italy) from 2 to 7 November 1988, in the framework of the International School of Materials Science and Technology. The workshop has brought together scientists from thirteen countries. Most participants are currently working on defect problems in either silicon submicron technology or in quantum wells and superlattices, where point defects, dislocations, interfaces and surfaces are closely packed together. Thus the scope of the workshop was twofold: first, to assess recent data on point and extended defects and their relationship, in elemental as well as in compound semiconductors; second, to address recent advances in techniques devised for defect investigation. The main concern of these lectures (as well as of subsequent discussions and round-table sessions) was the present range covered by these techniques in providing structural, chemical, electrical and optical information, and the ability of these techniques to see single defects on atomic resolution. There has been exciting progress during the last years - such as the scanning tunnel microscope (STM) or the major improvement of the point-to-point resolution of 1.5Å in high-resolution electron microscopy (HREM) - which allow for more precise insight into the electronic and atomic structure of defects, their dynamics and the interaction between defects of different dimensions.

All the contributors presented only recent and original results, although in the runstream of previous research. Thus a distinguishing feature of this workshop is the paper originality combined with extensive references to previous work. This should give further insight into defect physics and hopefully trigger new progress on the basis of what we know today. In accordance with the scope of the workshop, the book is divided in two parts. The first part highlights recent progress in understanding the structure and properties of defects, their role on electrical properties and their interactions, discussed at the light of various physical processes and phenomena. The second part describes in detail techniques devised for defect characterization. They include the electron beam induced current (EBIC) and cathodoluminescence (CL) modes of scanning electron microscopy, STM, lattice imaging by HREM, scanning tunnel potentiometry, light beam induced current (LBIC) imaging and recent development in deep level transient spectroscopy (DLTS). Some examples are given of ad-hoc techniques designed for a thorough study of certain observed properties. The contributions here collected give altogether a vivid account of rapidly advancing topics and respective methods of investigation.

We acknowledge the sponsorship of the European Physical Society, IBM Italy, NATO Scientific Affairs Division, the Italian National Research Council (CNR), the Italian Ministry of Scientific and Technological Research and the Sicilian Regional Government. We thank the Director Prof. A. Zichichi and all the staff members of the Majorana Centre for the excellent organization and stimulating atmosphere enjoyed at the Centre. Finally we thank the School Director, Prof. Minko Balkanski, for his interest, and all the lecturers and scholars for their solicit response and active participation. All these elements have contributed to a very successful workshop and, hopefully, to a significant proceedings book.

Giorgio Benedek, Anna Cavallini and Wolfgang Schröter

Erice, 7 November 1988

CONTENTS

STRUCTURE AND PROPERTIES OF POINT DEFECTS

IN SEMICONDUCTORS

L. C. Kimerling

AT&T Bell Laboratories
Murray Hill, New Jersey 07974

INTRODUCTION

Junction spectroscopic methods utilize voltage modulation of the depletion width of a semiconductor junction to isolate and detect electronic transitions at defect states which are positioned in the band gap. Measurement structures may employ p-n junctions, Schottky-barriers or metal-insulator semiconductor junctions. The key feature of the approach is the separation of the carrier capture and emission processes by control of the Fermi level with junction bias. The semiconductor junction provides an ideal sample system for the control of defect charge state, the measurement of defect density, and the observation of the spectrum of defect state energies. The primary limitation of this approach is that it provides no chemical signature. The energy position and capture cross sections of a defect state yield no direct aid in identification.

This report will review the status of new applications of junction spectroscopy as a method for the determination of microscopic structure and properties. These procedures are reviewed as four case studies which illustrate Polarized Excitation Photocapacitance, Stress and Electric Field Modulated DLTS, Spatially Resolved DLTS, and Charge State Control of Structure.

CAPACITANCE TRANSIENT SPECTROSCOPY

A semiconductor junction structure under reverse bias may be considered as a parallel-plate capacitor. The plate separation is equivalent to the depletion width (W) and the junction capacitance is a measure of the associated charge separation. Electronic transitions which correspond to the charging or discharging of the defect states within the depletion region are monitored directly as changes in the junction capacitance. A schematic summary of the measurement methodology is shown in Figure 1.

Schottky Barriers and p^+n Junctions

The depletion width W is determined by the applied *reverse* voltage V_R, the barrier height V_B, and the free carrier concentration, N_D.

$$W = \left[\frac{2\epsilon_s(V_R + V_B)}{q\,N_D} \right]^{1/2} \tag{1}$$

The junction electric field varies linearly through the depletion width in uniformly doped material with a maximum value, $E_{\max}$, at the junction and a zero value at W.

$$E_{\max} = [2q/\epsilon_s(V_R + V_B)N_D]^{1/2} \qquad (2)$$

The capacitance, C_s, of a junction of area A is given by

$$C_s = \frac{\epsilon_s A}{W} = A\left[\frac{\epsilon_s q N_D}{2(V_R + V_B)}\right]^{1/2} \qquad (3)$$

In equations (1), (2), and (3), ϵ_s refers to the relative dielectric constant of the semiconductor $(\epsilon_s = \epsilon'_s\epsilon_0 = \epsilon'_s \cdot 8.86\times10^{-14}\text{F/cm}$ and q refers to the unit electrical charge $(= 1.6\times10^{-19}$ Coul.).

The junction depletion region vanishes under *forward* bias as an exponentially varying spatial distribution of minority carriers is injected into the substrate. The average concentration of minority carriers existing within a diffusion length (L_p) of the junction is given by

$$p = J/q(L_p/D_p) \qquad (4)$$

where J is the injected current density and D_p is the minority carrier diffusion constant. Schottky barrier structures allow, however, only majority carrier flow.

MOS Structures

The ideal MOS structure is considered as two capacitances in series

$$C = \frac{1}{C_o} + \frac{1}{C_s} \qquad (5)$$

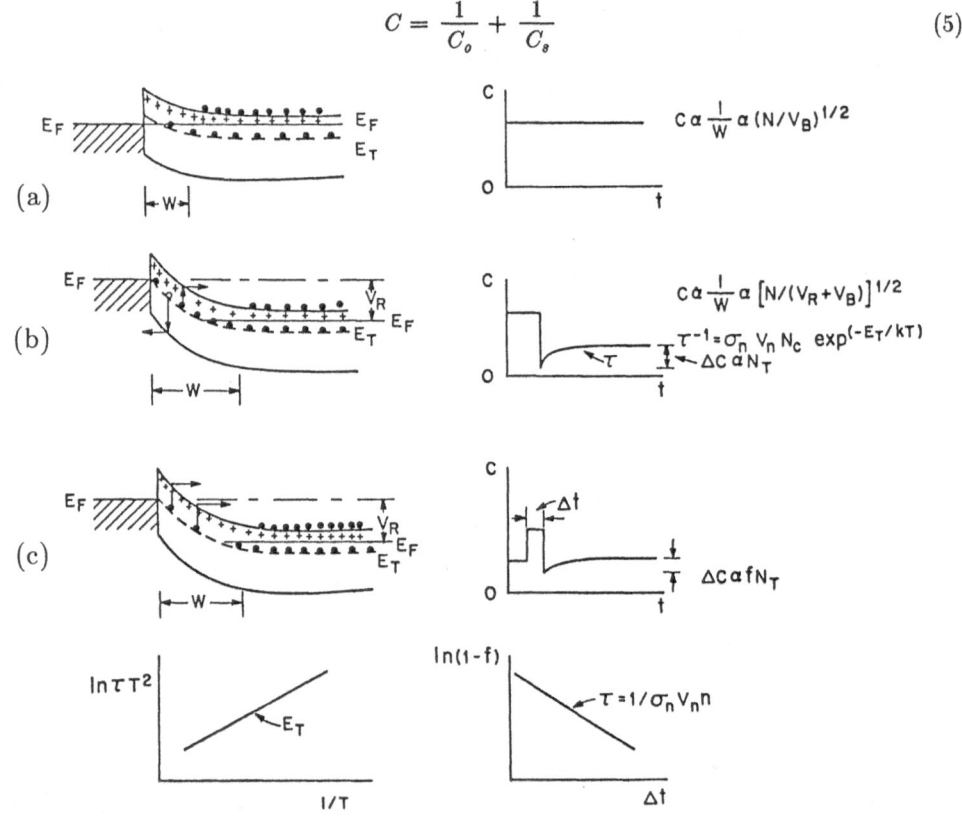

Fig. 1. Capacitance transient analysis of a Schottky barrier on n-type semiconductor structure: (a) zero bias, (b) zero to reverse bias (c) reverse to zero to reverse bias. $V_B =$ junction potential, $n =$ carrier conc., $C =$ capacitance, $\tau^{-1} = e_n =$ electron emission rate, $\sigma_n =$ capture cross section, $v_n =$ thermal velocity, $N_c =$ effective density of state in CB, $f =$ fraction of traps filled in time Δt.

The oxide capacitance, C_o, is determined by the oxide thickness as

$$C_o = \frac{\epsilon_o A}{t_o} \qquad (6)$$

In equation 5, C_g is associated with the depletion width capacitance generated by the applied gate voltage V_g and any trapped charge residing in the oxide or at the interface.

Defect State Parameters

The activation energy for carrier emission from a defect state to a band (E_T) is determined from the temperature dependence of the associated capacitance transient time constant (τ), and represents, exactly, the enthalpy of ionization at the measurement temperature, which is the approximate energy position of the state in the band gap. Activation energies are reported with the prefix H or E to denote hole emission to the valence band or electron emission to the conduction band, respectively. A prefix M is added when the state is a member of a family of metastable structural alternatives of a single defect system. Capture cross sections are determined by directly monitoring the decrease in emission signal (1-f) as the filling time (Δt reduced bias) is decreased. (An example for determining σ_{maj}, the majority carrier capture cross section is shown in Fig. 1c.) The depth distribution of a defect state is measured as the change in the magnitude of the capacitance transient (ΔC) as the state is filled at different reduced biases.

A defect state spectrum (deep level transient spectrum DLTS)[1] is generated by repetitive pulsing of the junction bias as the sample temperature is scanned. A peak in the transient correlator output is observed when the decay rate of the capacitance transient is coincident with the instrumental rate window. A complete review of experimental approaches to capacitance transient spectroscopy is given in reference 2. A typical measurement system for a junction structure is shown in Fig. 2. For this system the emission rate window is given by 0.4·[lock-in frequency]. Fig. 3 gives an example of the detection of Ag and Ni in a silicon p^+n junction using reference spectra (dashed lines).

Capacitance Transient Spectroscopy requires the filling and emptying of traps within a junction depletion region. Therefore, devices such as Schottky barriers, p-n junctions, MOS structures, field effect transistors, LED's and semiconductor lasers are proper test structures. Sample configurations varying from silicon wafers to packaged devices have been accommodated in measurement apparatus. Defect state concentrations and spatial profiles are easily determined in Schottky barriers or asymmetric junctions where the spread of the depletion region is clearly defined, but interpretation becomes difficult in complex device geometries.

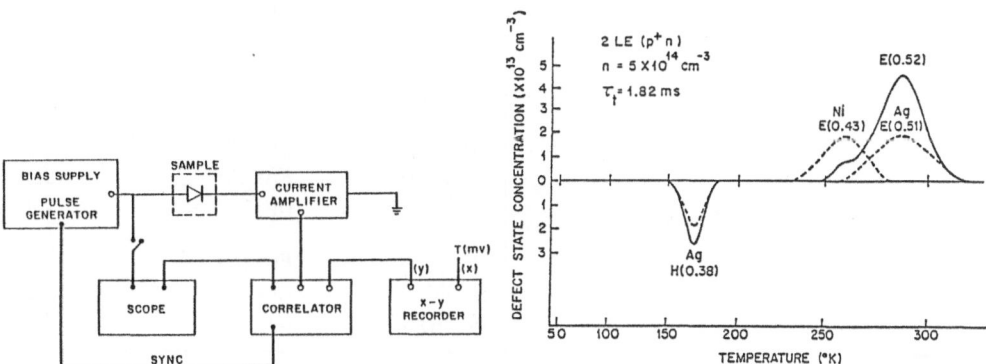

Fig. 2. Capacitance transient spectroscopy system.

Fig. 3. Detection of contamination in a silicon diode.

The limit of detection of capacitance transient spectroscopy is approximately 10^{-5} times the background free carrier concentration. Therefore, the method is least effective when applied to heavily doped materials. Furthermore, use of samples with lower doping avoids the production of large electric fields in the junction region, which can modify the observed activation energies of some defects.[3]

Reference Spectra

An unknown peak from a DLTS scan can be compared with reference data by two methods. First, a reference Arrhenius plot of $T^2\tau$ vs. $1/T$ can be constructed using the point given by the temperature of the known peak (T) at a time constant (τ) and a slope given by the activation energy (E_T) listed in Table 1. Alternatively, the temperature at which a signal from a listed defect should occur using any given instrumental time constant can be determined by iteration. A simple computer program can be used to set the ratio R to a value of unity.

Table 1. Electronic Parameters of Some Common Defects and Impurities in Silicon

Defect	E_T (eV)[a]	$T_{1.8\text{ msec}}$ (K)[b]	Defect	E_T (eV)[a]	$T_{1.8\text{ msec}}$ (K)[b]
Cr [0/+]	E(0.22)	108	O donor	E(0.07)	Below freeze out
(Cr·B) [0/+]	H(0.28)	123		E(0.15)	58
Mn [−/0]	E(0.11)	68	Dislocation	E(0.38)	225
Mn [0/+]	E(0.42)	216		H(0.35)	206
Mn [+/++]	H(0.25)	207			
			Glide debris	E(0.63-0.68)	288
Fe$_i$ [0/+]	H(0.43)	267	(V) [=/−]	E(0.09)	76
(Fe$_i$·B$_s$)[0/+]	H(0.10)	59	(V) [+/++]	H(0.14)	79
(Ni$_i$·B$_s$)[0/+]	H(0.14)	88	(V·V)[=/−]	E(0.23)	142
(Ni·?)	H(0.43)	257	(V·V)[−/0]	E(0.40)	241
			(V·V)[0/+]	H(0.21)	131
(Cu$_i$·B$_s$)[0/+]	H(0.22)	112			
(Cu·?)	H(0.41)	242	(V·0)[−/0]	E(0.17)	98
Au [−/0]	E(0.53)	288	(P·V)[−/0]	E(0.43)	215
Au [0/+]	H(0.35)	173	(As·V)[−/0]	E(0.47)	235
			(Sb·V)[−/0]	E(0.44)	224
			(Sn·V)(+/++)	H(0.07)	69
			(Sn·V)(0/+)	H(0.32)	192
			(Al$_i$)[+/++]	H(0.23)	203
			(Al·V)	H(0.52)	282
			(B$_i$[0/+]	E(0.13)	87
			(B$_i$[−/0]	E(0.45)	223
			(B·V)	H(0.32)	190
			(C$_i$)[−/0]	E(0.12)	64
			(C$_i$)[0/+]	H(0.27)	165
			(C$_i$O$_i$)[0/+]	H/0.36)	206

[a] E_T activation energy for electron (E) or hole (H) emission to respective band edge.
[b] $T_{1.8\text{ msec}}$ temperature at which the carrier emission time constant is 1.8 msec.

$$R = \frac{\tau_1 T_1^2 \exp^{-E_T/kT_1}}{\tau_2 T_2^2 \exp^{-E_T/kT_2}} \tag{7}$$

where the subscript 1 refers to the values of Table 1 and the subscript 2 to the values for the particular run. T_2 is varied until the ratio $R = 1$.

In our laboratory, a catalog of defect spectral parameters is kept on computer file. Entries can be searched by specifying material characteristics, processing data, and/or defect state parameters. A positive defect identification requires determination of E_T and σ_{maj}, specification of hole or electron trap, and consideration of the processing steps and their sequence.

STRESS AND ELECTRIC FIELD MODULATION

Applied external stress can produce two responses in junction spectroscopy: one, atomic reorientation of an anisotropic defect structure (or rebonding among equivalent Jahn-Teller distortions) to minimize total energy with respect to the projected stress field and two, modification of the defect ground state electronic energy level due the local and long range deformation potentials. Watkins and Corbett[4] have pioneered the application of uniaxial stress perturbations to probe the properties of defects in semiconductors in conjunction with electron paramagnetic resonance measurements. Meese, Farmer and Lamp[5] have identified both phenomena in the first uniaxial stress modulated DLTS study of the deep acceptor state of the A-center (V·O) in silicon.

Defects within the space charge region are subject to strong electric fields during a DLTS measurement. These fields can interact with the carrier emission process for electrons excited form donors and holes excited from acceptors.[3,6] This coupling arises from the positionally integrated voltage drop across the excited carrier and its oppositely charged defect site. The result is a lowering of the carrier binding energy which exhibits a $E^{1/2}$ dependence on electric field. If the defect site is anisotropic, one expects a superimposed effect which is smaller in magnitude (due to the limited range of interaction) but asymmetric in response to field orientation. This interaction is equivalent to a linear Stark effect in which local inversion symmetry is removed, for example, by a structural dipole. A rough estimate of the barrier lowering is given by

$$\Delta E = q\vec{E}\vec{d} \tag{8}$$

where $\vec{d}$ is the induced orbital eccentricity of the bound carrier and $\vec{E}$ is the electric field or potential drop across the defect. Martin, et al.[7] have considered the role of variations

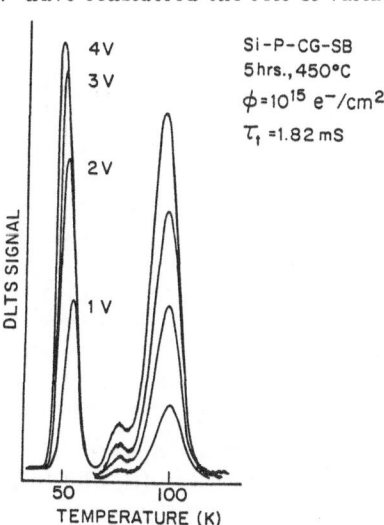

Fig. 4.   Capacitance transient spectrum of n-type silicon following 450 °C heat treatment + 1-MeV electron bombardment. At constant reverse bias, the filling pulse is reduced, and regions of lower electric field are sampled. The $E(0.15)$ oxygen donor peak (50 °K) shows a marked shift in position while the $E(0.18)$ A-center, acceptor peak (100 °K) does not.

5

in the structure of the local binding potential on the isotropy and rate of carrier emission.

The Oxygen Donor in Silicon

Two defect states, $E(0.07)$ and $E(0.15)$ have been identified by DLTS as oxygen donor levels in silicon.[3] This identification was made as follows: 1) the states were observed only in silicon which contained interstitial oxygen, ($[O_i] > 10^{17}\,cm^{-3}$); 2) the states appear during 450 °C anneal and disappear during anneal above 550 °C; and 3) a ($\sim E^{1/2}$) barrier lowering was observed, confirming the shallow donor nature of the center. Figure 4 shows the shift of the $E(0.15)$ donor peak to lower temperatures as the electric field is increased while the $E(0.18)$ acceptor peak of the A-center does not shift.[8] Table 2 lists the observed anisotropy in emission rate and shifts in energy position with electric field direction for the $E(0.15)$ donor state. The superposition of an isotropic, long range Poole-Frenkel (P-F) interaction (as manifested in the differences in τ with respect to the $<111>$ extrapolated zero field behavior) and an anisotropic field-

Table 2. Emission Time Constants of $E(0.15)$ at $T = 59$ °K

Direction	$<$Electric Field$>$, (V/cm)	$T^2\tau$, (°K^2 sec)	ΔE_E, (eV)
$<111>$	0 (extrap.)	4000	0
$<111>$	2.4×10^4	10.1	3.0×10^{-2}
$<110>$	2.4×10^4	8.9	3.1×10^{-2}
$<211>$	2.4×10^4	6.4	3.26×10^{-2}
$<100>$	2.4×10^4	3.7	3.54×10^{-2}

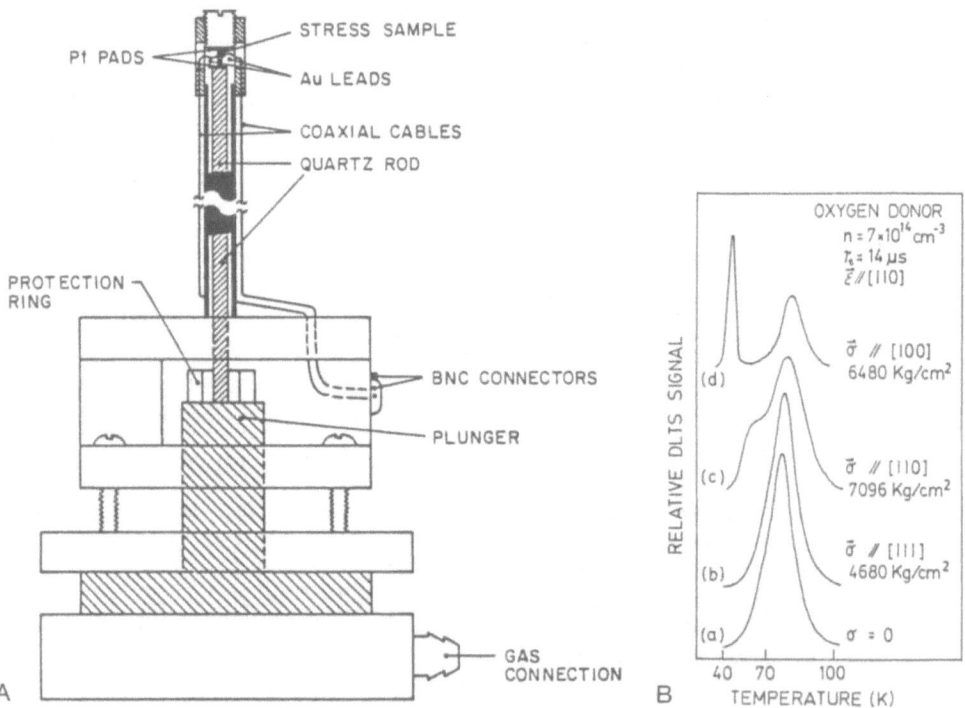

Fig. 5. (A) Schematic diagram of the apparatus used for uniaxial stress perturbed DLTS studies. (B) DLTS spectra of the oxygen donor peak, $E(0.15)$, under (a) no applied stress, and under uniaxial stress parallel to the (b) [111], (c) [110], and (d) [100] axes.

6

modulated interaction is clear. The approximate magnitudes of the barrier lowering interactions ΔE_E are $\sim 10^{-2}$ eV for $P\text{-}F$ and $\sim 10^{-3}$ eV for the anisotropic (Stark) component.

The splitting of the $E(0.15)$ state under uniaxial stress[9] is illustrated in Figure 5b. (A schematic of the DLTS stress apparatus is shown in Figure 5a). The maximum response exists for a $<100>$ stress. The splitting is linear in applied stress and consists of both a shift in ground state energy and in the "center of mass" energy of the defect. The deformation potential of the center of mass and the defect are approximately equal, falling in the range of 10-15 eV. The relative strengths of the ground state splitting, 2:1 for $<100>$, 1:2 for $<110>$, and none for $<111>$ are consistent with a $<100>$ primary axis and D_{2d} symmetry. The motion of the center of mass energy (weighted by the degeneracy fo each splitting) can be explained by stress coupling to the six conduction band minima.[9]

Interpretation of the Uniaxial Stress Perturbation

Figure 6 shows the three types of transitions which must be considered in the interpretation of the stress perturbed DLTS data of a donor state. Note that the *thermal* emission process which is monitored by DLTS is a statistical process and, therefore, is distinct from optical excitations (which are adiabatic and must conform to selection rules) because the lowest lying conduction band minima will always be selected. Hence, the final state of all electronic transitions is $1'$. The interpretation of the ground state splitting is dependent on the nature of the ground state wavefunction.

Shallow States. A *shallow, effective-mass state* has wave functions which are composed of contributions from the band extrema. For a shallow donor in silicon there are six such conduction band minima along the six $<100>$ directions. The defect state is delocalized and the stress coupling is long range, reflecting, primarily, the deformation potential of the band structure. The motion of the DLTS peaks can fall into one of three categories.

Category 1. If the ground state is degenerate, transitions $1 \rightarrow 1'$ will be observed. The lowest lying ground state and conduction band minima are selected. For level separations which are greater than 2 kT no peak splitting is observed. (Otherwise, a significant Boltzmann population of both levels will occur and some evidence of the higher energy level will be seen.) A slight shift in peak position may occur when the effective-mass make-up of the ground state deviates from ideal.

Category 2. If the ground state is a singlet with the T_d symmetry of the lattice, only the coupling of the conduction band is observed. The ground state is isotropic with respect to applied stress, but the conduction band responds in an anisotropic manner due to the $<100>$ origins of the conduction band minima. *A shift in peak position to lower temperature (transition $3 \rightarrow 1$) is observed.*

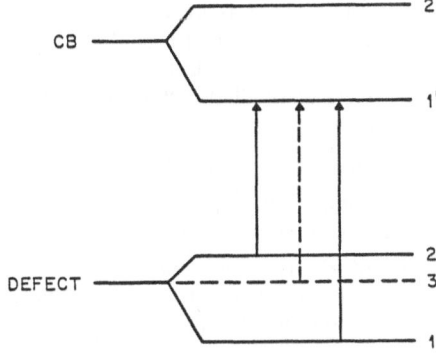

Fig. 6. Schematic diagram of the possible electronic transitions in a stress perturbed DLTS experiment observing electron transitions to the conduction band. The conduction band splits $(1',2')$ under long range deformation potentials and the defect splits $(1,2,3)$ according to its atomic symmetry and the character of its ground state.

Category 3. If the ground state is a singlet with lower than T_d symmetry splitting is observed due to the orientational degeneracy and transitions 1 → 1' and 2 → 1' occur. Both the ground state and the conduction band couple to the external stress. The relative intensities of the split peaks correspond to projections of the stress on the ground state symmetry. As stated above and as illustrated in Figure 5, the relevant symmetry is that of the band structure.

Deep States. Stress coupling for a *deep state* is short range and reflects the symmetry and nature of localized bonds. The symmetry is generally low (category 3) due to Jahn-Teller distortions or other mechanisms which lower total energy and maximize the strength of the bonds between the imperfection and the host. Since the wave function is not simply composed of band extrema, splitting is a unique function of the defect stress tensor which will clearly reflect the orientational degeneracy.

Category 3. For low symmetry with either a singlet or multiplet ground state splitting is observed according to transitions 1 → 1' and 2 − 1'. This result was first observed with DLTS by Meese, et al. for the A-center in silicon.[5]

The oxygen donor defect exhibits a classic, helium-like ionization spectrum which closely matches the ground state energies predicted by effective mass theory for a shallow state. If one takes this observation as a starting point, critical insight into the nature of the defect is evident from the data shown in Fig. 5b. These conclusions are summarized below.

Since splitting is observed, the defect state has a Shallow State (3) classification. The relative intensities of the stress split peaks are independent of temperature, verifying that *the ground state is a singlet.*

The preferred orientation of each oxygen donor is fixed during heat treatment and possesses a random distribution. No atomic stress alignment could be produced at temperatures up to 350 °K. The intensity ratios of the split components show that the oriented populations are distributed equally about the six <100> directions.

The defect state wave function is derived from a single pair of conduction band valleys and the local atomic symmetry is tetragonal (D_{2d}) or lower. This electronic structure is unique with respect to substitutional, effective mass donors (P, As, etc.) which possess T_d site symmetry and a $1s(A_1)$ ground state. The shallow, effective mass character of the wave function requires that it be composed of functions from the conduction band minima. These minima in silicon are located at six valleys in the six <100> directions. The stress results show that the oxygen donor selects only one <100> axis and, hence, one pair of valleys for its wave function. The atomic symmetry which forces this selection, therefore, must either possess a <100> axis or exhibit a projected interaction which isolates a particular <100> axis. These results are equivalent to the behavior of the TD° and TD^+ IR spectra[10] and the NL-8 EPR spectra[11] of the oxygen donor.

Interpretation of the Electric Field Perturbation

Anisotropic field emission phenomena are not understood. Martin et al.[7] have attempted a numerical model of the role of binding potential anisotropy in field emission effects. Within the qualitative bounds of their approach, the ordering and magnitude of the effects of the Table 2 are consistent with the existence of a <100> oriented dipole at the defect.

One can estimate the magnitude of such an effect by equation 1, $\Delta E = q\vec{d}\vec{E} = 10^{-3}$ eV, where $\vec{d}$ is an oriented dipole separation of 1Å and $\vec{E}$ is the average electric field of 10^5 V/cm. The energy scale of the observed effects is given by $\ln[\tau(\vec{E})/\tau(0)_{111}]$ which yields values in the range of 10^{-3} eV.[9] The shallow, extended nature of the defect state wave function may require modification in the details of this localized potential approach.

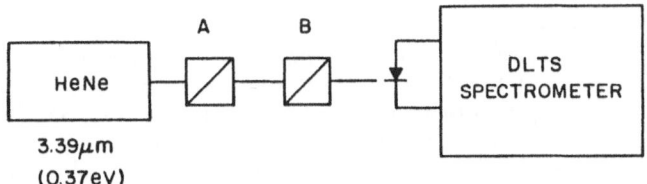

HeNe

3.39μm
(0.37eV)

Fig. 7. Schematic diagram of PEP
measurement equipment. A and B
are polarizers.

POLARIZED EXCITATION PHOTOCAPACITANCE (PEP)

Photocapacitance measurements are well recognized as a powerful method for the
determination of the form of the cross section for optical excitation. Clear advantages
exist over absorption and photoconductivity techniques because the junction
measurement allows preparation and isolation of a given transition. Since the optical
transition is adiabatic, the photon polarization directly probes the atomic structure of
the defect. The full sensitivity of this method is possible only when the defects can be
prepared in a state of preferred alignment. An example is given for the case of the
divacancy donor state H(0.21) in silicon.

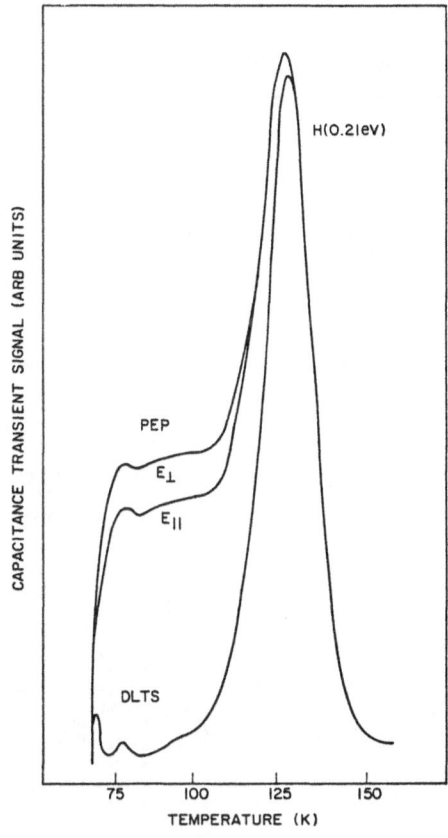

Fig. 8. DLTS spectrum and PEP spectra
of the H(0.21) hole trap
(divacancy) introduced by 1-MeV
electron irradiation.

9

The Divacancy in Silicon

EPR,[12] IR absorption[13] and photoconductivity[14] studies have yielded extensive insights into the nature of the divacancy structure and physical properties. DLTS has provided the best results regarding the defect state energy positions. In particular, the state H(0.21) has been assigned to the donor charge state using standard analytical procedures described earlier.[15] Microscopic confirmation of this identity was completed with the PEP techniques described below.[16]

Figure 7 shows a schematic diagram of the PEP apparatus used in the divacancy study. A He-Ne laser tuned to the 3.39 μm line was employed to excite within the 3.9 μm divacancy adsorption band. Polarizer A is oriented at 45 ° to the sample stress axis. Polarizer B is rotated at $\pm$45 ° with respect to polarizer A and selects equal intensities of light polarized perpendicular and parallel to the applied stress.
The sample was stressed along [110] and illuminated along [001]. A uniaxial stress was applied at 160 °C and maintained while cooling to room temperature to produce an aligned defect population. A set of PEP spectra are shown in Figure 8. The anisotropy in the optical cross section for the emptying of state H(0.21) is evident in the difference in the signal levels with the photons polarized $\perp$ and $\parallel$ to the applied stress. The absence of any background above the DLTS signal level at high temperatures ($T > 150$ °K) confirms that the excitation wavelength excites only from H(0.21).

A summary of results is given in Table 3. The ratios of the optical cross sections are in agreement with calculated values for a model of the defect which predicts a

Table 3. Dichroic Ratios for Several Stress and IR Beam Directions

Stress direction	IR beam direction	Magnitude of stress (kg/cm^2)	$\sigma_\perp^0/\sigma_\parallel^0$ Exp.	$\sigma_\perp^0/\sigma_\parallel^0$ Cal.
[110]	[001]	2800	1.32	1.52
[110]	[110]	2800	1.16	1.27
[112]	[111]	2000	1.14	1.15
[001]	[110]	2100	1.03	1.00

Fig. 9. Comparison of defect state spectra for deformed silicon: (a) initially dislocation free and (b) initial dislocation density $\simeq 10^5$cm^{-2}.

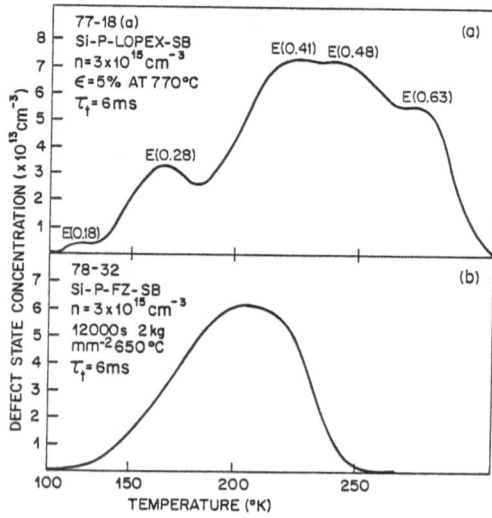

<110> oriented transition moment along a bond linking the atoms at the ends of the divacancy axis. The absence of a stress response along <001> and the presence of a response along the other directions confirms the <111> atomic axis of the defect. Finally, the disappearance of the stress-induced dichroism upon anneal in the 100-150 °C temperature range yields atomic reorientation kinetics of $\tau^{-1} = 1.3 \times 10^{12}$ exp $(-1.22 \text{ eV}/kT)$ which are in agreement with the EPR and IR determinations. The identification of H(0.21) as a defect state of the divacancy is thus complete. The body of junction spectroscopy evidence now includes data of the introduction rate dependence on electron bombardment energy, the independence of introduction rate on impurity background, the annealing recovery kinetics, the determination of local atomic and electronic structure, and the measurement of local reorientation kinetics. The junction spectroscopy tools are now proven and available to derive this same information for an unknown defect and directly yield assignments of structure and identity.

DEFECT STATE MICROSCOPY

Determination of the defect spatial distribution or microstructure is frequently a key element in identification and phenomenology. Literally, the previous discussions refer to structures on an invisible Angstrom scale, whereas this discussion addresses *microstructures* on a visible micrometer scale. Typically, an uneven spatial distribution of defects suggests stray impurity contamination, striation patterns related to crystal growth, and distributions created by process steps such as implantation or diffusion. A classic defect system with microstructure is the dislocation.

Dislocations in Silicon

Figure 9a shows a typical DLTS spectrum which is produced by deformation of silicon.[17] Even though dislocations are observed in the microstructure, it is unlikely that they are responsible for all of the states. Heat treatment at 900 °C removes all states but one, $E(0.37)$, but the microstructure continues to exhibit a high dislocation density. Scanning DLTS measurements in regions of high and low dislocation density reveal that the $E(0.63)$ levels dominate in the region of slip bands, while the $E(0.37)$ signals are most prominent in regions of isolated dislocations.[19] Figure 10 shows a schematic diagram of a Scanning Electron Microscope DLTS system.

Figure 9b shows the defect state spectrum which results from mild deformation of n-silicon which contained $\sim 10^5 \text{cm}^{-2}$ dislocations before deformation and exhibited no slip bands following deformation. The $E(0.37)$ state dominates. Figure 11 shows a plot of defect state density against dislocation density for a series of the mild deformations. Since the dislocation density is nonuniform, it is critical that junction techniques on the same junction be used to measure all relevant densities. SEM-Charge Collection Microscopy provided the images on the right which were used to count dislocations. The

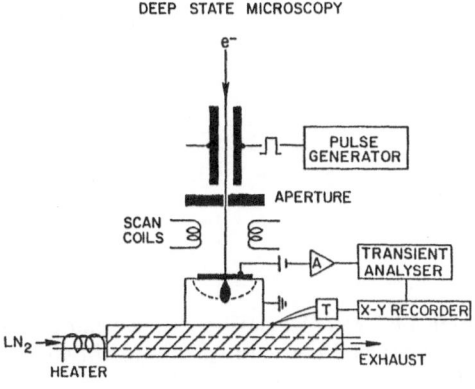

DEEP STATE MICROSCOPY

Fig. 10. Schematic diagram of Deep State Microscopy apparatus which employs a pulsed electron beam excitation in a scanning electron microscope.

same junctions were used for the SEM-CCM and DLTS measurements. The data correspond to one electrically active site for every 200Å of dislocation length. The $E(0.37)$ state has been assigned to kink sites. The recoverable deep states are identified as representing point defect debris which are created during cross slip of dislocation tangles when a dense network is present.

CHARGE STATE CONTROL OF STRUCTURE

The junction structure provides means for direct control of defect charge state within the depletion region. The bonding of deep states and, hence, their structure and physical properties are directly related to the local charge state. Thus, the charge state-structure interaction may be explored in detail.

One of the most exciting frontiers in semiconductor materials science has been the discovery of metastable defect associates, their reactions and structural transformations. In particular, observation of structural metastability by a variety of capacitance transient measurements and the relation of these phenomena to interstitial defects has focused renewed experimental and theoretical attention on the interstitial.[19] Unlike substitutional imperfection, interstitial defects are not confined to host lattice symmetry and often possess multiple atomic configurations with similar total energies. The search for these configurations has led to the detection of a complex hierarchy of defect reactions and an initial glimpse into the relationship between reaction kinetics and structure.

The Iron-Boron Pair in Silicon

The properties of interstitial iron in the silicon lattice have been extensively studied through characterization of its association reaction with boron acceptor impurities[20]

$$(Fe_i)^+ + (B_s)^- \rightleftarrows (Fe_iB_s)^0 \qquad (9)$$

The attractive interaction is electrostatic and, thus, can be controlled by the charge state of either constituent.

The donor state of iron H(0.43) is a deep state in the lower half of the gap. At room temperature, the Fermi level in neutral material is below the defect state, and it is charged positively. Within a depletion region, the defect is empty of holes and is neutral. Figure 12 schematically depicts the charge state controlled stability for the associates. Figure 13 shows a diagram of the binding potentials and interaction energies associated with the (Fe_iB_s) pair as derived by junction spectroscopy. The charge state control of

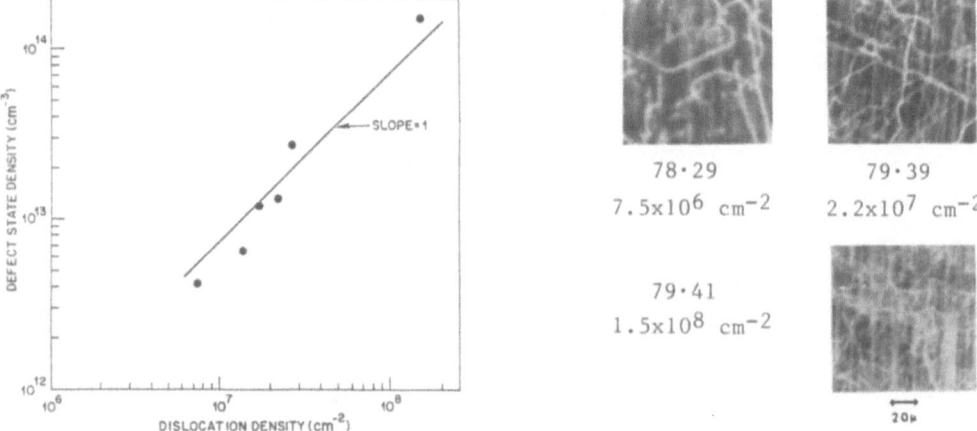

Fig. 11. Defect state concentration as a function of local dislocation density under the Schottky barrier in samples similar to those of Figure 6b. EBIC (SEM-Charge Collection) Micrographs of some samples are also shown.

Fig. 12. Influence of (Fe$_i$) charge state on the (Fe$_i$B$_s$) ion pair stability. Under reverse bias, the charge state is (Fe$_i$)^0 within the depletion region. At zero bias the Fermi level is positioned below the H(0.43) donor state of iron and the charge state is (Fe$_i$)$^+$. The shaded areas denote Fermi level positions for which the association reaction will occur or nearest neighbor pairs will remain bound.

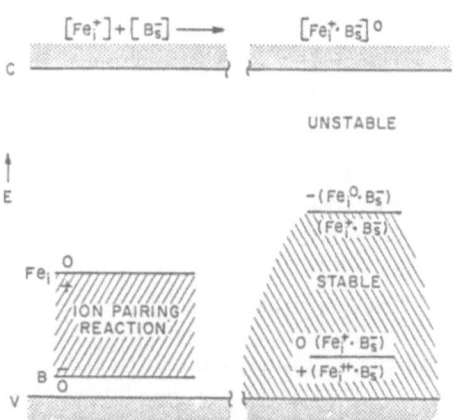

the structure is evident, because E_B (the tilt in potentials) is dependent on the presence of an electrostatic attraction.

In the iron-acceptor pair system[20,21,22] the local pair structure depends on the charge state of the interstitial iron (Fe$_i$)$^{0/+/++}$. The system exists in two structural states with (Fe$_i$) located at either the nearest- or second-nearest tetrahedral interstitial position relative to the substitutional acceptor. Preparation of the defect in either of the two structures is accomplished by cooling the sample with the defect in the proper charge state and trapping the configuration in a metastable minimum. Figure 14 shows the results for (Fe$_i$Ga$_s$) and (Fe$_i$In$_s$). For (Fe$_i$B$_s$), (Fe$_i$Al$_s$) and (Fe$_i$Ga$_s$), the 1st neighbor position is stable and the 2nd neighbor position is metastable. For (Fe$_i$In$_s$) the two structures are bistable, depending on charge state.

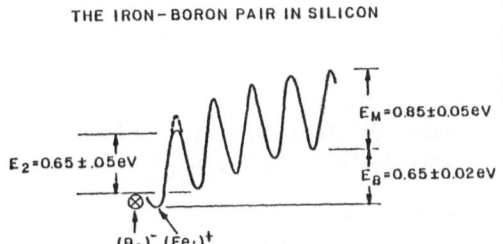

THE IRON–BORON PAIR IN SILICON

$E_M = 0.85 \pm 0.05\,eV$

$E_2 = 0.65 \pm .05\,eV$

$E_B = 0.65 \pm 0.02\,eV$

$(B_s)^- \; (Fe_i)^+$

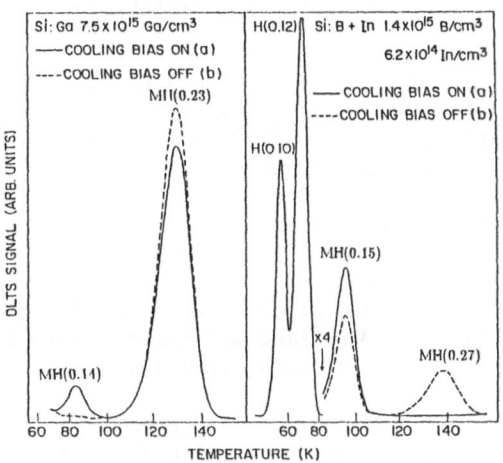

Fig. 13. Schematic diagram of the lattice potentials for (Fe$_i$)$^+$ near (B$_s$)$^-$. E_m is the free (Fe$_i$)$^+$ migration enthalpy observed in the association reaction. E_B is the (Fe$_i^+$B$_s^-$) pair binding energy derived from equilibrium studies. E_2 is the single hop barrier to second neighbor (Fe$_i$)$^+$ for nearest neighbor pair formation.

Fig. 14. DLTS spectra for Fe-contaminated, silicon doped with Ga or B+In ($\tau = 6$ ms). MH(0.23)/MH(0.27) and MH(0.14)/MH(0.15) represent the first and second neighbor (Fe$_i$Ga$_s$)/(Fe$_i$In$_s$) ion pair configurations, respectively.

SUMMARY

The perfect spectroscopy measures the properties of a system which are critical to its application. Junction spectroscopy has traditionally fulfilled this specification. The examples presented in the review highlight the emerging role of junction spectroscopy as a laboratory for microscopic science. The identity of spectral features can be derived with confidence; local defect structures can be determined; and new physical properties can be discovered. In an applied sense, these methods are becoming standard tools for for monitoring materials processing steps and evaluating device performance in semiconductor materials.

The perturbation methods present an ultimate application of the semiconductor junction, but one which will not likely reach the factory floor. Special samples must be prepared to assure uniform, axial fields and detailed consideration of all physical interactions is required for interpretation. However, these methods provide direct support for more routine uses of the measurement. The unique ability of junction spectroscopy to contribute in both roles places it in the active center of semiconductor materials research.

REFERENCES

1. D. V. Lang, *J. Appl. Phys.* 45:3023 (1974).
2. G. L. Miller, D.V. Lang, and L. C. Kimerling, *Ann. Rev. Mater. Sci.* 7:377 (1977).
3. L. C. Kimerling and J. L. Benton, *Appl. Phys. Lett.* 39:410 (1981).
4. G. D. Watkins and J. W. Corbett, *Phys. Rev.* 121:1001 (1961).
5. J. M. Meese, J. W. Farmer and C. D. Lamp, *Phys. Rev. Lett.* 1:1286 (1983).
6. J. Frenkel, *Phys. Rev.* 54:647 (1938).
7. P. A. Martin, B. G. Streetman and K. Hess, *J. Appl. Phys.* 2:7409 (1981).
8. J. L. Benton, L. C. Kimerling and M. Stavola, *Physica* 116B:271 (1983).
9. J. L. Benton, K. M. Lee, P. E. Freeland and L. C. Kimerling, *J. Electronic Materials* 14b:647 (1985).
10. M. Stavola, K. M. Lee, J. C. Nabity, P. E. Freeland and L. C. Kimerling, *Phys. Rev. Lett.* 54:2639 (1985).
11. K. M. Lee, J. M. Trombetta and G. D. Watkins, in: "Microscopic Identification of Defects in Semiconductors," N. M. Johnson, S. G. Bishop and G. D. Watkins, eds., Matr. Res. Soc. Proc., Vol. 46, MRS, Pittsburgh (1985), p. 263.
12. G. D. Watkins and J. W. Corbett, *Phys. Rev.* 138A:543 (1965).
13. L. J. Cheng, J. C. Corelli, J. W. Corbett and G. D. Watkins, *Phys. Rev.* 152:761 (1966).
14. A. H. Kalma and J. C. Corelli, *Phys. Rev.* 173:734 (1968).
15. L. C. Kimerling, in "Radiation Effects in Semiconductors 1976," Inst. Phys. Conf. Ser. 31, edited by N. B. Urli and J. W. Corbett, Institute of Physics, London, 1977 p. 221.
16. M. Stavola and L. C. Kimerling, *J. Appl. Phys.* 54:3897 (1983).
17. L. C. Kimerling and J. R. Patel, *Appl. Phys. Lett.* 34:73 (1979).
18. L. C. Kimerling, J. R. Patel, J. L. Benton and P. E. Freeland, in "Defects and Radiation Effects in Semiconductors 1980," Inst. Phys. Conf. Ser. 59, edited by R. R. Hasiguti, Institute of Physics, London, 1981 p. 401.
19. L. C. Kimerling, M. T. Asom, J. L. Benton, P. J. Drevinsky and C. E. Cafer *in:* "Proceedings of the 15th International Conference on Defects in Semiconductors, Budapest, 1988," Materials Science Forum, Trans-Tech. Pub., Switzerland (1988).
20. L. C. Kimerling and J. L. Benton, *Physica*, 116B:297 (1983).
21. A. Chantre and D. Bois, *Phys. Rev.*, B31:7979 (1985).
22. A. Chantre and L. C. Kimerling, *Mater. Sci. Forum.*, 10-12:387 (1986).

CONDUCTIVITY OF GRAIN BOUNDARIES AND DISLOCATIONS
IN SEMICONDUCTORS

R. Labusch and J. Hess
Institut für Angewandte Physik
der T.U. Clausthal

Introduction

In spite of the high perfection that can be achieved in the production of monocrystaline Silicon, the working horse of the semiconductor industry, many devices are and will be made in the future of
compound as well as of elemental semiconductor material which is not
so perfect, either for economic or technological reasons. This
material contains dislocations and grain boundaries (GB-s). Consequently, the investigation of the electronic properties of these
defects will continue to be of interest from a practical point of
view.

Beyond this, dislocations and GB-s can be also of fundamental interest as one- and two-dimensional electronic systems. Two-dimensional
conductivity at surfaces has lead to the famous von Klitzing effect
and some one-dimensional conductors exhibit transport by charge density waves (CDW), a phenomenon related to superconductivity, and a
Peierls phase transition due to a strong coupling between Bloch-waves
and standing lattice waves.

All these interesting effects can in principle be studied in GB and
dislocations as well and these extended defects are potentially even
more ideal than other known systems of reduced dimensionality. Thus,
for instance, localised states in one dislocation core can be perfectly decoupled from other dislocation cores, while in so-called
"one-dimensional" crystalline conductors there is always a small but

finite overlap between Bloch functions and also a Coulomb interaction
between CDW-s on different chains.

Furthermore, the investigation of the conductance along extended defects can provide also information on their other electronic properties which is difficult to obtain by other means

In this work new results on the two-dimensional conductivity of grain
boundaries in Ge are compared with measurements of the conductance
across the boundary. Former discrepancies between the two sets of
results are resolved. Direct evidence for one-dimensional conduction
along dislocations in Ge is obtained from DC-measurements for the
first time. A Peierls transition and a Peierls gap of 22 meV are observed. The I-V-characteristics are strongly nonlinear and formally
fit the theoretical expression for charge density waves, although
with a negative threshold field. No evidence has been found so far
for one- or two-dimensional conduction at extended defects in Si.

Conduction perpendicular and parallel to GB-s

Most experiments on the conductivity along GB-s have been performed
so far in Ge-bicrystals. In earlier measurements low angle GB-s with
tilt angles up to 15° and a (100) tilt axis were investigated in n-Ge
[1,2,3]. The results were interpreted by G. Landwehr et al in terms
of an effective mass calculation based on the following model [3]: Associated with the GB is a sheet of acceptor states with a strongly
localised charge. This charge is neutralised by an equal number of
holes in two dimensional bands that split from the valence band. The
hole states are localised with a half width of roughly 40 $\overset{o}{A}$. The
screened potential in the neutral state which is obtained from a Hartree calculation has the same range. The density of bound holes is
high enough for complete degeneracy. In n-type material the number
of holes is of course somewhat reduced because electrons are trapped
in the GB states, but the degeneracy persists.

This model was confirmed by all transport experiments, including
Schubnikov-de Haas measurements that allowed to determine the effective mass of the bound holes. It is represented in its simplest
possible form in figure 1. The horizontal axis in this figure is a
coordinate perpendicular to the GB. The vertical bar whose width is

about 40 Å is meant to be the 2-dimensional GB-band. The three parts of the figure stand for n-type material in which the GB has a nega- tive charge and is surrounded by a potential barrier, for the neutral GB, and for a GB in p-type material, respectively.

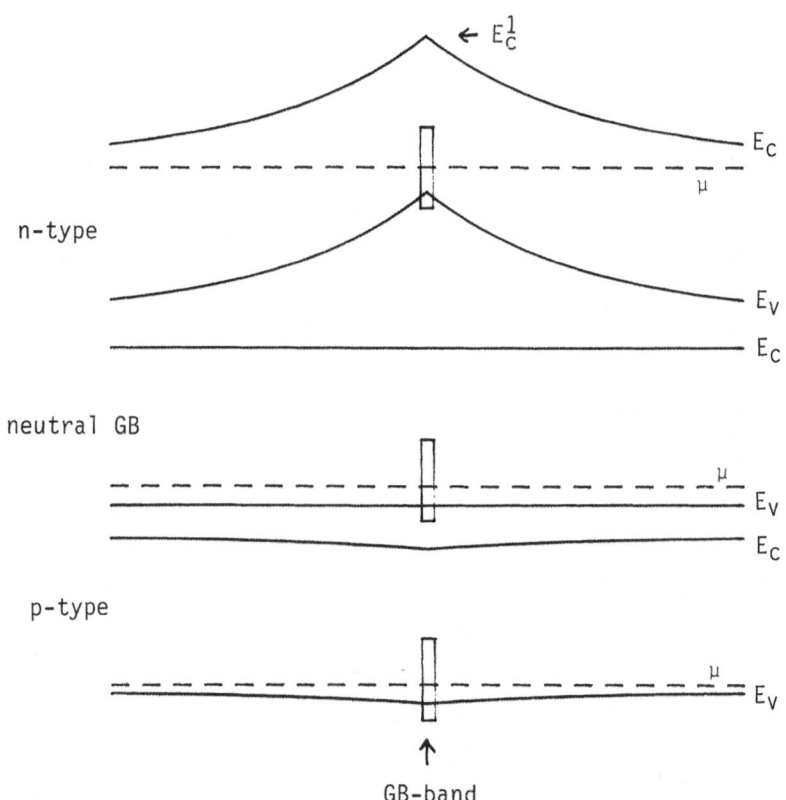

Figure 1.   Schematic diagram of GB-states and of the potential barrier at a GB.

It is possible that the Fermi level in the neutral state is so close to the valence band that there is no barrier at all or even a small potential of the opposite sign in p-type material. The latter is ex- pected if the Fermi level of the bulk is higher than the occupation limit of the neutral GB. We notice that the range of the potential barrier in n-material is given by the screening length which depends on doping but is typically of the order of several thousand angstroms, while the Hartree potential that binds the holes in the GB-band has a range of about 40 Å.

More recently, we have extended the conductivity and Hall effect measurements to large angle GB-s of the Σ11-structure in n- as well as in p-Ge [4]. The results are essentially the same as in previous measurements: We find two-dimensional quasi metallic conduction with a carrier density of 5 to 6 times 10^{12} cm^{-2}. Examples are given in figures 2 and 3 where we have plotted the resistivity or the resistance versus 1/kT. Other specimens gave very similar results.

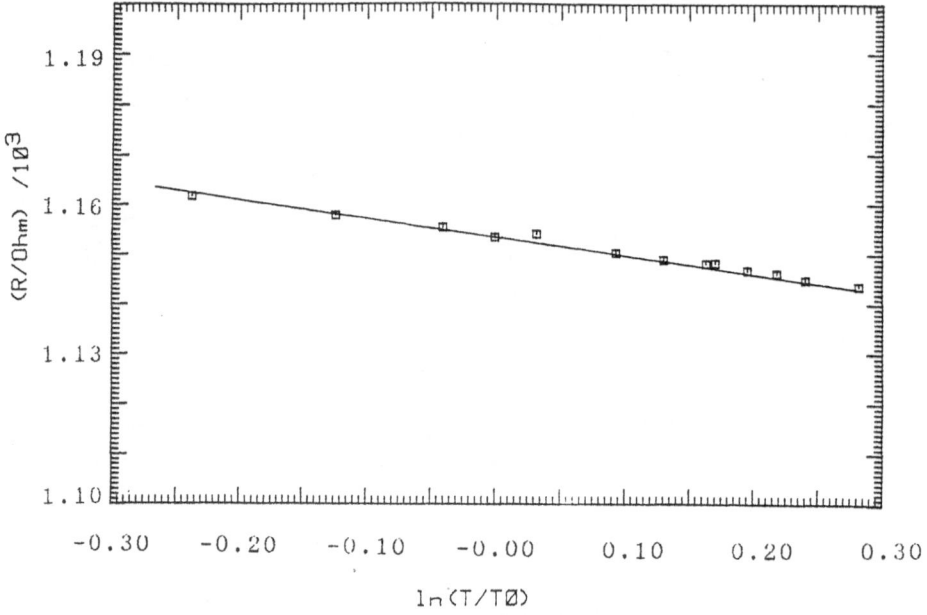

Figure 2. 2-dimensional conduction at GB-s in n-Ge. The carrier density is $6 \cdot 10^{12}$ degenerate holes per cm^{-2}. Theoretically, the resistivity is expected to be linear in ln(T).

In the case of the p-type specimen it is impossible to avoid current flow through the bulk which is electrically parallel to the GB. Therefore the GB-conductivity shows up only at very low temperatures where the bulk conduction is frozen out. The bulk resistance, measured next to the GB at a distance of less than 1 mm is given by the filled dots. It has the expected exponential temperature dependence while the GB resistance becomes almost constant at low temperatures. In n-Ge it is possible to avoid conduction through the bulk by using alloyed Au-In contacts which are rectifying on n-material but ohmic on the p-type GB-conductor. The alloying takes place at a temperature of only 356° C where In and Au are practically immobile, so that there is no danger of contaminating the GB.

For the same type of GB ($\Sigma 11$) there are also many measurements available of the conduction <u>across</u> the GB [5]. As we have mentioned before, the GB traps electrons and has a negative charge in n-Ge and is consequently surrounded by a long range potential barrier which is repulsive to electrons. At low temperatures where all screening is done by charged donors, the potential of a GB at x = 0 is given by

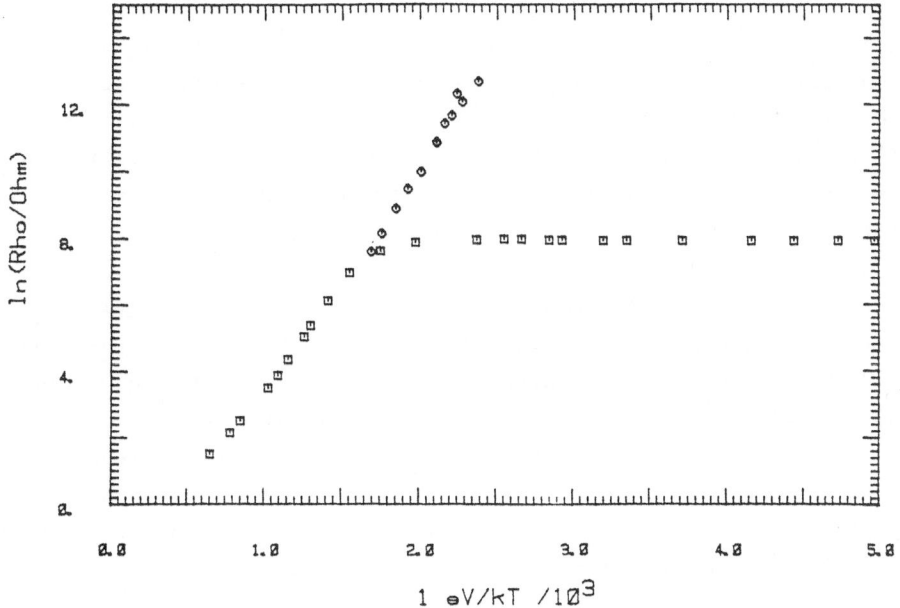

Figure 3. Resistance of a GB in p-Ge. The bulk resistance, measured with the same contact geometry, is given by the filled dots.

(1) $\Phi(x) = \Phi_0 \cdot (d-abs(x))^2/d^2$

where d is the thickness of the space charge layer:

(2) $d=(2\varepsilon\varepsilon_0\Phi_0/q\cdot N_D)^{1/2}$

q being the elementary charge, ε the dielectric permittivity, N the donor concentration and Φ_0 the barrier height. Furthermore, the relation between Φ_0 and the number of trapped electrons at the GB is given by the charge balance equation

(3) $2N_D \cdot d = (2\varepsilon\varepsilon_0\Phi_0 2N_D/q)^{1/2}$

At higher temperatures these relations have to be modified but can still serve as a rough estimate. Accurate expressions have been given in ref. 5. The conductance across the barrier can be estimated in different ways. Wu et al used an expression obtained from the thermal emission theory:

(4) $S = A^* \cdot q \cdot (T/k_B) \cdot \exp(-(E_c - \mu + q\Phi_0)/k_B T)$

where A^* is a modified Richardson constant, E_c the lower edge of the conduction band and μ the position of the Fermi level. Since in this theory it is assumed that the electrons cross the barrier in free flight while actually, in the range of interest, the mean free path is often shorter than the barrier width d, one might prefer a different approach, which is very simple.

In the limit of a short mean free path, the local conductivity in the barrier is simply given by $\sigma_B \cdot \exp(-q\Phi(x)/k_B T)$ where σ_B is the conductivity in the bulk. Therefore the resistance of the barrier is $R = (1/\sigma_B) \cdot \int_{-d}^{+d} \exp(q\Phi(x)/k_B T)\, dx$. As the largest contributions to R come from the vicinity of the top of the barrier, we can replace $\Phi(x)$ by $2\Phi_0 \cdot (1 - abs(x)/d)$ and extend the integral from $-\infty$ to $+\infty$ in a good approximation. This yields $(2dk_B T/q\Phi_0) \cdot \exp(-q\Phi_0/k_B T)$ for the integral and, for the conductance per unit area,

(5) $S = (\sigma_B q\Phi_0/2dk_B T) \cdot \exp(-q\Phi_0/k_B T)$

Numerically the preexponential is not very different from the factor obtained by the thermal emission theory but its temperature dependence is different, and this will lead to small difference of the activation energies which are obtained from an evaluation according to eq. (4) and (5) respectively. In the range of temperatures in which S has been measured, σ_B and Φ_0 are almost constant or even decreasing with temperature so that the preexponential in eq. (5) is proportional to a positive power of 1/T while it is proportional to T in eq. (4). The truth will be somewhere between these extremes.

X. J. Wu, V. Szkielko, and P. Haasen measured the conductance perpendicular to grain boundaries in n-Ge and found an Arrhenius law with an activation energy close to 0.7 eV. They determined the barrier height and the position of a GB-level at 0.2 eV above the edge of the conduction band from an analysis based on the assumptions that the thermal emission formula is quantitatively correct and that the GB-states are exclusively acceptor levels. The latter seemed to be justified because they observed no barrier in p-type material. For a small number of trapped electrons and for temperatures between 200 and 400 K where the measurements had been done, the GB-level is then well above the Fermi level: $(E_T-\mu) = kT \cdot \ln(N_T/n_T)$, where N_T is the number of GB traps and n_T the number of trapped electrons. Consequently, the position of the Fermi level comes out higher than it would if the analysis were based on the model of Landwehr et al.

This analysis and its results contradict not only the competing model described in the introduction but also our own measurements of transport along the same type of GB in several ways:

1. If the GB has only acceptor levels at or above 0.2 eV, there should be no GB-conductivity in p-type material at low temperatures, in contrast to the experiment. On the other hand, if we assume, in order to save the model, that the GB-conduction in p-material is carried by a donor level, or rather a donor band, below 0.2 eV, the analysis would have to be modified significantly because, at temperatures between 200 K and 400 K, the number of holes trapped in this band would not be negligible, also in n-type material.

2. In an acceptor or in a donor band, the carrier density would be equal to the number of trapped electrons or holes, respectively. At low temperatures where most conductivity measurements have been done, the Fermi level must be close to the lower edge of the donor band in n-type (or close to the upper edge of the acceptor band in p-type) material on the one hand while, on the other hand, outside the range of the GB-potential it is close to the edge of the conduction band of the bulk. Therefore the height of the barrier is nearly equal to E_c-E_T of the neutral GB, almost independent of the doping concentration (although different in n- and p-type material). As a consequence, according to eq. (3), the carrier density in the GB should be proportional to the square root of the doping concentration while, actually, it turns out to be always p-type and nearly independend of doping for concentrations which differ by more than two orders of magnitude. For comparison, the maximum number of trapped electrons determined by Wu et al was $4 \cdot 10^{12}$ cm^{-2} for a doping concentration of $4.7 \cdot 10^{16}$ cm^{-3} and less than 10^{11} cm^{-2} for a doping of $2 \cdot 10^{13}$ cm^{-3}, while in our specimens the carrier density in the GB was 5 to $6 \cdot 10^{12}$ cm^{-3} for doping concentrations between $5 \cdot 10^{13}$ and $1.5 \cdot 10^{16}$ cm^{-3}.

Looking for an alternative analysis that could lift these discrepancies, we start now from the assumption that, instead of an acceptor level, a two dimensional band is associated with the GB which, in its neutral state, contains a large number of free holes at its upper edge so that the Fermi level is well within the band, in a position E_o above the valence band of the bulk.

If this GB-band traps electrons, the Fermi level is shifted to a higher position within the band, but the shift is quite small: The density of states in a two dimensional parabolic band with the effective mass $m^* \simeq 0.3\ m_o$ is $N_{GB} = m^*/\pi\hbar^2 \simeq 5\cdot10^{13}eV^{-1}cm^{-2}$. With this value the shift, even in the worst case (as quoted above), must be quite small and its dependence on n can be linearised.

The energy difference between E_c^l, the local edge of the bulk conduction band at the top of the barrier, and the position of the Fermi level is now $(E_c^l-\mu) = (E_c-E_o-\alpha\cdot n_T)$. At the same time, the electron density, and therefore also the conductance S, at E_c^l is proportional to $exp(-(E_c^l-\mu)/k_BT)$ so that the experimental activation energy of S is

(6) $\quad Q = (E_c-E_o-\alpha\cdot n_T)$.

In a rigid band model we have $\alpha = 1/N_{GB}$. Since the bound hole states at the GB depend on the actual potential, the value of α is somewhat modified. With increasing n_T ,the energy levels are shifted down towards the local edge of the valence band because the binding potential gets stronger and therefore the bound holes become more localised. This cancels part of the shift of the Fermi level which is due to the change of occupation with electrons. However, the effect is only of the order of $q\Phi_o\cdot w/d$ where w is the half width of the bound eigenfunctions and d the width of the potential barrier. With $w = 40\ \AA$ and more than thousand $\AA$ for d this is only a few percent of the barrier heigth and the term $\alpha\cdot n_T$ remains negligible for most practical purposes.

In figure 4 we have plotted the data of one n-type specimen with a doping concentration of $1.5\cdot10^{15}$ cm^{-3} that we have measured with particular care, together with other specimens from reference [5]. We notice that all the data are in agreement within the experimental uncertainties. In all cases the curves bend upward at low temperatures, probably due to leakage currents along the specimen surface [6]. Disregarding this temperature regime, we find a thermally activated

regime that, without correction of the data, is limited by another
bending around at high temperatures. Responsible for the latter is
the bulk resistance of the specimen which is not negligible any more
so that the measured conductance is now $S = (S_{GB}^{-1} + S_B^{-1})^{-1}$ where S_B is
the bulk conductance of the specimen whose magnitude and temperature
dependence are known.

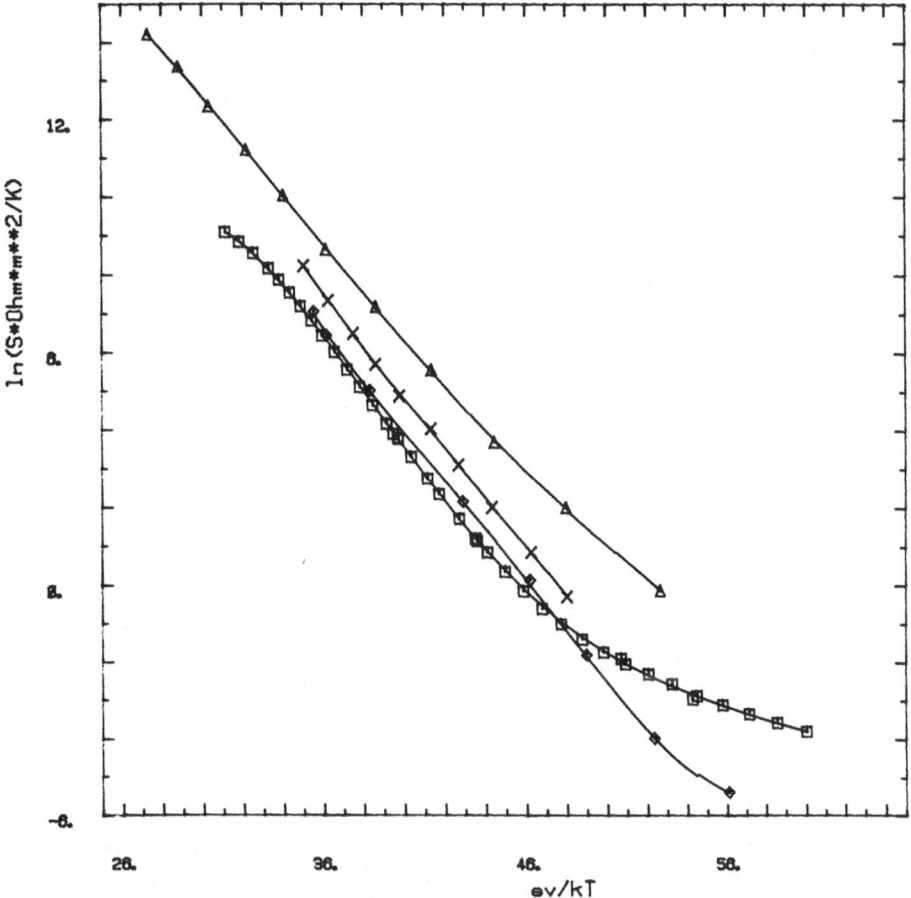

Figure 4.   Conductance across GB-s in Ge specimens with doping
concentrations between $2 \cdot 10^{13}$ and $4.7 \cdot 10^{17}$ cm^{-3}. Squares are
our own measurements, $N_D = 1.5 \cdot 10^{15}$ cm^{-3}. The other data were
taken from reference 5 .

Figure 5 shows our own data after correction for this effect. The
activation energy can be obtained from this plot with high accuracy.
Its value is Q = 0.760 eV. If the temperature dependence of the preexponential according to the thermionic formula is taken into account

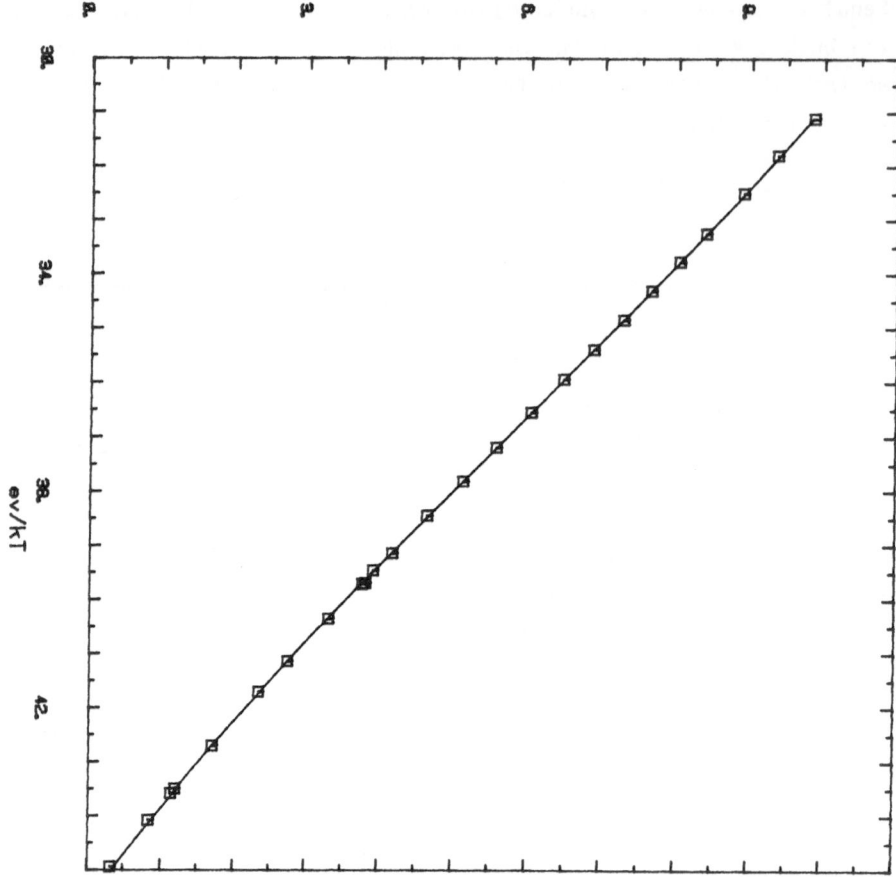

Figure 5. Conductance S across a GB in Ge after correction for the bulk resistance. The slope of lnS(1/kT) is 0.760 eV.

the activation energy comes out somewhat lower: Q = 0.733 eV. Since Q is the extrapolated value at T = 0, it has to be compared with the extrapolated bulk gap whose value is 0.785 eV [7]. Therefore E_0, the Fermi level of the neutral GB is only between 0.025 eV and 0.052 eV above the edge of the valence band. Consequently, a barrier in p-Ge, if it exists at all, could only be detected at very low temperatures.

The experimental value of the preexponential in the thermal emission formula is $1.76 \cdot 10^{14}$ $\Omega^{-1}m^{-2}$. Theoretically we expect a factor of $5 \cdot 10^{12}$ $\Omega^{-1}m^{-2}$ from the thermal emission probability and another factor of $\exp(4.4 \cdot 10^{-4}eV/k_BT)$ from the temperature dependence of E_g. Together this is $8.25 \cdot 10^{14}$, or about five times higher than the experimental value. Scattering at the GB and reflection due to the mismatch of the Bloch waves in the two crystals may be responsible for at least part of the difference, but we notice also that our local conductivity formula (5) yields a lower value of the preexponential and a better fit. We also notice that the use of the theoretical thermal emission probability in the calculation of the barrier from experimental data would lead to an error of about $1.6k_BT$ in the barrier height.

The situation in Si is very different from Ge. GB-s of the $\Sigma 25$-type ((100) tilt axis, 16.3^O tilt angle) have been investigated in great detail by G. Petermann [8,9] and Petermann and Haasen [10]. The most striking difference to Ge in the conductance at U = 0 is that in Si the activation energy depends on doping while in Ge it is virtually constant (see figure 4). From this dependence and also from an analysis of the I-V-characteristics, the authors obtained a continuous density of states with a broad maximum near the middle of the gap. Its order of magnitude is 10^{11} to 10^{12} $cm^{-2}eV^{-1}$ which is about two orders of magnitude lower than the estimated value for a parabolic band of free carriers.

In a preliminary experiment we tried to measure the conductivity along a GB of the same type in n-type material with a doping concentration of $10^{13}cm^{-3}$, using alloyed Al-contacts which are rectifying on n- and ohmic on p-type material. No conduction could be measured at liquid nitrogen temperature. With the sensitivity of our instruments we would have detected a GB conductivity if it were higher than 10^{-15} Ω^{-1} per square. We therefore conclude that the GB-states are localised.

Conduction along dislocations

Conduction along dislocations has been observed directly until now only in CdS. Although the results were interpreted as one-dimensional conductivity in dislocation states, it could not be completely excluded that the conduction took place in a thin cylinder of segregated impurities or intrinsic defects around the dislocation [11]. In compound semiconductors both, intrinsic defects and impurities, are always potentially present in sufficient concentrations to mask the true dislocation states. In the elemental semiconductors the conditions are more favourable concerning impurities but, so far, only AC-conductivity measurements on dislocation networks are available [12,13]. These are easier to perform than DC-measurements, because there are no problems with the electric contacts to individual dislocations but, on the other hand, they have some disadvantages:

There is no way to distinguish between different types of dislocation and also, even more important, no way to distinguish unambiguously between true one-dimensional conductivity along dislocations and polarisation of defects.

Nevertheless, the AC-measurements in Ge and Si strongly suggest that, at least in Ge, one-dimensional conduction does exist.

To investigate this topic further we have therefore started experiments in Ge which are aimed at measurements of the conduction along single dislocations. Our starting material was n-type with $1.5 \cdot 10^{13}$ donors per cm^3. We introduced long straight dislocations by the well known technique of expanding half loops from scratches on the (111)-surface of a <123>-oriented specimen. It is known that most loops which expand under compression from a scratch which is perpendicular to the trace of the main glide plane, have a Burgers vector parallel to the surface and consist of two 60^0-segments which intersect the surface and one connecting screw segment parallel to an below the surface ^{14}. However, the two most extended groups of half loops that we found after compression at 370^0 C and under a load of 60 MPa happened to lie in the unexpected glide plane so that it was not certain from the beginning whether the two dislocation arms intersecting the surface are both 60^0- or one 60^0- and one screw type. All our experiments, so far, were performed with these two groups of loops which had diameters of up to 500 microns.

A thin slice containing the loops at one surface, was cut from the specimen and was further thinned by grinding and polishing from below to a thickness of approximately 100 microns. At this stage we could see etch pits on the back side of the platelet which apparently belonged to the same two groups of dislocations that had been observed on the front surface.

From the conductivity along grain boundaries, including low angle GB-s which could be described as dense arrays of dislocations (although not of the types that are introduced by loop expansion), we infered that, most likely, the conduction along dislocations, if it existed at all, would be p-type. Therefore our strategy was to apply contacts to both sides of our specimen which are ohmic on p-type and rectifying on n-type material, so that, at least at low temperatures, there would be no or only a negligible bias current through the bulk. Suitable for this purpose are the same alloyed Au-In contacts which were used already in the GB-experiments. The contact spots on the upper surface had dimensions of 50µ·500µ and covered one group of 40 dislocation segments each. Furthermore, a reference contact of the same size was applied in a dislocation-free area to check the magnitude of the bias current. The back side was furnished with similar contact spots and then fastened to a small aluminium block with silver epoxy, while the spots on the upper surface could be contacted separately with thin elastic wires.

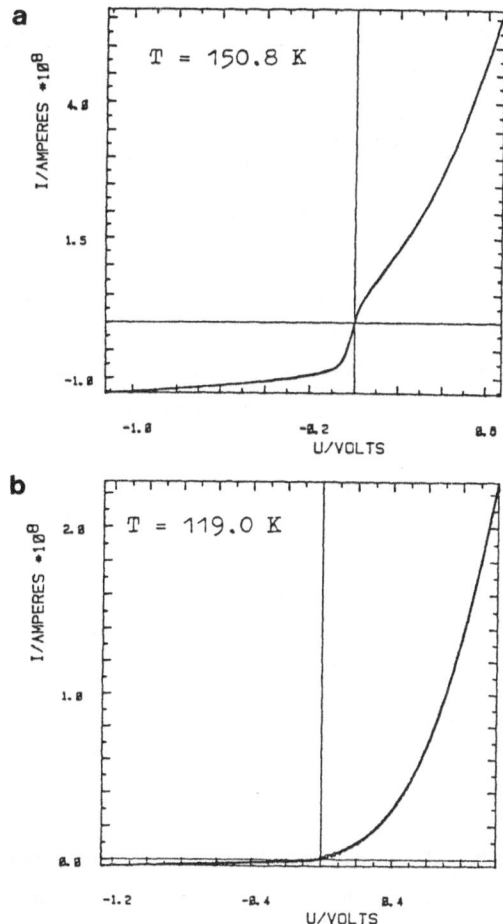

Figure 6. I–V–characteristic of a pair of rectifying contacts on Ge at 151 K (a) and at 119 K (b). One contact is alloyed AuIn, the other alloyed AuIn plus silver epoxy.

Figure 6 shows I–V–curves for the reference contact at 151 K and 119 K. These curves are typical of a pair of two asymmetric diodes in opposite direction in series. The observed asymmetry must be expected because the back side contact consists actually of two diodes in parallel, one for the Ge-AuIn and one for the Ge-silver epoxy. Notice that, for abs(U) ≤ 0.6V, the highest current is less than

$6 \cdot 10^{-8}$ A at 151 K and less than $6 \cdot 10^{-9}$ A at 119 K. Below 110 K the contact resistance became so high that meaningful measurements were not possible any more.

From the first measurements it became clear that the two arms of the half loops, called A and B in the following, have different electrical properties:

The contacts to B-dislocations had practically the same I-V-characteristics as the reference contact (see figure 6). In contrast, the contacts on A-dislocations are almost ohmic with a small asymmetry between about 70 K and 150 K, and develop a nonlinear but symmetric shape below 70 K. Typical examples are shown in figure 7. At all temperatures below 150 K the resistance for $U = 0$ is by a least two orders of magnitude lower and the current at $abs(U) = 0.6$ V by two orders of magnitude higher than the corresponding values of the reference contact or of the B-dislocations. below 27 K the nonlinearity of the I-V-curves becomes weaker with decreasing temperature. However, we have the suspicion that in this regime where the resistance is very high, the dislocation effects may be masked by bias currents along the specimen surface (see also below and figure 10).

Each curve in figure 7 represents 600 data pairs. In some of the curves, particularly in those at the lowest temperatures there was some random noise but because of the large number of single measurements this could be largely removed by a spline procedure. The worst example is given in figure 7d. Only splined curves will be used in the following.

As a preliminary hypothesis, based on various theoretical models of the dislocation core [15], we assume that B-dislocations are screws and A-dislocations are of the 60°-type, although at present we have no solid experimental evidence to prove this conjecture.

Discussion of the I-V-curves of A-dislocations

A plausible explanation for the asymmetry between 90 K and 150 K seems to be the assumption that the dislocation itself is a nearly ohmic resistor with a symmetric I-V-curve which is slightly biased by the same asymmetric pair of opposite diodes which is observed with the reference contact. At 120 K the slope of the I-V-curve for large negative U is negligible. Therefore the bias can be reconstructed by

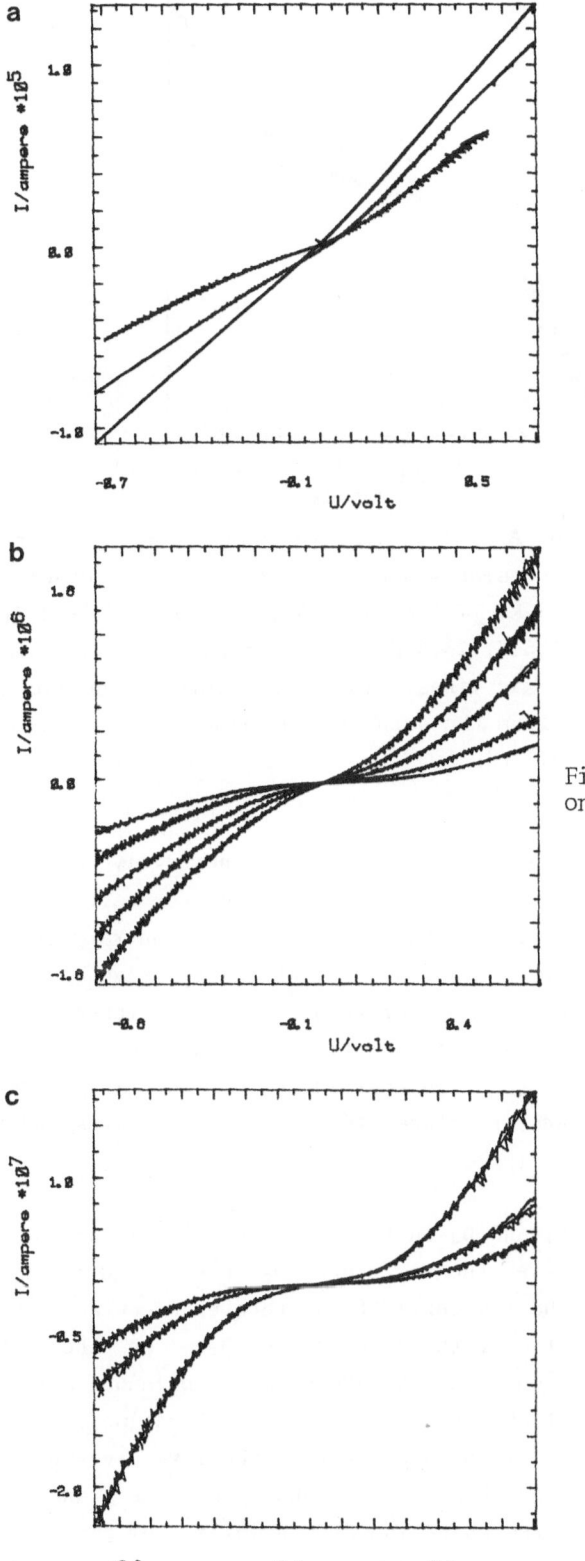

Figure 7. Caption
on next page.

29

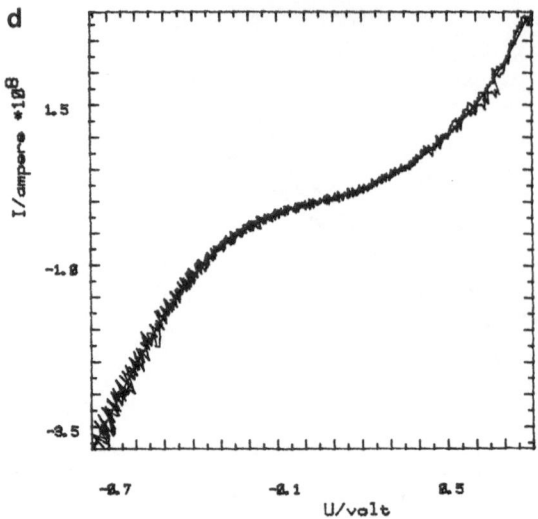

Figure 7. I-V-curves of A/dislocations (presumably 60°) at different temperatures. The slopes are decreasing with temperature. a: 123 K, 94.5 K, 77.5 K. b: 54,4 K, 50,6 K, 47.2 K, 42,8 K. c: 35.1 K, 31,6 K, 28,5 K. d: 26.5 K. The raw data are connected by zig-zag lines. The smooth curves, sometimes hardly distinguishable from the data, are obtained by a third order spline over a small window.

subtracting from the experimental data a perfectly symmetric I-V-curve which coincides with the measured curve for voltages less than −0.2 volt. The result is shown in figure 8. Contrary to our expectation it turns out that the bias current for positive U is by about two orders of magnitude higher than for the reference contact so that we have to look for an alternative explanation. In fact there is still another possible current path, shown in figure 9, which involves charge transfer between the dislocation states and the bulk conduction band.

The two surface diodes D_1 and D_2 in figure 9 are the same as for the reference contact. Of these, the lower one, D_2, has a much higher reverse current so that the I-V-curve of the reference contact is essentially the characteristic of the upper diode, D_1. On the other hand, for the dislocation contact there is a continuous array of diode-like connections between the bulk and the dislocation resistor. We suggest that the asymmetric bias current in the I-V-curves between 60 K and 120 K is carried by these connections. The same transitions play also a role in the recombination of nonequilibrium carriers at dislocations in deformed material and have been investigated and dis-

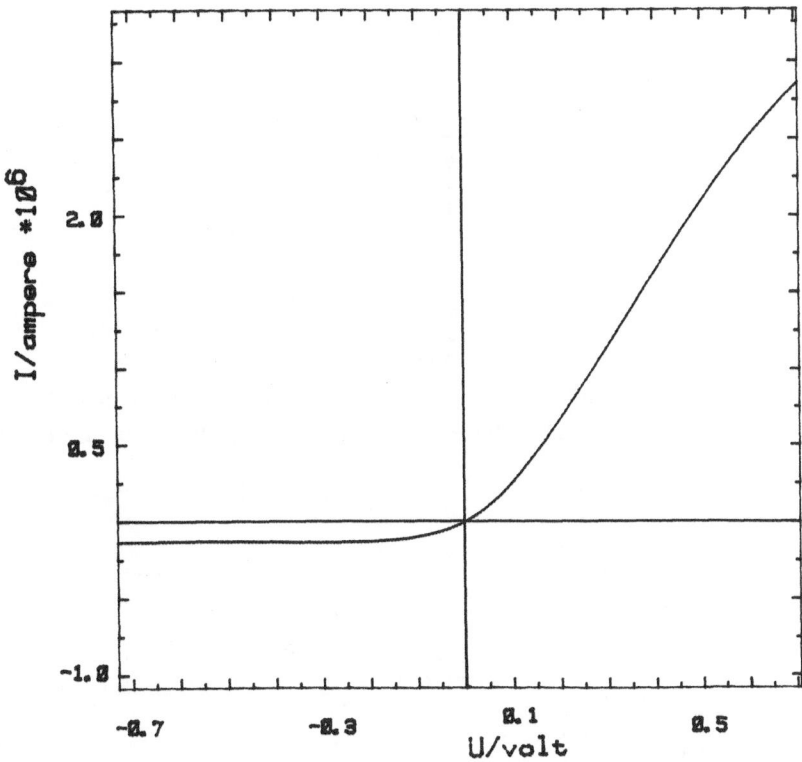

Figure 8. Asymmetric bias current at 123 K. A symmetric a
curve was subtracted from the curve in figure 7a which has the
same slope for U < -0.25 V and is linear for abs(U) < 0.25 V.

cussed in great detail [16]. In the transition, the electrons have to
overcome a potential barrier which is given by the electrostatic
potential of the negatively charged dislocation. According to W.
Schröter [17] the effective height of this barrier can be considerably
reduced due to tunneling. Nevertheless we find that our bias currents are much higher than one would expect from the observed recombination rates in comparable deformed specimens (1 μA bias in our
experiment corresponds to a transition rate of more than
$3 \cdot 10^{18}$ cm^{-3}s^{-1} for a dislocation density of 10^6 cm/cm^3). Therefore
it will be necessary to invoke a modification of the dislocation
potential. Such a modification could for instance come about near
the surface, through the presence of the contact spots.

At low temperatures, the bias which is thermally activated is expected to disappear and, consequently, the symmetric I-V-curves below
60 K are interpreted as conduction by the bound dislocation states
alone.

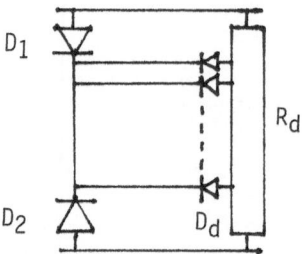

Figure 9. Replacement circuit for rectifying contacts at the ends of a negatively charged dislocation with a conducting core.

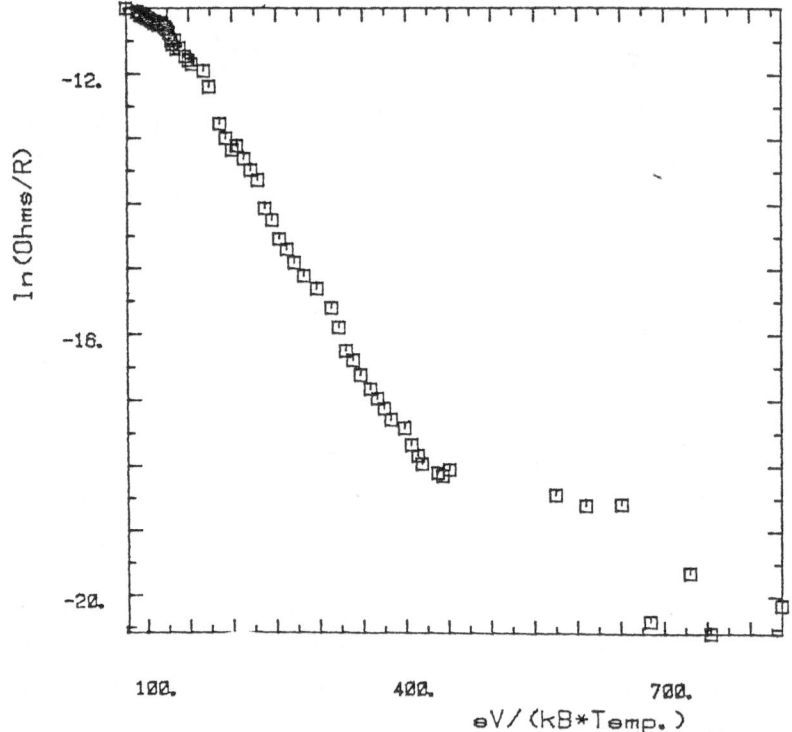

Figure 10. Conductance of A-dislocations at zero applied voltage in an Arrhenius plot. The slope of the linear regime is −0.022 eV.

Figure 10 shows a logarithmic plot of the conductance at $U = 0$ versus $1/k_B T$. All measurements up to 150 K have been included in this plot without correction for the contribution of the asymmetric bias which is negligible at $U = 0$. The result is a straight line except at the lowest temperatures (below 25 K) where the resistance is so high that other bias currents may come in. From the linear part we obtain an activation energy $Q_P = 0.022$ eV. A dependence like this is expected for one-dimensional quasi-metallic conductors which undergo a Peierls transition at low temperatures. Q_P is half of the Peierls gap. Theoretically the Peierls transition is expected at $T_P = 0.57 \cdot Q_P/k_B$ [18]. Some one-dimensional conductors show a maximum of the conductivity at the transition temperature but in our case we are unable to observe this because the bias current through the bulk becomes too big above 150 K.

Considering all our experimental evidence and theoretical expectation we conclude that we have detected one-dimensional conduction along one type of dislocations in Ge. If we assume that the 40 dislocations in one group have all the same conductance and if we take into account the segment length of 100 microns, the conductance of a single dislocation is given by $s = s_0 \exp(-Q_P/kT)$ with $s_0 = 6.9 \cdot 10^{-11}$ Ω^{-1}m.

Theory predicts the possibility of electric transport by charge density waves (CDW) [19]. These are many-body states that come about through the coupling of Bloch waves near the Fermi level with travelling lattice waves. They resemble the superconducting state in so far as there is no effective scattering mechanism for their propagation. However, in real "one-dimensional" crystals these waves seem to be always pinned, either by impurities or by an incompatibility of their wavelength with the lattice periodicity. This has the consequence that CDW-s contribute to the current only if the applied field is greater than a threshold field E_{th} which is necessary for unpinning. The theories predict a dependence of the conductivity on the applied field of the form [20,21]

$$(9) \quad \sigma = \begin{array}{ll} \sigma_0 & \text{for } E < E_{th} \\ \sigma_0 + \sigma_1 \cdot \exp(-E_0/(E-E_{th})) & \text{for } E > E_{th} \end{array}$$

E_{th} is of the order of 100 V/m. This dependence has indeed been found in some 1-dimensional crystals [22,23]. However, a plot of our data according to this formula shows a somewhat strange agreement: σ_1 was determined from an extrapolation of $\ln(S(U)-S(0))$ versus $1/U$ to $1/U=0$ which is rather straightforward and accurate. In figure 11 we have then plotted our data in the form $-1/\ln((\sigma-\sigma_o)/\sigma_1)$ versus applied voltage for temperatures between 26.5 K (lowest curve) and 47.2 K (highest curve). The data in a very small regime (abs(U) < .04 Volt) where $S(U)-S(0)$ goes to zero and its logarithm diverges, were ommitted from the plot. According to equation (9) the resulting curves should be straight lines, which they are indeed in a very good approximation. However, all intercepts are positive which means that the threshold fields are negative. At first sight this seems to be physically meaningless.

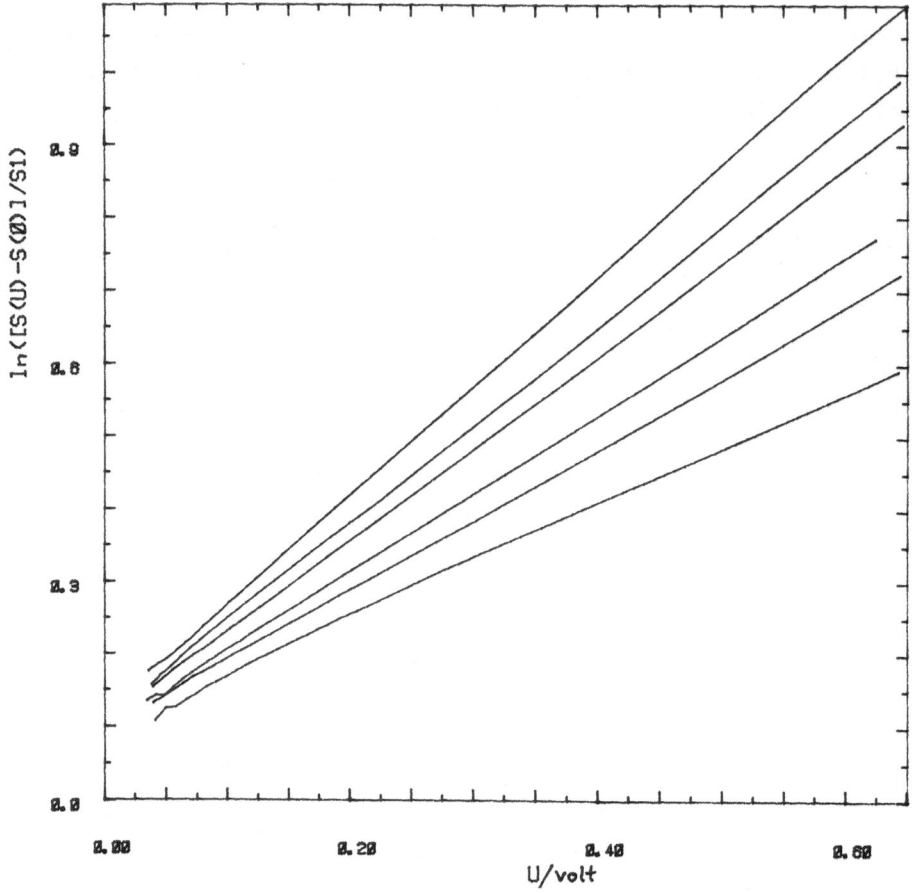

Figure 11. Deviations of the conductance from its value at zero applied voltage, plotted according to the theory of CDW-s.

Further possibilities that we have checked are local heating of the dislocation core and injection of carriers into the dislocation bands from the surface contacts. Of these, the heating effect is easy to calculate with the heat transfer coefficient between dislocation and bulk as a free parameter, but the curves obtained in this way have the wrong shape. Injection would qualitatively lead to a nonlinear I-V-dependence, but a quantitative analysis of this effect is very difficult and would require additional knowledge about the excitation spectrum of the dislocation states and about the influence of the intrinsic field in the depletion zone under the surface contacts on the Peierls transition. This information is not yet available.

In summary, at present we cannot provide a solid physical model to explain the nonlinearity and consider this question as a challenge for further theoretical work.

Outlook

The electronic properties of dislocations in Ge and Si have been studied for many years in specimens containing not only different types of dislocation but also deformation produced point defects and very short or curved dislocation segments. This creates ambiguities in the interpretation of experimental data.

Measurements of the AC-conductivity in deformed Ge show a temperature dependence which is much weaker than that of our dislocations which are 100 micrometers long and presumably very straight because they were generated at a rather low temperature. The reason for the difference is most likely the contribution of much shorter segments to the AC-conductivity in the deformed specimen. For short segments the lattice distortion which is associated with the Peierls transition continues beyond the segment length and requires additional positive elastic energy while the negative electronic energy contribution per unit length remains the same. Therefore the Peierls transition will be shifted to a lower temperature or even be suppressed altogether. The AC measurements provide only averages over a wide range of segment lengths, while our new measurements give us for the first time the opportunity to investigate a single type of well defined dislocations. With some additional effort it should even be possible to do measurement on one single dislocation rather than on a group of 40.

The excitation spectrum of disloctions which has also been a subject of investigation with ambiguous results can be studied in the future via measurements of the 1-dimensional photo conduction under illumination with photon energies lower than the band gap of the bulk. It seems possible that this holds not only for A- but also for B-dislocations, which show no one-dimensional conduction in the dark but may do so under illumination.

Unfortunately the situation seems to be somewhat less favourable in Si. AC-measurements in deformed material show a very small conductivity and, as mentioned before, a small angle grain boundary consisting of edge dislocations in the <100> direction had no measureable conductivity at all. Only if the dislocations in the deformed material were loaded with atomic hydrogen in a plasma, the dislocation conduction became comparable with that in Ge, although it was still lower. Of course the dislocations are subject to a drastic change by this treatment which probably saturates all dangling bonds in the core and leaves only levels which are bound by the deformation potential of the stress field. But, on the other hand, these modified dislocations could be interesting objects in their own right.

Acknowledgement: The authors are very grateful to V. Szkielko and G. Petermann of the Institut für Metallphysik in Göttingen who supplied the bicrystals for our GB measurements. J. Lüdecke performed the measurements of the conductance perpendicular to a GB in Germanium in the course of his diploma thesis.

References

1. B.M. Vul and E.I. Zavaritskaya, Inst. Phys. Conf. Series No. 43, London 1979, p 421

2. G. Landwehr and P. Handler, J. Phys. Chem. Solids 29, 891 (1962)

3. G. Landwehr, E. Bangert and S. Uchida, Solid State Electronics 28, 171 (1985)

4. R. Labusch and H. Thümmel, Iswestja Akademia Nauk SSSR, Seria Fizitsetskaja 4, moskow 1987, page 798

5. X.J. Wu, V. Szkielko, and P. Haasen, J. de Physique suppl. 10, vol.43, C1-135.

6. G. Petermann, Diploma thesis, Göttingen 1983

7. F.J. Morin and J.P. Maita, Phys. Rev. 94, 1525 (1954)

8. G. Petermann, Ph. D. thesis, Göttingen 1987

9. G. Petermann, phys. stat. sol. (a), 106, 535 (1988)

10. G. Petermann and P. Haasen, Proc. MRS meeting, Boston Dec. 1987

11. G. Döding and R. Labusch, phys. stat. sol.(a) 68, pages
    143 and 469 (1981)

12. V. Kveder, R. Labusch, and Yu. A. Ossipyan, phys. stat.
    sol.(a) 92, 293 (1985)

13. Yu. A. Ossipyan, Crys. Res. Technol. 16, 239 (1981)

14. H. Schaumburg, phys. stat. sol. 40, K1 (1970)

15. R. Labusch, Physica 117/18B, 203 (1983)

16. D. Mergel and R. Labusch, phys. stat. sol.(a) 41, 431
    and 42, 165 (1977)

17. W. Schröter, phys. stat. sol. 31, 177 (1969)

18. D. Allender, J.W. Bray and J. Bardeen, Phys. Rev. B9,
    119 (1974)

19. H. Fröhlich, Proc. R. Soc. A 223, 296 (1954)

20. P.A. Lee and T.M. Rice, phys. Rev. B 19, 3970 (1979)

21. J. Bardeen, Phys. Rev. Lett. 42, 1498 (1979) and Phys.
    Rev. Lett. 45, 1978 (1980)

22. R.M. Flemming, C.C. Grimes, Phys. Rev. Lett. 42, 1423
    (1979)

23. P. Monceau, J. Richard, M. Renard, Phys. Rev. B 25, 31
    (1982)

24. J. Richard, P. Monceau, M. Renard, Phys. Rev. B25, 948
    (1982)

POINT DEFECTS IN GaAs

E.R. Weber, K. Khachaturyan, M. Hoinkis and M. Kaminska

Department of Materials Science and Mineral Engineering
University of California, Berkeley, CA 94720, USA, and
Center for Advanced Materials, Lawrence Berkeley Laboratory
Berkeley, CA 94720, USA

INTRODUCTION

In the last five years, rapid progress has been made in the identification of electrically active defects in GaAs and related compounds. Part of this progress is due to the application of spectroscopic experimental techniques sensitive to defect symmetries, such as electron paramagnetic resonance, which has been successfully used to identify a large number of defects in silicon. State-of-the art calculations of low-symmetry defect configurations have also been instrumental in this progress, demonstrating that metastability of defects in covalent crystals can be due to isolated point defects which undergo a symmetry-breaking lattice relaxation. In this paper we discuss the microscopic identity and new results concerning the electronic properties of some of the most important defects in GaAs and related compounds.

DEEP/SHALLOW DONORS IN COMPOUND SEMICONDUCTORS

In some binary and many ternary compound semiconductors deep donor states are observed in dopants, which are expected to create shallow donor levels, as they replace host atoms from the adjacent column of the periodic table. The most famous examples of such defects are the "DX-centers" in AlGaAs[1]. Such deep donor states of usually shallow donor atoms have been well known for more than twenty years. Examples include IV_{III} or VI_V donors in III/V semiconductors, such as S in GaSb[2], and S, Se and Te in GaAsP[3], or III_{II} donors in II/VI semiconductors, such as Ga, In or Tl in ZnS.[4] However, the most technologically relevant donors of this kind are IV_{III} or VI_V donors in AlGaAs. Lang et al.[5] investigated the properties of Si, Se and Te in AlGaAs in detail. Deep donor states are found for all donors in n-$Al_xGa_{1-x}As$ for x>0.2. This deep state can be transformed into a shallow state upon illumination, which cannot be reversed at low temperatures. After removal of light, persistent photoconductivity is observed at low temperatures. The large difference between the photoionization threshold energy and the thermal level position, as measured e.g. by Hall effect, implies that the deep ground state in this case is accompanied by a lattice relaxation. The barrier against re-capture of the photo-excited carriers into the lattice-relaxed ground state can only be overcome by warming up the sample so that thermally activated capture is possible. Lang et al. proposed that such large lattice relaxation indicates a defect complex (donor "D" with an unknown partner "X"). Today it is widely accepted that this behavior is characteristic of the isolated donor atom in the specific compound band structure. This conclusion is mainly based on observation of such shallow/deep transitions even in n-GaAs under hydrostatic pressure,[6, 7] similar to those observed previously in in GaAs:S,[8] GaSb:Se, Te,[2] and n-CdTe.[9]

However, it is difficult to understand how an isolated, substitutional donor can undergo a lattice relaxation upon the capture of a single carrier. Some authors suggested that the lattice relaxation might be small[10, 11] ("inner crossing," see Fig. 1a), but a recent detailed analysis of the stress coefficients of both capture and emission processes led to the conclusion that the deep donors in GaAs and AlGaAs undergo a large lattice relaxation[7] ("outer crossing," see Fig. 1b). In the following it will be shown, that this is understandable if such donors have a negatively charged ground state, capturing two electrons instead of one as in the case of "normal" shallow donors.

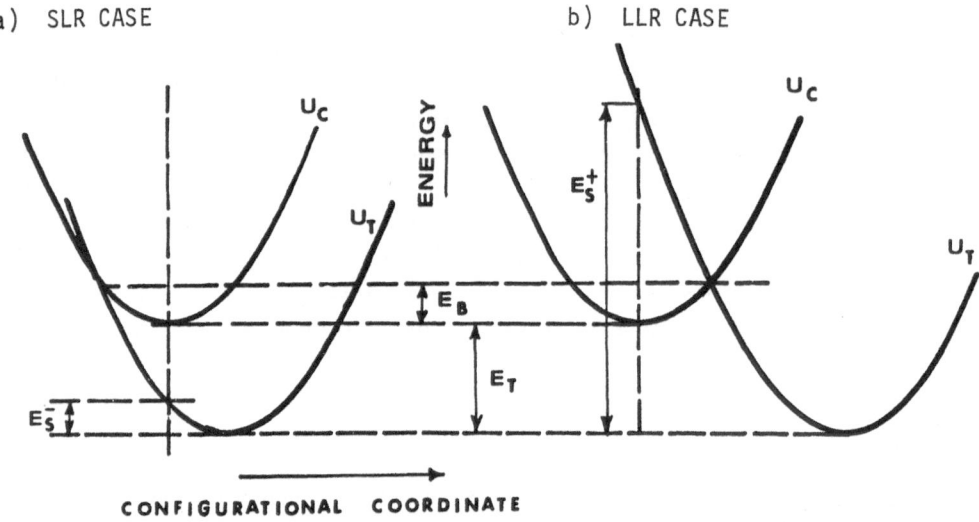

Figure 1. Schematic configuration coordinate diagrams of a defect with a) small, b) large lattice relaxation of the ground state.[7] Note that both diagrams can explain the same capture barrier E_B and thermal energy difference E_T between relaxed and unrelaxed state.

NEGATIVE-U MODEL OF METASTABLE SHALLOW DONORS

It is well known that lattice relaxation can lead to electron pairing[12], resulting in a two-electron D⁻ ground state with a negative Hubbard electron correlation energy U:

$$2D^0 \rightarrow D^+ + D^-$$ (1)

Whether or not shallow/deep donors will undergo this disproportionation reaction can be theoretically predicted by comparing the lattice relaxation energy needed to produce the experimentally-observed persistent photoconductivity with the lattice relaxation energy required for negative U pairing.

Toyozawa described the lattice-relaxed state as extrinsic self-trapping of a carrier with lattice relaxation. The necessary condition for this process is[13]

$$-e^2/(\varepsilon_o a) - W + T_{kin}(a) < \quad E_{el-ph} \tag{2}$$

Negative electron correlation energy can be observed if the lattice relaxation energy gained by the capture of two carriers is larger than the electronic repulsion between these carriers[12]

$$1/2e^2/(\varepsilon_o <r_{el-el}>) < \quad E_{el-ph} \tag{3}$$

In eqs. (2) and (3), $e^2/(\varepsilon_o a)$ and W are Coulombic and short-range parts of the donor impurity potential, $e^2/(\varepsilon_o <r_{el-el}>) \approx e^2/\varepsilon_o a$ is the averaged electron-electron repulsion energy in the localized state, and $T_{kin}(a)$ is the kinetic energy in the localized state.

We have shown recently[14, 15, 16] that if (i) the Bohr radius r_b of the metastable hydrogenic state is much greater than the interatomic distance a, and (ii) the defect is an isolated substitutional donor with electronegativity close to that of the host lattice atom for which it substitutes (W is small), then (2) is more restrictive than (3), i.e., if the lattice relaxation energy E_{el-ph} is large enough to result in extrinsic self-trapping (2), it should be also large enough to cause negative U.

This allows us to propose that **all impurities with a shallow hydrogenic state ($r_b >> a$) and weak short-range potential W, which show extrinsic self trapping (as evidenced e.g. by persistent photoconductivity) are expected to be negative-U centers.** This conclusion applies particularly to DX-centers in AlGaAs. Figure 2 shows a schematic configuration coordinate diagram of a negative-U donor with a lattice- relaxed negatively charged ground state.

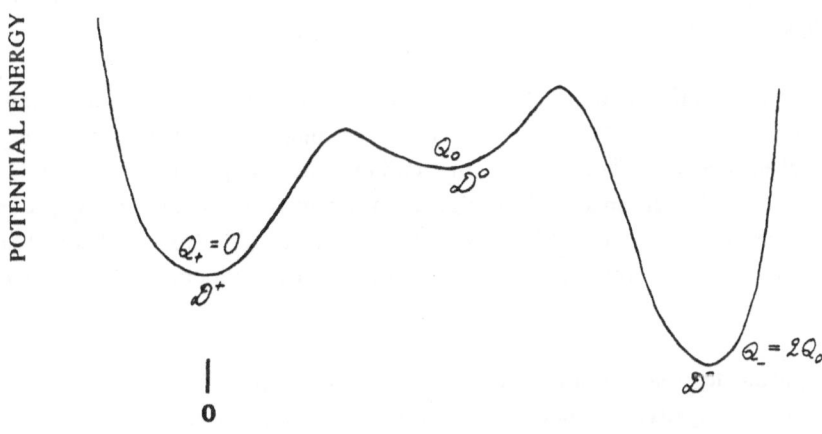

DEFECT CONFIGURATION COORDINATE Q

Figure 2. Configuration coordinate diagram of of a negative-U donor impurity.

A number of investigators have investigated the mechanism of the thermally activated carrier capture into the deep, lattice-relaxed state.[5, 17, 18, 19, 20] Figure 2 shows that the capture barrier is related to the energy of an intermediate state, which is connected with the neutral donor state. The energy of this state may appear to be related to one of the minima of the conduction band, i.e., the L-minimum in the direct bandgap region and the X-minimum in the indirect bandgap region, as recently postulated,[21] though it will generally be derived from a mixture of the conduction band minima. In this way, the negative-U model easily reconciles opposing interpretations of experimental results.[18, 20, 21]

EVIDENCE FOR THE NEGATIVE-U MODEL

Räuber and Schneider seem to have been the first to experimentally observe negative-U behavior in a semiconductor[4]. They detected no EPR signals due to Ga, In or Tl donors in ZnS after cooling down in the dark. Subsequent photo-excitation, however, did produce such signals. They explained their observations on the basis of co-existing equal concentrations of monovalent and trivalent donors after cooling down in the dark (corresponding to positively and negatively charged group III atoms on group-II sites, eq. (1)), whereas they found that the neutral, paramagnetic divalent charge state was metastable, populated only after photo-excitation.

We performed a detailed EPR study on MBE-grown 10μm thick $Al_{0.36} Ga_{0.64}As$ doped with $10^{17}cm^{-3}$ Si. Five samples were sandwiched together for increased sensitivity. Though the total number of expected spins is well above the sensitivity limit of the EPR spectrometer, and donor EPR signals in semiconductors are generally easy to observe, no EPR signal due to neutral donors could be detected after cooling down in the dark. Illumination with white light resulted in the formation of a high concentration of persistent free carriers. These carriers decrease the quality factor Q of the microwave resonator, so that it could not be decided whether neutral, metastable donor states can be populated.[22] A metastable EPR and ODMR signal in the indirect bandgap region of n-AlGaAs has repeatedly been reported,[23, 24, 25, 26] and interpreted to be the hydrogenic state related to the X-minimum, the lowest minimum of the conduction band in indirect AlGaAs.

The lack of observation of an EPR signal of neutral deep donors after cooling down in the dark in ZnS,[4] AlGaAs,[22, 25, 26] and ZnCdTe[22] is a strong argument in favor of negative-U behavior of these shallow/deep donors. In fact, in his original paper on negative-U centers, Anderson[12] used the absence of EPR signals of defects in semiconducting glasses as an argument for a negative-U model for these defects. In the case of ZnS, Räuber and Schneider reached essentially the same conclusion.[4] We discuss further experimental evidence for this model below.

The fact that an electron pair rather than a single electron is involved in capture and emission processes in the negative-U model changes the relationship between emission and capture barriers on one hand, and the thermal depth of a donor, on the other hand. In the positive-U case with single carrier capture and emission, the difference between the emission energy E_e and capture barrier E_c, as measured e.g. by DLTS, corresponds to the thermal depth of the donor evident from Hall effect measurements. Since the binding energy of an electron pair is twice the thermal depth of the donor, the following holds in a negative U model

$$E_e - E_c = 2E_{th} \tag{4}$$

This relationship has been confirmed in an experimental example of $Al_{0.35}Ga_{0.65}As:Si$:[19]

carrier emission barrier $E_e = 0.4eV$, capture barrier $E_c = 0.2eV$, corresponding to a level position $E_{th} = 0.1eV$. The same was found to hold for ZnCdTe,[22] with carrier emission barrier $E_e = 0.29eV$, capture barrier $E_c = 0.13eV$, and the level position was determined by Hall effect to be $E_{th} = 0.07eV$. These results were obtained in both cases with E_{th}/k_B as the slope of the Arrhenius plot of the carrier concentration.

It was recently shown that for a negative-U donor the slope of an Arrhenius plot of the carrier concentration log n vs. $1/T$ is always E_{th}/k_B, both when $n > N_A$ and $n < N_A$ (N_A: concentration of acceptors).[15, 27] This is in contrast to the positive-U case when the slope is E_{th}/k_B if $n < N_A$ but is $E_{th}/(2k_B)$ if $n > N_A$. To the authors' best knowledge, no change of slope of such plots usually derived from Hall effect measurements has ever been reported. Usually, this problem was attributed to an assumed strong compensation[28], though this assumption should not hold over as wide a concentration range as experimentally observed.

Experimental evidence in favor of the negative-U model comes also from measurements of carrier mobility in samples containing deep/shallow donors. The low-temperature mobilities after cooling down in the dark are very small, and frequently an increase in mobility is observed when deep donors in AlGaAs[3, 28, 29] or ZnCdTe[30] are ionized by light or temperature increase. This mobility behavior cannot satisfactorily be explained by a positive U, D^0 ground state model. In the negative-U model, the ionization process $D^- \longrightarrow D^+ + 2e^-$ does not change the total number of scattering ions. The observed increase in mobility after ionization may be due to the increase in screening caused by greater free electron concentration.

Finally, recent pseudopotential calculations by Chadi and Chand[31] showed that neutral Si_{Ga} and S_{As} donors in GaAs are stable, whereas for negatively-charged centers a metastable state exists with a large motion of the Si atom (for Si_{Ga}) or a Ga-neighbour (for S_{As}) in the antibonding direction. Chadi and Chand predict that in GaAs subjected to hydrostatic pressure and in AlGaAs this metastable state becomes the ground state, which would essentially describe a negative-U behavior as discussed above.

INFLUENCE OF DEEP/SHALLOW INSTABILITIES ON DEVICE PERFORMANCE

The DX-centers found in $n-Al_xGa_{1-x}As$ for Al concentrations larger than 22% are at present the technologically most relevant deep/shallow donor instability in semiconductors. According to the model discussed above, a DX-center does not contain an unknown "X" component. It is, rather, an isolated IV_{III} or VI_V donor, which can capture an electron pair and reach a new, lattice-relaxed ground state. This conclusion has important technological consequences: these defects can no longer be regarded as unintended deep level defects, as they constitute the basic n-doping of AlGaAs layers. The presence of the deep donor ground state can be observed in fast modulation-doped FETs (MODFETs), which make use of a two-dimensional electron gas at the AlGaAs/GaAs interface, the carriers for which are supplied from a n-AlGaAs layer. At room temperature, carrier capture and emission related to the deep donors results in slow transients, and at low temperatures, which should offer the fastest MODFET performance, carrier freeze-out can prevent device operation completely.

The lattice-relaxed deep donor level is even present as a resonant state in the conduction band in $Al_xGa_{1-x}As$ with x<0.22[32] and in n-GaAs.[33] It can be populated by high doping, when the Fermi level shifts into the conduction band. The negative-U character of the deep level will result in an effective Fermi level pinning at this energy, which might explain why the maximum obtainable electron concentration in GaAs does not exceed about $10^{19}cm^{-3}$, independent of the donor used. A similar explanation has been proposed which does not take into account the negative-U character of this level.[33] Another consequence of the existence of the deep donor level in the conduction band for x<0.22 is the increase of the apparent deep level concentration with doping observed in n-$Al_xGa_{1-x}As$ near x=0.2.[34] For x=0.195, all Si donors were found to be in the shallow state up to a concentration of $10^{17}cm^{-3}$, whereas for higher dopant concentrations an increasing population of the deep donor state was observed. This effect can be ascribed to the shift of the Fermi level with higher doping concentration. Therefore, caution is required in designing devices in which the deep donor state in the AlGaAs layer is avoided by a reduction of Al concentration below 20%: at the consequent high carrier concentration, the unwanted effects of the deep donor state will be noticeable.

EL2 IN GaAs

Another defect of great technological relevance is EL2 in GaAs. This intrinsic midgap donor is found in As-rich bulk or epitaxial material. It determines the Fermi level in GaAs, if its concentration is larger than the concentration of acceptors, which in turn has to be larger than the concentration of residual shallow donors. With the Fermi level pinned at midgap, GaAs can reach resistivities well above $10^6 \, \Omega cm$, so that it is semi-insulating (s.i.). Details of the defects involved in the complex compensation mechanism of s.i. GaAs have been discussed elsewhere.[35]

The presence of EL2 defects in s.i. GaAs results in an optical absorption in the energy region from 0.75eV up to the GaAs band gap with a characteristic band peaked around 1.2eV.[36] Careful optical absorption measurements revealed an additional, narrow zero-phonon line at 1.04eV followed by phonon replicas related to an EL2 intracenter transition.[37] The magnitude of the absorption band is mainly determined by the concentration of *neutral* EL2 defects, but it is nevertheless frequently used for the determination of *total* EL2 concentrations. More precise measurements are possible taking into account the absorption cross-sections of EL2 in various charge states.[38] Further details of the electronic properties of EL2 can be found in recent reviews.[39, 40, 41]

The stoichiometry dependence of the EL2 concentration suggests that native defects typical for As-rich GaAs, such as V_{Ga} or As_{Ga} antisite defects, are constituents of EL2. Electron paramagnetic resonance measurements with monochromatic illumination (photo-EPR) indeed showed a midgap donor level at $E_v+0.75eV$ for As_{Ga} antisite defects.[42] However, a group-V atom on a group-III site is expected to be a double donor, and the photo-EPR measurements showed a second donor level of As_{Ga} at $E_v+0.52eV$. The relation between the As_{Ga} double donor and the EL2 donor was solved with the identification of a second donor level for EL2 as well,[43, 44] and today it is generally accepted, that an As_{Ga} antisite defect is the main constituent of EL2.

One of the most characteristic properties of EL2 is its metastability. That is, illumination of s.i. GaAs with light corresponding to the EL2 absorption band at a temperature below 130K leads to optical quenching of the EL2 absorption spectrum,[36] which can be recovered by warming up above 130K. This optical quenching and the thermal recovery of the EL2 level can also be observed electrically by photocapacitance measurements[45] and by photo-EPR measurements of the As_{Ga}^+ spectrum.[46] A detailed investigation of the photoquenching transients of the EPR signal showed that the As_{Ga} defect had first to be brought into the neutral charge state before it could be quenched into the metastable state. This behavior can be attributed to a transition of the EL2 defect upon illumination into a state with large lattice relaxation.

The observation of a large lattice relaxation was for a long time considered to mean that EL2 is a complex defect. The EPR measurements of As_{Ga}^+ show an isotropic four-line quadruplet with a linewidth of about 250G. These lines are not sensitive to small perturbations in the proximity of the As_{Ga}^+ defect, such as another defect outside the first nearest neighbour shell of atoms. Based on rather indirect evidence, several complex models were suggested for EL2, such as $V_{As}V_{Ga}As_{Ga}$[47] or $As_{Ga}-As_i$[48] complexes. Optically detected electron nuclear double resonance (OD-ENDOR) measurements have been interpreted in favor of the As_{Ga}-As_i model.[49]

However, recent self-consistent calculations of the electronic structure of the As_{Ga}-As_i defect showed clearly that this complex of an As_{Ga} double donor with an As_i single donor is a *triple donor* in GaAs, with the first level very close to the conduction band, and the second donor level at midgap.[50] These calculations could explain the metastability of EL2 by motion of the As_i atom. However, having the *second* donor level of EL2 at midgap contradicts the widely observed compensation mechanism of GaAs.[35] This mechanism requires as high an EL2 concentration as possible. In the case of the As_{Ga} - As_i model a *second* donor midgap level would dominate the Fermi level only if the crystal contained acceptors in a concentration exactly *between once and twice* the EL2 concentration, which seems to be quite unlikely for unrelated defects. On the other hand, high EL2 concentrations would make the material n-type in contradiction with experimental obeservation. A further argument against the As_{Ga}-As_i complex as a model for EL2 is the observed thermal stability of EL2 up to about 1000°C. Two donors cannot attract each other by Coulomb forces, and it is very unlikely that lattice deformation can provide a bond strong enough to keep As_i near the As_{Ga} antisite. Further arguments in this discussion were presented in a recent review by Baraff.[51]

On the other hand, the uniaxial stress measurements of Kaminska et al.[52] showed that the zero phonon line of EL2 corresponds to a transition from an A_1 ground state to a T_2 excited state, implying T_d symmetry for the ground state. This experiment was recently reproduced with the same result.[53]

Moreover, two independent pseudopotential calculations[54, 55] showed convincingly that an isolated As_{Ga} antisite defect can undergo a transition into a metastable state upon optical excitation, which can be described as a reaction:

$$As_{Ga} + h\nu \quad \rightarrow \quad V_{Ga} + As_i \tag{5}$$

According to these calculations, the photon needs to have a minimum energy of about 1 eV in order to transfer an electron of $As_{Ga}°$ into the excited state. In the excited state, there is a probability for the process (5) to take place. Upon thermal activation, As_i is recaptured by V_{Ga}, effectively reversing this process. This mechanism, which had been proposed before,[56] provides an explanation of the metastability of EL2 in Kaminska's model which is analogous to that of the deep/shallow donor metastability. In addition, it has the great advantage of being the simplest model under consideration for EL2.

The discussion of the correct model for EL2 is not yet closed. The OD-ENDOR results[49] and a number of other experimental observations[51] are still not well explained. The near future might bring the final answer to the question of the identity of EL2.

Anion antisite defects in GaAs seem to play other important roles. They appear to be involved in the mechanism of Schottky barrier formation[42, 46, 57, 58] and very recently were detected for the first time in high concentration in low-temperature grown As-rich MBE layers (Fig. 3),[59] which show very promising device isolation properties.[60]

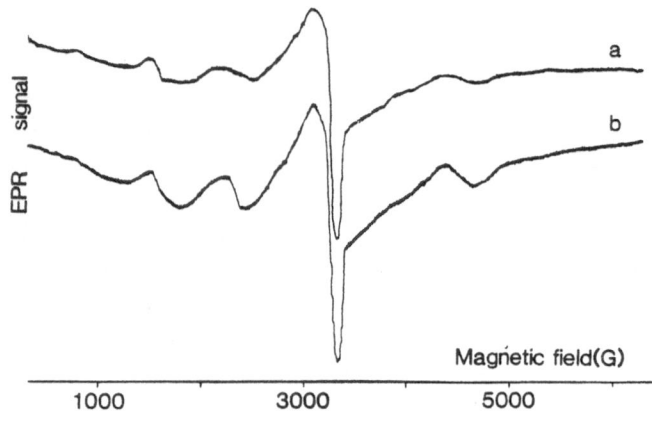

Figure 3. EPR spectrum of As_{Ga}^+-defects in MBE GaAs/GaAs(100) grown under As-rich conditions at 200°C (a) after cooling down in the dark, (b) after illumination. The strong line near 3200G is cavity background.[59]

BE 1 IN GaAs

The last part of this chapter introduces a defect which is not yet very widely recognized, but might be of considerable interest in the near future. EPR measurements of semi-insulating GaAs frequently show a very strong spectrum after photo-quenching of the As_{Ga}^+-spectrum (see Fig. 4).[39] This spectrum was labeled GaAs-BE1 and seems to be identical with the FR1 spectrum described by Baeumler et al.[61]

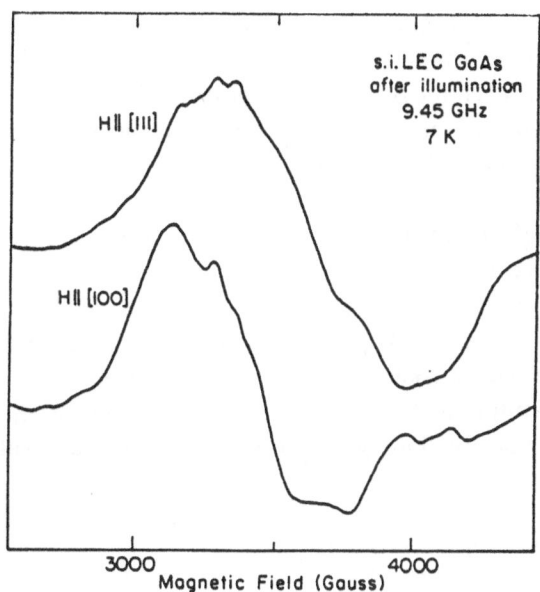

Figure 4. EPR spectrum of GaAs-BE1 in two orientations of the crystal to the magnetic field.[39]

The BE1 spectrum shows a distinct, but hitherto unresolvable anisotropy.[62] The signal disappears after warming the sample to 90K, well before the re-appearance of the As_{Ga}^+-spectrum, which excludes any direct correlation of these two defects. The strength of the EPR signal is due in part to the fact that this signal is an electric rather than a magnetic dipole transition. Comparison of a number of as-grown and annealed GaAs crystals shows that this spectrum is largest after a quench from high temperatures, whereas it disappears during slow cooling down or upon annealing at 450°C. The appearance in quenched and in annealed samples is similar to that of a vacancy-related defect found in high concentrations ($10^{17} cm^{-3}$ range) in GaAs.[63] The microscopic nature of this defect is not yet clear; however, it is worth noting that there seems to be a correlation between the occurence of a strong BE1 spectrum in s.i. GaAs substrate material and poor device performance of this material.[64] Further work on this "dark horse defect" is clearly required.

CONCLUSION

A number of isolated donor impurities in III-V and II-VI semiconductor compounds an alloys which would be expected to create shallow donor levels have deep ground states because of extrinsic self-trapping, as evidenced by effects such as persistent photo-conductivity at low temperatures. Based on experimental evidence and simple theoretical considerations we propose that the extrinsic self-trapping is due to capture of two carriers, which can lead to a strong, symmetry-breaking lattice relaxation. This applies specifically to DX centers in AlGaAs. This conclusion is supported by experimental observations and the recent pseudopotential calculations of Chadi and Chang[31] for the case of deep/shallow donors in GaAs. Previous work on donors in ZnS[4] leads essentially to the same conclusion. We expect that these are only a few examples of a whole class of defects in III/V and II/VI semiconductors showing extrinsic self-trapping and persistent photoconductivity.

The metastable EL2 defect in GaAs might also be an isolated anion antisite defect undergoing a lattice relaxation in the neutral state, i.e., when filled with two carriers, after additional optical excitation. However, the final word on the detailed model for EL2 is not yet spoken. Besides their involvement in the EL2 defect, anion antisite defects are ubiquituous native defects, which are of relevance in a wide range of GaAs applications.

BE1 in GaAs is a defect which deserves further attention, as it seems to be at present unnoticed in high concentrations in state-of-the-art GaAs.

These examples show considerable progress in the detailed understanding of the nature of defects in GaAs and related compounds and of their fundamental properties. Especially encouraging for the future is the interplay between experiment and theory, which now permits the calculation of even low-symmetry defect configurations with great reliability.

ACKNOWLEDGEMENT

This work was supported in part by the Director, Office of Energy Research, Office of Basic Energy Sciences, Materials Science Division of the U.S. Department of Energy under Contract DE-AC03-76SF00098 and by the Air Force Office of Scientific Research.

REFERENCES

1. D. V. Lang and R. A. Logan, *Phys. Rev. Lett.* **39**, 653, 1977.
2. B. Kosicki, W. Paul, A. Strauss and G. W. Iseler, *Phys. Rev. Lett.* **17**, 1175, 1966.
3. M. G. Craford, G. E. Stilman, J. A. Rossi and N. Holonyak, *Phys. Rev.* **168**, 867, 1968.
4. A. Räuber and J. Schneider, *Phys. Stat. Solidi.* **18**, 125, 1966.
5. D. V. Lang, R. A. Logan and M. Jaros, *Phys. Rev.* **B19**, 1015, 1979.
6. M. Mizuta, M. Tachikawa, H. Kukimoto and S. Minomura, *Jpn. J. Appl. Phys.* **24**, L143, 1985.
7. M. F. Li, P. Y. Yu, E. R. Weber and W. L. Hansen, *Appl. Phys. Lett.* **51**, 349, 1987.
8. W. E. Howard and W. Paul, 1960, unpubl. in H. Ehrenreich, *Phys. Rev.* **120**, 1951, 1960.
9. A. G. Foyt, R. E. Halsted and W. Paul, *Phys. Rev. Lett.* **16**, 55, 1966.
10. H. P. Hjalmarson and T. J. Drummond, *Appl. Phys. Lett.* **48**, 656, 1986.
11. J. C. M. Henning and J. P. M. Ansems, *Semicond. Sci. Technol.* **2**, 1, 1987.
12. P. W. Anderson, *Phys. Rev. Lett.* **34**, 953, 1975.
13. Y. Toyozawa, *Physica.* **116B**, 7, 1983.
14. K. Khachaturyan, E. R. Weber and M. Kaminska, LBL-report 24994, March 1988.
15. K. Khachaturyan, E. R. Weber and M. Kaminska, Proc. 15th Internat. Conf. on Defects in Semiconductors, Budapest 1988, in print.
16. K. Khachaturyan, E. R. Weber and M. Kaminska, 1989, to be publ.

17. T. N. Theis, T. F. Kuech, L. F. Palmateer and P. M. Mooney,
    Inst. Phys. Conf. Ser. **74**, 241, 1984.

18. D. V. Lang, in: "Deep Centers in Semiconductors," ed. S.T. Pantelides (Gordon and Breach, New York 1986) p. 489.

19. P. M. Mooney, E. Calleja, S. L. Wright and M. Heilblum, in: "Defects in Semiconductors," ed. H.J.v. Bardeleben (Trans Tech, Aedermannsdorf 1986) p. 417.

20. T. N. Theis, *Inst. Phys. Conf. Ser.* **91**, 1, 1987.

21. M. F. Li, W. Shan, P. Y. Yu, W. L. Hansen, E. R. Weber and E. Bauser, Proc. 15th Internat. Conf. on Defects in Semiconductors, Budapest 1988, in print.

22. K. Khachaturyan, M. Kaminska, E. R. Weber, P. Becla and R. A. Street, 1988, to be publ.

23. C. Weisbuch and C. Herman, *Phys. Rev.* **B15**, 816, 1977.

24. E. Glaser, T. A. Kennedy and B. Molnar, Proc. Third Internat. Conference on Shallow Impurities, Linkopping 1988, in print.

25. P. M. Mooney, W. Wilkening, U. Kaufmann and T. F. Kuech, 1989, to be publ.

26. H. J. v. Bardeleben, D. Stievenard and J. C. Bourgoin, 1989, to be publ.

27. G. Baraff, unpubl.

28. N. Chand, T. Henderson, J. Klem, D. Masselink, R. Fischer and H. Morkoc, *Phys. Rev.* **B30**, 4481, 1984.

29. A. K. Saxena, *Appl. Phys. Lett.* **36**, 79, 1980.

30. B. C. Burkey, R. P. Koshla, J. R. Fisher and D. L. Losee, *J. Appl. Phys.* **47**, 1095, 1976.

31. D. J. Chadi and K. J. Chang, *Phys. Rev. Lett.* **61**, 873, 1988.

32. T. N. Theis, B. D. Parker, P. M. Solomon and S. L. Wright, *Appl. Phys. Lett.* **49**, 1542, 1986.

33. P. M. Mooney, T. N. Theis and S. L. Wright, *Inst. Phys. Conf. Ser.* **91**, 1988.

34. T. Ishikawa, T. Yamamoto and K. Kondo, *Jpn. J. Appl. Phys.* **25**, L484, 1986.

35. E. R. Weber and M. Kaminska, in: "Semi-Insulating III-V Materials," ed. G. Grossmann and L. Ledebo (Adam Hilger, Bristol and Philadelphia 1988) p. 111.

36. G. M. Martin, *Appl. Phys. Lett.* **39**, 747, 1981.

37. M. Kaminska, M. Skowronski, J. Lagowski, J. M. Parsey and H. C. Gatos, *Appl. Phys. Lett.* **43**, 302, 1983.

38. P. Silverberg, P. Omling and L. Samuelson, in: "Semi-insulating III-V Materials," ed. G. Grossmann and L. Ledebo (Adam Hilger, Bristol 1988) p. 369.

39. E. R. Weber and P. Omling, in: "Festkörperprobleme/ Advances in Solid State Physics XXV," ed. P. Grosse (Vieweg, Braunschweig 1985) p. 623.

40. G. M. Martin and S. Makram-Ebeid, in: "Deep Centers in Semiconductors," ed. S.T. Pantelides (Gordon and Breach, New York 1986) p. 399.

41. M. Kaminska, *Phys. Scr.* **T19B**, 551, 1987.

42. E. R. Weber, H. Ennen, U. Kaufmann, J. Windscheif, J. Schneider and T. Wosinski, *J. Appl. Phys.* **53**, 6140, 1982.

43. J. Lagowski, D. G. Lin, T. P. Chen, M. Skowronski and H. C. Gatos, *Appl. Phys. Lett.* **47**, 929, 1985.

44. J. Osaka, H. Okamoto and K. Kobayashi, in: "Semi-insulating III-V Materials," ed. H. Kukimoto and S. Miyazawa (North-Holland, Amsterdam 1986) p. 421.

45. G. Vincent and D. Bois, *Sol. State Commun.* **27**, 431, 1978.

46. E. R. Weber and J. Schneider, *Physica.* **116B**, 398, 1983.

47. J. F. Wager and J. A. V. Vechten, *Phys. Rev.* **35**, 2330, 1987.

48. H. J. v. Bardeleben, P. Stievenard and J. C. Bourgoin, *Appl. Phys. Lett.* **47**, 970, 1985.

49. B. K. Meyer, D. M. Hofmann, J. R. Niklas and J. M. Spaeth, *Phys. Rev.* **B36**, 1332, 1987.

50. G. A. Baraff, M. Lannoo and M. Schlüter, *Mat. Res. Soc. Proc.* **104**, 375, 1988.

51. G. Baraff, Proc. 19th International Conference on the Physics of Semiconductors, Warsaw 1988, in print.

52. M. Kaminska, M. Skowronski and W. Kuszko, *Phys. Rev. Lett.* **55**, 2204, 1985.

53. K. Bergmann, P. Omling, L. Samuelson and H. G. Grimmeiss, in: "Semi-insulating III-V Materials," ed. G. Grossmann and L. Ledebo (North Holland, Amsterdam 1988) p. 397.

54. D. J. Chadi and K. J. Chang, *Phys. Rev. Lett.* **60**, 2187, 1988.

55. J. Dabrowski and M. Scheffler, *Phys., Rev. Lett.* **60**, 2183, 1988.

56. W. Kuszko, P. J. Walczak, P. Trautman, M. Kaminska and J. M. Baranowski, *Materials Science Forum.* **10-12**, 317, 1986.

57. W. E. Spicer, T. Kendelewicz, N. Newman, R. Cao, C. McCants, K. Miyano, I. Lindau and E. R. Weber, *Appl. Surf. Science.* **33/34**, 1009, 1988.

58. E. R. Weber, W. E. Spicer, N. Newman, Z. Liliental-Weber and T. Kendelewicz, Proc. 19th Internat. Conf, on the Physics of Semiconductors, Warsaw 1988, in print.

59. M. Kaminska, Z. Liliental-Weber, E. R. Weber, T. George, J. B. Kortright, F. W. Smith, B. Y. Tsaur and A. R. Calawa, 1989, to be publ.

60. F. W. Smith, A. R. Calawa, C. L. Chen, M. J. Manfra and L. J. Mahoney, *IEEE Electron. Dev. Lett.* **EDL-9**, 77, 1988.

61. M. Baeumler, U. Kaufmann and J. Windscheif, in: "Semi-insulating III-V Materials," ed. H. Kukimoto and S. Miyazawa (North Holland, Amsterdam 1986) p. 361.

62. M. Hoinkis and E. R. Weber, in: "Semi-insulating III-V Materials," ed. G.Grossmann and L. Ledebo (Adam Hilger, Bristol and Philadelphia 1988) p. 43.

63. J. S. Dannefaer, P. Mascher and D. Kerr, in: "Defect Recognition and Image Processing in III-V Compounds," ed. E.R. Weber (Elsevier, Amsterdam 1987) p. 313.

64. M. Hoinkis, E. R. Weber and R. Zuleeg, 1989, to be publ.

CHANGES OF ELECTRICAL PROPERTIES OF SILICON CAUSED BY PLASTIC DEFORMATION

Helmut Alexander

Abteilung für Metallphysik im II. Physikalischen Institut
Universität Köln
D-5000 Köln 41

INTRODUCTION

When the interest in the electrical properties of dislocations in semiconductors started some 30 years ago plastic deformation was just a technique to introduce a dislocation density into the crystal sufficient high for producing measurable effects.
Mainly by EPR spectroscopy it became clear that standard deformation procedures generate a surprisingly large number of thermally stable point defect clusters in the bulk of the crystal /1/. In some cases it was possible to show that the physical properties of these clusters are identical irrespective of their origin from neutron irradiation or from plastic deformation. So it is now generally accepted that deformation induced changes of crystal properties cannot be just attributed to dislocations; they rather have to be carefully traced back to one of three species of defects: point defects in the bulk, point defects near or within dislocations and pure dislocations. Obviously this is a difficult task. The situation became even more intricate when we learned, also from EPR, that dislocations of a certain type are not characterised by a given density of paramagnetic centers per unit length: rather this number develops more or less proportional to the distance the dislocation has moved and depends on the circumstances of the deformation, as stress, temperature, protecting gas etc..From that we have to accept that physical properties of dislocations are not intrinsic but depend on history. In other words: to compare different crystals one has to know their complete previous processing. As a cosequence as many as possible investigations should be done on one and the same crystal.
In a certain sense today we feel further away from the ultimate aim to understand the properties of pure dislocations and their response to outer variables than we did decades ago. But, of course, we are much more aware of facts and much nearer to actual processes which also are of some importance to production and use of semiconductor devices.
As follows from the foregoing it is not very reasonable at the present state to speak on properties of dislocations alone. We first have to understand what is going on during plastic deformation.

ELECTRIC PROPERTIES INFLUENCED BY PLASTIC DEFORMATION

Activities by which the defects mentioned in the introduction influence the electrical properties of a crystal can be classified in the following manner: the defects can act as
1) Structural dopants

i. e. they can be donors and/ or acceptors with a certain level in the
energy gap,
2) Recombination centers
influencing the life time of (minority) carriers,
3) Scattering centers
determining the mean free path of charge carriers, i.e. their mobility
4) Probably certain dislocations introduce onedimensional bands into the
energy gap, so providing quasimetallic contributions to the conductivity.
5) Additionally to these direct influences there are indirect ones:
modification of the spatial distribution and charge state of impurity
and dopant atoms or incorporation of new impurities from outside etc..

Clearly it is impossible to summarise in short the understanding we have
reached in all these fields. Instead I will tackle some particular cases
from which we learn something on basic problems. First I have to review
what we know from EPR and DLTS spectroscopy. Then I will discuss in which
way straight dislocations cause low temperature microwave conductivity.

EPR AND DLTS SPECTROSCOPY

Electron paramagnetic resonance (EPR) of defects in silicon is an
inexhaustible source of knowledge. Also on plastic deformation it yields
next to electron microscopy the most clear-cut results.Detailed analysis
of the spectra of crystals deformed under different conditions brought to
light the following facts:
Socalled standard deformation by single slip (T = 650°C, $\varepsilon \approx$ 5%, resolved
shear stress τ= 30 MPa) produces paramagnetic centers the density of
which is of the order 10^{16} cm^{-3}. The majority (65 - 80%) are point
defect clusters of high thermal stability /2/. The remaining centers by
their anisotropy are related to the dislocation geometry /3/. The
distinction between the two classes of paramagnetic defects , for a long
time being issue of an obstinate debate /4/, can be made by several
methods: Taking advantage of the spin lattice relaxation time T_1 being
different by orders of magnitude Kisielowski-Kemmerich /5/ succeeded to
separate the spectra completely. This technique allows to demonstrate the
differences of anisotropy most clearly /2/. On the other hand using
unusual passage conditions he could show both parts of the spectrum,
usually recorded at low temperature and at room temperature, respectively, at 15 K simultaneously /6/. So any idea of a phase transition of
the paramagnetic system around 50 K is out of discussion. Moreover, it is
possible by a two-step deformation with an annealing between the two
deformations to suppress to a large extent the formation of point defect
centers which produce the "high temperature" part of the EPR spectrum
/7/. As EPR shows the dislocation related centers are not influenced by
the annealing.
As mentioned before the point defect centers produced by plastic
deformation at 650°C are stable up to 700°C whereas radiation defects
disappear around 400°C. Deformation at lower temperature (390°C) in fact
generates at least one defect (Si-P1) well known from neutron irradiation
/8/. This is further proof for the attribution of the room temperature
spectrum to point defects.
Considering the dislocation-related low temperature spectrum the first,
very important conclusion is:the core of the partial dislocations must be
reconstructed with the exception of about 3% of the sites (relating spin
density to the total density of partial dislocations, see below). The
spectrum consists of three distinct parts: a wide line called Si-Y /9/,
several narrow lines, two for each activated slip system, called Si-K1
/10/ and finally a series of pairs of lines (10 G wide) belonging to spin
S $\geq$ 1 centers (Si-K2) /10 - 12/. Generations of research students and
post-docs have collected a mass of data on these spectra which indicated

that Si-K1 is related to the 30°-partials of screw dislocations /3/ and that illumination with light and doping influence the ratio of intensities of K1 and K2 in a characteristic manner /13/. At the time being Kisielowski-Kemmerich applying LCAO-type molecular orbital analysis is outlining structural models for all these centers /14/. Two findings yield the key for the models: the g-tensor of both Si-Y and Si-K2 are very similar and indicate the orbital of the unpaired electron to be parallel to that «110» direction [0,-1,1] which is perpendicular to the burgers vector of the dislocations. On the opther hand the hyperfine lines of Si-K1 showed that the two lines are due to dangling bond orbitals parallel to two «111» directions: the normal on the primary glide plane [1,-1,1] and on the cross glide plane [1,1,-1], respectively. The simplest model distinguishing just these two dangling bond directions is a vacancy in the core of the 30°partials of screw dislocations /3/. Actually Kisielowski-Kemmerich /14/ attributes Si-Y to such vacancies with threefold coordinacy: the third electron which is unpaired resides in a substituting bond parallel to [0,-1,1]. Si-K2 represents chains of such vacancies which are coupled by dipole-dipole interaction. Several characteristics of Si-K1 may well be explained treating a four-center "molecule" consisting of the vacancy neighbouring a reconstruction defect. Relating now the density of unpaired electrons to the density of screw dislocation partials the ratio becomes larger: about 40% of the sites are occupied by vacancies (Si-Y centers) and 5% by Si-K1 centers after standard deformation. This conclusion is supported by high resolution electron microscopy /15/. These results, as satisfactory they are, open new problems. For instance: why behaves the 30°partial in a screw dislocation different from a 30°partial belonging to a 60°dislocation? And what about the atoms removed from the core of (screw) 30°partials? This latter process can be looked at as local transition from the glide set to the shuffle set structure /16/ and may assist kink motion /17/, However, the point defects left by moving dislocations are, as far as we know from EPR, also of vacancy type and not interstitials /8/. It is well known and supported by recent diffusion experiments /18/ that dislocations in Si are effective sinks for self-interstitials. Probably the two puzzles are interrelated: self-interstitials probably are attracted mainly by 60°dislocations filling vacancies also in their 30°parrtial. So only screws hold the vacancies uncompensated which may be necessairy for kink motion. A further important result concerns the generation of the paramagnetic centers in the dislocation core: Their number is not simply related to the dislocation length, but to the area swept by moving dislocations (i.e. shear strain) since the last annealing treatment /2/. This means: they are not intrinsic part of the core of screw dislocations, but they are produced during dislocation motion. This becomes particularly clear if one compares spectra of standard deformed crystals with spectra of crystals deformed at 420°C with higher shear stress. The dislocation density is higher in the last named case by a factor 2 to 3, but the strain ε is only 2%. The spin density of both parts of the EPR spectrum (point defect and dislocation related, respectively) follows ε ! Just the composition of the parts may change: Si-K1 decreases more than Si-Y /19/. Although both classes of centers are produced parallel to strain the former belief /20/ that both are generated and destroyed always simultaneously was disproved by the observation that annealing at 800°C between two deformation steps may suppress to a large extent the formation of point defects during the second deformation without changing number and character of dislocation centers /7/.As these crystals deformed at low temperaturte and high stress (LT/HS) contain straight dislocations parallel to the «110»Peierls valleys and a much higher portion ($\approx$50%) screw dislocations it is interesting to note that the density of Si-K1 ($\approx$ 4 10^{-4}) and Si-Y ($\approx$1.5 10^{-2}) is much smaller here than in standard deformed crystals (Si-K2 is even under the detection limit /21/). Probably this is related to the reduced relative mobility of screws under LT/HS conditions which is the reason for the enhanced screw density.

Modelling paramagnetic centers must take into account results of annealing experimants /4,22/. Using the capabilities of relaxation spectroscopy /5/ it became possible to follow the annealing of each defect type separately /22/. However, obviously it is possible that a defect changes its spectroscopic "finger print" (e.g. T_1) during annealing /22/. In general the different parts of the EPR spectrum anneal out. between 600 and 725°C more or less simultaneously with the only exception of part of Si-Y which was called Si-O in /4/. This signal resembles to the EPR of amorphous silicon /4,22/; especially it shows increasing line width between 180 K and room temperature /22/. The meanig of these results is not yet clear. Apparently the vacancy content of screw 30°partials is strongly reduced between 600 and 725°C. From the activation energy of the annealing process (3.1 - 3.6 eV) Weber /22/ concludes that self-diffuision by a vacancy mechanism /23/ is involved.

The electrical activity of the deformation induced defects can be tested by photo-EPR of deformed n-type Si ($\approx 10^{16} \cdot cm^{-3}$ P atoms). Any decrease of the EPR signal of (neutral) P^0 atoms at 13 K indicates acceptor activity of the defects. In fact the point defect spectrum of a standard deformed specimen and the Si-P1 spectrum of a neutron irradiaded crystal equally disappear in the dark and the P^0 signal at the same time is strongly reduced /3, 24/.Illumination (0.6 eV ≤ hv ≤ 1.24 eV) restores both the point defect and P^0 spectra. This indicates that most of the point defect clusters are acceptors in the lower half of the gap.

If one is interested in the dislocation related centers one uses crystals deformed at 800°C (doping 10^{15} cm^{-3}, strain 1.6%). From the P^0 spectrum one deduces that 3 10^{14} cm^{-3} electrons must be captured by defects. Those crystals are not only free of EPR active centers (with the exception of Si-O) but DLTS shows that the total density of deep (≥ 50 meV) traps does not exceed 2 10^{13} cm^{-3}.(That means:dislocations do not show up as deep trap levels!) Over-band-gap light restores 7 10^{13} cm^{-3} electrons back to the P-atoms. So two questions arise: which traps not accesible to DLTS can trap 2.8 10^{14} cm^{-3} electrons? and why can light bring only 25% of them back to the phosphorous? - To answer the first question one needs a detailed analysis of the properties of an extended chargeable defect with its screening charge in the depletion region of a Schottky barrier /25/. The second problem may be easier: as one learns from the rapid relaxation of the photo-effect after ending the illumination one is measuring a dynamical equilibrium between capture of holes on the negatively charged dislocation and recombination in the dislocation core.

Although understanding of the photo-EPR of dislocation centers is far from being complete it is now clear how to isolate this effect by avoiding the production of point defects and using the charge state of P-atoms as a convenient probe for electron capturing by dislocations.

Deep level transient spectroscopy (DLTS) has been applied to deformed silicon by a number of groups (for references see /26/). Using Schottky barrier probing of the upper and lower half of the band gap has to be done with n-type and p-type material, respectively. Although more traps are located in the lower half, n-type Si has been investigated more thoroughly. Typical spectra are shown in fig.1. All lines are broadened and exhibit non-linear filling characteristics /26/. On account of the strong temperature dependence of the filling behaviour for part of the centers TSCAP (thermally stimulated capacitance) is not a reliable method for determining absolute concentrations /27/.

Logarithmic filling generally is ascribed to the Coulomb barrier around multichargeable extended defects. The best candidate for this character of course is the dislocation. However, also point defects sitting within the space charge around a dislocation should behave in the same manner. Omling et al. /26/ could show for two deformation centers (E 0.27 eV and

E 0.55 eV, see table 1) that for very short filling pulses ($t_P \leq 10^{-7}$s) the filing is a linear function of t_P becoming logarithmic afterwards. This is proof for the barrier model. There are defects the filling characteristic of which is even more complicated , neither logarithmic nor linear. Especially to fill the prominent level E 0.55 eV (table 1) is the more difficult the lower the temperature /27/.

Very recently Feklisova et al./28/ have demonstrated that (dislocation?) centers in very lightly doped and weakly deformed Si arrive in two different states at low temperature, depending on reverse bias applied during cooling or not. The authors explain this finding by metastability of those centers and speculate about reconstruction of dislocation cores. Nitecki and Pohoryles /25/ stress the fact that dislocations which are not parallel to the plane of the Schottky barrier locally shortcircuit this barrier. Shikin and Shikina /29/ emphasise that in case of dislocations the influence of the deformation potential has to be taken into account to calculate the capture dynamics (not so much the energetic position of the traps).

From that it follows that the correct evaluation of DLTS spectra of defects in or around dislocations is by no means established. But the position of trap levels seems to be only weakly influenced by these difficulties. Table 1 contains best estimates of these positions for prominent traps. Since electron and hole traps never have been determined in one crystal it is unknown whether the electron trap E 0.55 eV and the hole trap H 0.63 eV are identical.

The value of DLTS for identifying defects greatly increases if it is pos sible to correlate defect levels with EPR spectra because these are high- ly specific.Problems which arise are summarised in a footnote to table 2. 2. One tryes to exploit the variation of the deformation mode resul - ting in different defect populations (table 2). The program has been re- alised but only to a small part.

The present state is as follows:
1) The shallow levels from table 1 (distance from the next band edge $\leq$ 200 meV) are absent when after a two-step deformation the point defect clusters Si- K3, K4 and K5 are lacking in the EPR spectrum. In fact Erdmann /30/ has shown by photo-EPR that these centers are influenced by photons in two energy ranges: 0.55 eV $\leq$ hv $\leq$ 0.60 eV and within 200 meV below band gap energy. So it is not surprising that near band edge (NBE) DLTS disappears with these defects. It should be examined whether midgap DLTS peaks also contain contributions from Si- K 3 to Si- K5.

2) Standard deformation results in about four times as many hole traps as electron traps. Since the room temperature EPR contains about the same number of centers as hole traps are present (EPR: 3.8 10^{15} cm^{-3}, DLTS: 3,2 10^{15} cm^{-3}) we think that the hole traps H.0.33, 0.39 and 0.49 eV are the (point defect type) EPR centers producing the room temperature EPR. Direct confirmation for this idea again comes from photo-EPR: monitoring the signal of neutral phosphorous Erdmann /30/ noticed that from the energy range (E_{gap} -0.5 eV) $\leq$ hv $\leq$ (E_{gap} - 0.34 eV) about 2.5 10^{15} cm^{-3} electrons are excited into the donors. In this energy range, therefore, that number of acceptors is located, which act as hole traps in DLTS. The related increase of the high temperature EPR signal can also be shown by photo-EPR.

3) Considering the doping influence on the spectra one finds that the EPR centers Si- K1 (and K 2) are much stronger influenced than the wide line Si -Y /31/. This means: the charge of the dislocations and, therefore, their Coulomb potential are more influenced by Si - K1 centers than by core vacancies. If this is correct the centers Si -Y should be not very affected in DLTS spectroscopy since here only rechargeable traps play a role. As a matter of fact the DLTS line E 0.48 eV fits this type of behaviour: in standard deformed crystals it is only weak, but on the other side only it remains after 850°C annealing - exactly as part of Si -Y does.

The DLTS line E 0.48 eV clearly represents a group of levels which can be demonstrated by variation of the rate window /32/.

4) Most difficult seems to be an identification of the centre causing the - sometimes prominent - DLTS line E 0.55 eV which due to its strange filling behaviour is measured by different authors in quite different positions. Its strength depends nonlinearly on strain and it is absent in two-step deformed specimen which are outstanding by nearly ideal single slip. Both findings point to dislocation interactions as source of E 0.55 eV traps. Kimerling et al. /33/ localised the E 0.55 eV centers in regions where EBIC contrast shows dark slip bands.

5) In two-step deformed crystals both levels near midgap (H 0.49 and E 0.55 eV) appear strongly reduced when compared with standard deformation (fig. 1).

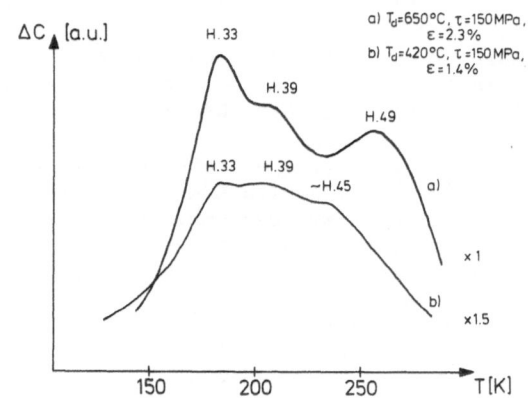

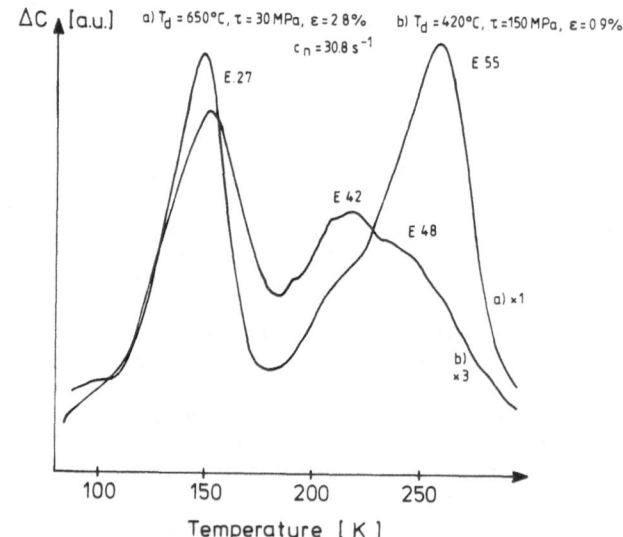

Fig. 1. DLTS spectra of deformed silicon
Above: p-type Si ($N_A-N_D= 3.3 \ 10^{16} cm^{-3}$). a) Standard deformed with 150MPa stress. b) Two-step deformed. Below: n-type Si ($N_D-N_A = 4 \ 10^{15} cm^{-3}$), a) Standard deformed, b) Two-step deformed.

Table 1. Best estimates of trap levels in deformed silicon.

Electron traps

E (0.55*-0.6) 0.48 0.42* 0.27 0.17 eV

Hole Traps

H 0.10 0.21 0.29 0.33*0.36 0.39 0.49 0.63 eV

---

* Levels prominent in standard deformed crystals
Underlined: levels found in our experiments

Table 2. Deformation procedures which lead to simple defect population.
(Strain is chosen so that with n-doping 5 10^{15}cm^{-3} EPR and DLTS
spectroscopy can be applied)*

Deformation mode	Procedure	Result dislocation centers	point defect
standard	650°C, τ=30MPa, ε=2.5%	+	+
high temperature	800°C, τ=8-12MPa, ε=1.6%	Si –O	–
two steps	800°C, τ=8 MPa, ε=0.7-1,6%		+
	420°C, τ=200-250MPa, ε=1.2%	straight disl. (K1/Y reduc)	(without K3,4,5)
three steps	800°C, τ=8 MPa, ε=0.7-1,6% 800°C 16 h anneal 420°C, τ= 200 MPa, ε≤0.1%	straight dis.	–

---

*) For comparison of EPR with DLTS doped material has to be used the
doping level clearly exceeding the trap density. To restrict the
conductivity to a range where EPR is feasible the deformation
(i.e.trap density) should be weak. Then EPR is applicable even at room
temperature to specimens thinner than the skin depth.
  In order to compare EPR and DLTS quantitatively we assume that the
formation of all defect types does not depend on doping below n,p ≈
5 10^{15}cm^{-3}, so that EPR from undoped Si can be used. This assumption
seems to be reasonable since the Fermi level at the deformation temperature is not changed by this doping.

MICROWAVE CONDUCTIVITY OF DISLOCATIONS

Considering the core of a partial dislocation as a structure being periodic in one dimension one is lead to postulate the existence of one (or several) one-dimensional band of electronic states which can be situated in the energy gap of the matrix crystal. The (unrealistic) model of a periodic array of dangling bonds would result in a half filled band, while a perfectly reconstructed core is equivalent to a full band separated from an empty band. Metallic cnductivity could be expected in the first case whereas in the second case doping by electrons or holes should lead to band conductivity. Early attempts to demonstrate dc conductivity had no clear result; remembering deviations from perfect periodicity as core point defects, kinks and jogs, suggests to search for high frequency conductivity /34/.

Here it is shortly reported on recent results comparig curved with straight dislocations /35/. We used three step deformed (table 2) FZ-Si with n-type doping. From the complex permittivity of 2 - 3 mm diameter discs , measured at a frequency of 10 GHz, the microwave conductivity (MWC) σ_w was calculated /36/. Fig. 2 shows σ_w of a sample only predeformed and annealed at 800°C, which contains irregularly shaped dislocations, edge character being dominant. The disc is cut parallel to the primary glide plane, the two curves correspond to two orientations of the electrical MW field: parallel and perpendicular to the burgers vector of the dislocations. One notices that σ_w below 30 K becomes nearly temperature independent and so by far exceeds σ_w of the starting material. The MWC parallel to edge dislocatios is two to three times larger than parallel to screw dislocations. This anisotropy reverses sign during the LT/HS deformation, the third step of a three step deformation, although the strength of the anisotropy is not very well reproducible.In fact in these crystals the proportion of screws is greatly enhanced /37/. Fig, 3 presents σ_w in a disc cut parallel to the (011) plane, i.e. perpendicular to the primary burgers vector (three step deformed). Turning the electrical field from orthogonal to parallel to the main glide plane (1.-1,1) σ_w increases from 10^{-5} to $7 \cdot 10^{-4} (\Omega cm)^{-1}$. This result should not be mixed up with a similar anisotropy of the electron mobility found by Eremenko et al. /38/ : there it a matter of bulk conductivity (T $\geq$ 100 K) and the effect is restricted to Czochralski grown Si. Under these circumstances impurities collected into the glide planes by moving dislocations act as scattering centers. In our case the density of point defects after deformation is comparatively small and we are watching an effect of extra conductors and not of scattering.

In a crystal of higher doping level (5 10^{15} cm^{-3}) the (weak) anisotropy after three step deformation reversed sign during short (5 min) annealing at the deformation temperature (420°C)(fig 4); in fact this type of annealing relaxes to a large extent the LT/HS dislocation morphology /39/

Comparing the MWC parallel to screw dislocations for three different doping levels (2 10^{13} : 1.5 $10^{-5} (\Omega cm)^{-1}$, 5 10^{14}: 3 10^{-3} $(\Omega cm)^{-1}$, 5 10^{15}: 3 10^{-3} $(\Omega cm)^{-1}$) one is lead to the conclusion that a certain doping of the dislocations is necessairy to obtain pronounced MWC. First experiments with p-type Si confirm that also doping by holes is effective.

Illuminating the disc with monochromatc light results in a small (60%) increase of σ_w for photons of 0.5 to0.66 eV and s total break down of the MWC under band gap light (fig, 5). The first effect is due to increase of the dislocation doping by exciting electrons from midgap acceptors via the cpnduction band into the dislocations. (The same is observed watchinmg the transition of P atoms from the positive into the neutral state). The breakdown is due to decrease of the dislocation charge by holes from electron- hole pairs.

There are indications that screw dislocations contribute to the MWC more than proportional to their length. Dividing σ_w by the length of screw dislocations one obtains $\sigma_w/N_s = 1,5 \cdot 10^{-11}$ cm/Ω (LT/HS deformation, N_D-

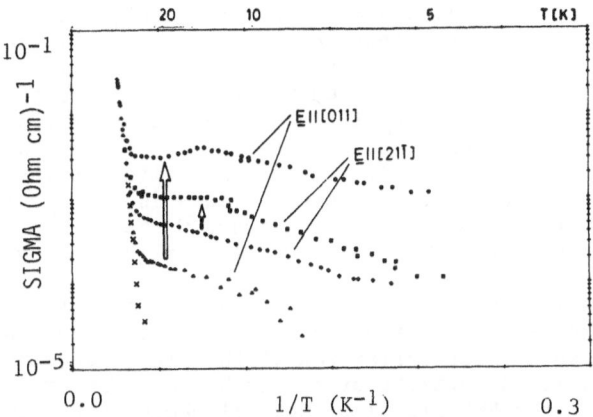

Fig. 2. MWC of deformed n-type Si ($N_D-N_A = 5 \cdot 10^{14}$ cm^{-3}). Lower curves: First two steps of a three-step deformation. Upper curves: Three-step deformed. Crosses: Underformed.

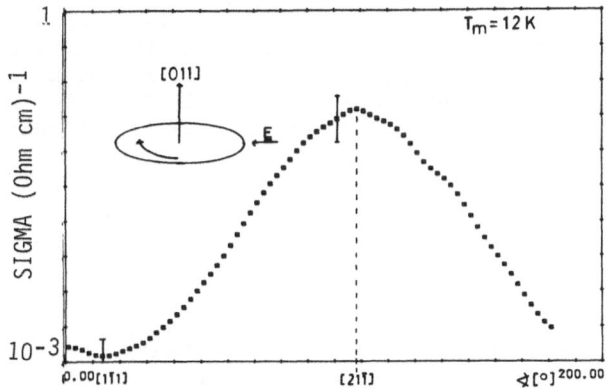

Fig. 3. MWC of deformed n-type silicon ($N_D-N_A= 5 \cdot 10^{14}$cm^{-3}), three-step deformed, but only 30 min annealed. The plane of the disc is (o11), the x-axis marks the angle between the E field and the normal on the glide plane [2,1,-1] is the direction of edge dislocations.

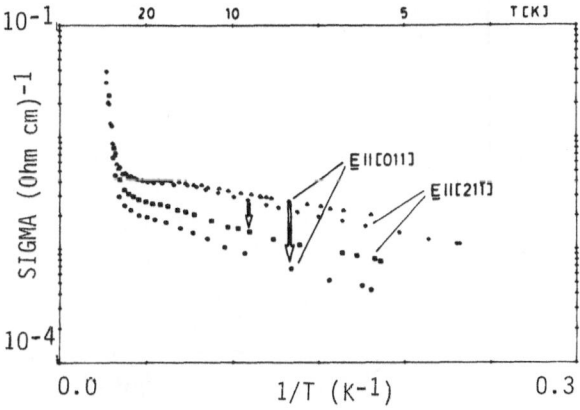

Fig. 4. MWC of deformed n-type silicon ($N_D-N_A= 5 \cdot 10^{15}$cm^{-3}).
Upper curves: Three-step deformed
Lower curves: After additional 5 min. anneal at 420°C.

N_A= 5 10^{14} cm^{-3}, T = 10 K). (Taking the total dislocation density instead of the screw density changes this number by a factor of two).
Under certain conditions /34/ this value can be identified with$((\sigma_B/N)+(2\pi/3)\,S)$, where σ_B is the bulk conductivity, to be neglected at 10 K, and S = L/R the specific conductance of a dislocation segment of length L. Typical one-dimensional conductors out of the class of polymers exhibit similar values of S /34/. Comparison with previous measurements on dislocated semiconductors by other groups is difficult because of different deformation and doping conditions: Germanium: S = 2.5 10^{-9}.... 1.4 10^{-10} cm/Ω /40/ and 1.6 10^{-12} cm/Ω /41/, Silicon: σ_w = 4 10^{-2} (Ωcm)$^{-1}$ /42/.
Further analysis in order to determine L and the effective volume of the dislocation segments needs measuring the frequency dependence of σ_w which is under way. Then the depolarisation of the finite segments can be taken into account. A rough estimate of the number of carriers involved can be made as follows: the EPR signal of neutral phosphorous atoms at 13 K for three step deformed crystals shows that about 2.5 10^{14} cm^{-3} electrons are captured by dislocations. Dividing σ_w by this number one arrives at an average carrier mobilty $\mu \approx$ 75 (cm^2/V s). Below of 10 K the MWC clearly is thermally activated with small activation energy (0.2 - 2 meV). Relatively low mobility and temperature dependence point to some kind of localisation for which several effects can be responsible. In this connection it is noteworthy that in deformed n-type Si a new type of EPR signal is observed which is influenced as well by the magnetic as by the electric component of the MW field /43/. This "combined resonance" /44/ seems to indicate that one is concerned with partly delocalised electron spins.

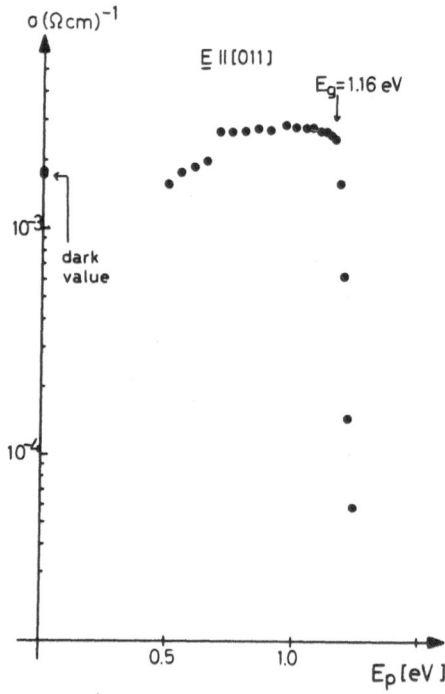

Fig. 5. MWC of deformed n-type silicon (N_D-N_A= 5 10^{14} cm^{-3}) under illumination, three-step deformed. x-axis: photon-energy.

CONCLUSIONS AND OUTLOOK

Investigations as described before made clear the proportion of dislocation related and not related defects and showed how to manage rather simple conditions by variation of the deformation procedure.
It has to be emphasised that the majority of EPR and DLTS active centers anneal out around 750°C. Obviously in this temperature range the dislocation density is essentially not changed. (Straight dislocations become curved at much lower temperature). One should not draw the hasty conclusion that all those centers do not belong to dislocations: it is easy to imagine that the core fine structure (core reconstruction, core point defects etc.) is changed on a scale not attainable by today electron microscopy. Nevertheless deformation induced effects which do not anneal out are of particular interest. At present time we know of 4 of them:
1) Part of the EPR signal Si-Y (= Si-O).
2) MWC (with reduced anisotropy (fig.4),

3) DLTS level E 0.48 eV and a relatively broad band of hole traps,
4) Photoluminescence (PL) lines D3 and D4 /45/.
Probably these effects reflect particularly essential properties of the dislocations in silicon. In this respect PL deserves special interest: after two step deformation straight dislocations cause a new PL line D 5 with fine structure which in our interpretation brings to light that the two partials coupled together by a stacking fault ribbon are seats of donors and acceptors, respectively. This polarisation at least in case of screw dislocations must be the result of the previous motion of the dislocation. This polarisation anneals out but the PL band D5 converges to the known line D 4. By the way differences between the two 30°partials of screw dislocations under LT/HS conditions are demonstrated by different mobilities of these two partials /37, 46/. More effort should be devoted to scanning electron microscopy (EBIC mode) of widely separated partials.
From a theoretical point of view unified treatment of one-dimensional deformation potential valleys coinciding with band bending by the space charge around charged dislocations would be highly desirable. There are unsolved problems about recharging dislocations by DLTS pulses which possibly are related to high electrical fields of the order of $5 \cdot 10^5$ V/cm around charged dislocations with fillling factors of f = 5%.
Recent observations by EBIC and IRBIC /47/ microscopy indicate that the electrical field near dislocation groups /48/ and grain boundaries /49/ can act as generation center separating electrons from holes as was observed with stacking faults from oxidation /47/.

Acknowledgements

The author wants to express his gratitude to all former and present coworkers for ideas, experiments and discussions. Special thanks are due to Dr. H. Gottschalk and Dr. C. Kisielowski-Kemmerich.
The Deutsche Forschungsgemeinschaft gave continuous financial support.

REFERENCES

1 E. Weber, H. Alexander, Inst. Phys. Conf. Ser. 31 (1977) 266.
2 C. Kisielowski-Kemmerich, G. Weber, H. Alexander, J. Electr. Mater. 14a (1985) 387.
3 E. Weber, H. Alexander, J. Physique 40 (1979) C6-101.

4 V.A. Grazhulis, V.V. Kveder, Yu. A. Ossipyan, Phys. Stat. Sol. (b)$\underline{103}$, (1981) 519.

Yu. A. Ossipyan, Sov. Sci. Rev. Sect. A $\underline{4}$ (1982) 219.

5 C. Kisielowski-Kemmerich, J. Magn. Res. $\underline{66}$ (1986) 307.

6 H. Alexander, H. Gottschalk, C. Kisielowski-Kemmerich in: Dislocations in Solids, Tokyo Univ. Press 1985, 337.

7 C. Kisielowski-Kemmerich, J. Czaschke, H. Alexander, Mater. Sci. Forum $\underline{10-12}$ (1986) 745.

8 M. Brohl, C. Kisielowski-Kemmerich, H. Alexander, Appl. Phys. Lett. $\underline{50}$, (1987) 1733.

9 M. Suezawa, K. Sumino, M. Iwaizumi, Inst. Phys. Conf. Ser. $\underline{39}$ (1981) 407.

10 H. Alexander, M. Kenn, B. Nordhofen, E. Weber, Inst. Phys. Conf. Ser. $\underline{23}$ (1975) 433.

11 U. Schmidt, E. Weber, H. Alexander, W. Sander, Sol. State Comm. $\underline{14}$ (1974) 735.

12 L. Bartelsen, Phys. Stat. Sol. (b) $\underline{81}$ (1977) 471.

13 R. Erdmann, H. Alexander, Phys. Stat. Sol. (a) $\underline{55}$ (1979) 251.

14 C. Kisielowski-Kemmerich, to be published.

15 A. Bourret, J. Desseaux-Thibault, F. Lancon, J. Physique $\underline{44}$ (1983) C4-15.

16 H. Alexander, J. Physique $\underline{35}$ (1974) C7-173.

17 F. Louchet, J. Thibault-Desseaux, Rev. Phys. Appl. $\underline{22}$ (1987) 207.

18 N.A. Stolwijk, D. Grünbaum, M. Perret, M. Brohl, Proc. 15. ICDS Budapest 1988, in press.

19 C. Kisielowski-Kemmerich, PhD Thesis Univ., Köln 1985.

20 E.R. Weber, H. Alexander, J. Physique $\underline{44}$ (1983) C4-319.

21 M.N. Zolotukin, Sov. Phys. Sol. State $\underline{28}$ (1986) 1862.

22 G. Weber, Diploma thesis Univ. Köln 1985.

23 H. Siethoff, J. Physique $\underline{44}$ (1983) C4-217.

24 C. Kisielowski-Kemmerich, H. Alexander, to be published.

25 R. Nitecki, B. Phoryles, Appl. Phys. $\underline{A36}$ (1985) 55.

26 P. Omling, E.R. Weber, L. Montelius, H. Alexander, J. Michel, Phys. Rev. $\underline{B32}$ (1985) 6571.

27 E.R. Weber, C. Kisielowski-Kemmerich, H. Alexander, Isv. Acad. Nauk USSR $\underline{51}$ (1987) 644.

28 O.V. Feklisova, E.B. Yakimov, NA.A. Yarikin, Proc. 15. ICDS Budapest, 1988 i.p.

29 V.B. Shikin, N.I. Shikina, Phys. Stat. Sol. (a) $\underline{108}$ (1988) 669.

30 R. Erdmann, PhD Thesis Univ. Köln 1979.

31 C. Kisielowski-Kemmerich, B. Bollig. J. Palm, H. Alexander, unpubl. results.

32 C. Kisielowski-Kemmerich, E.R. Weber, to be published.

33 L.C. Kimerling, J.R. Patel, J.L. Benton, P.E. Freeland, Inst. Phys. Conf. Ser. $\underline{59}$ (1981) 401.

34 R. Labusch, Physica 117B + 118 B (1983) 203.

35 M. Brohl, M. Dressel, H.W. Helberg, H. Alexander, submitted to Phil. Mag.

36 A. Gleitz, H.W. Helberg, Phys. Stat. Sol. (a) $\underline{90}$ (1985) K209.

37 K.Wessel, H. Alexander, Phil. Mag. $\underline{35}$ (1977) 1523.

38 V.G. Eremenko, V.I. Nikitenko, E.B. Yakimov, JETP Lett. $\underline{26}$ (1977) 72.

39 H. Gottschalk, Proc. 10. Int. Congr. Electr. Micr., Hamburg 1982, 527.

40 Yu. A. Ossipyan, Cryst. Res. Technol. 16 (1981) 239.

41 M. Dressel, H.W. Helberg, Phys. Stat. Sol. (a) 96 (1986) K199.

42 V.A. Grazhulis, V.V. Kveder, Yu. Mukhina, Phys. Stat. Sol. (a) 44 (1977) 107.

43 J. Krüger, Dipl. Thesis Univ. Köln (1988).

44 V.V. Kveder, Yu. A. Ossipyan, A. Shalynin, JETP Lett. 40 (1984) 729; ibidem, 43 (1986) 255.

45 R. Sauer, C. Kisielowski-Kemmerich, H. Alexander, Phys. Rev. Lett. 57, (1986) 1472.

46 H. Alexander in: Dislocations 1984, P. Veyssiere, L.K. Kubin, J. Castaing eds., CNRS, p.283.

47 A. Castaldini, A. Cavallini, A. Poggi, E. Susi, Proc. 15. ICDS Budapest 1988, in press.

48 M. Kolbe, unpublished results.

49 O. Hollricher, unpublished results.

INTERNAL FRICTION DUE TO DEFECTS IN SEMICONDUCTORS

P. Haasen, U. Jendrich and D. Laszig

Inst. für Metallphysik, Univ. Göttingen and SFB 126
D 3400 Göttingen, FRG

INTRODUCTION

It is well established that the dependence of the dislocation velocity on applied stress σ and temperature T in Ge and Si can be described by the equation[1-4]

$$v = v_0 \cdot \left(\frac{\sigma}{\sigma_0}\right)^m \cdot \exp - \frac{H_v}{kT} \tag{1}$$

This behaviour can be explained under the assumption that the dislocations move by overcoming a high primary Peierls potential (H_{KPF}) by kink pair formation (KPF) and the secondary Peierls barriers (H_{KM}) by single kink migration (KM) along the dislocation line.

A clear method to resolve the two elementary steps of the dislocation movement, the KM and KPF, seems to be the internal friction. The height of the secondary Peierls potential can be measured by investigating the internal friction of deformed crystals at sufficient low temperatures, where no kink pair formation occurs. At these temperatures the only possible motion of the dislocations is the motion of geometrical kinks. A great number of internal friction data is available[7-9, 12-15] but unfortunately no unique picture was obtained.

Recent calculations of the kink enthalpy H_K[10] and H_{KM}[11] of the 90° partial dislocation in Si result in a rather small kink enthalpy $H_K \approx 0.4-0.5eV$ and a large kink migration enthalpy $H_{KM} \approx 1.3eV$, confirming indirect observations[5,6].

An internal friction maximum corresponding to the movement of geometrical kinks over a high potential can be detected only at high temperatures and at low frequencies. Those conditions were fulfilled in this work.

Measurements of the internal friction on deformed Ge at frequences from 0.2 to 20Hz at temperatures between 150 and 820K have shown two maxima α and β, occuring after deformation only. Their activation enthalpy H, prefactor f_o and relaxation strength Δ are: H_α = 1.108±0.01eV, $f_{o\alpha}$=0.14 bis $15\cdot10^9s^{-1}$ (increasing with increasing deformation), $\Delta_\alpha \approx 10^{-2}$; H_β=2.07±0.2eV, $f_{o\beta} \approx 10^{13}s^{-1}$, $\Delta_\beta \approx 3\cdot10^{-2}$). Both shift to lower temperatures with increasing deformation. They are attributed to the movement of geometrical kinks (α) and to kink pair formation (β) on single partial dislocations. Doping with $2 - 4\cdot10^{17}cm^{-3}$ Sb lowers the kink migration enthalpy H_α by 0.3-0.4eV. This indicates that the lowering of the activation enthalpy of the dislocation velocity by doping is due to that of the kink migration enthalpy H_α only.

First results are also presented on energy losses due to EL2 centers in GaAs. The measurements were performed at 100kHz on crystals of various orientations, dopings and with different heating/cooling rates.

EXPERIMENTAL

For the mentioned purpose an inverted torsion pendulum of the Max-Planck-Institut für Metallforschung in Stuttgart was used, where the specimen bar is clamped at both ends. The pendulum frequency could be adjusted by change of inertia weights in the range of 0.2 to 30Hz, depending on the dimensions of the samples too. The strain amplitude was controlled to a value of normally $\epsilon \approx 10^{-5}$ at the specimen surface. To study the amplitude dependence the strain was raised up to $\epsilon \approx 1.5 - 2\cdot10^{-4}$. The internal friction was measured by free decay of the oscillation, as the amplitude of 20 to 200 periods were detected, depending on the pendulum frequency.

Most specimens were cut with 7 x 7mm^2 cross section in <110>, some in <411> direction, out of Czochralski grown Ge single crystals. "Intrinsic" material ($\approx 10^{14}cm^{-3}$ p-doped) was sawn along the growth axis of the single crystals (≈ 40mm long), the Sb-doped specimens at right angles to this axes (≈ 20mm long) to get a homogeneous doping level along the specimen. These samples had <110> direction too.

These bars were polished mechanically and etched. All were deformed in compression in an Instron machine at 600°C under a forming gas atmosphere. The deformation velocity was normally 0.1mm/min, for the most heavily deformed specimens (Ge8.0 (see table 1) and all the doped specimens) it was increased up to 1mm/min to get a high dislocatin density. The shear stress at the end of the deformation of all the doped samples was chosen equal to that reached with Ge8.0. An approximately equal dislocation density can be expected then.

The deformed ingots were mostly sawn into 6 pieces, ≈ 2x3mm^2 cross section each. These pieces were ground to get parallel side faces. Some of them were thinned in the middle with a diamond wheel to achieve even lower pendulum frequencies. The ends covered with an acid resistant varnish, the thinned parts of these samples were etched in CP4 (2 parts HNO_3, 1 part CH_3COOH, 1 part HF plus 0.5ml Bromine per 100ml) to get a smooth surface.

There was no difference in the damping behaviour between etched and unetched specimens, so surface effects can be ruled out. But all the samples polished at the ends could not be fixed well in the holder. They produced a high, irreproducible background damping or fractured if clamped more tightly.

EXPERIMENTAL RESULTS

Intrinsic Germanium

In fig. 1 the internal friction above room temperature is shown. The two peaks named α and β shift to lower temperatures with increasing deformation and are absent in undeformed samples. The different height of the maxima in <110> and in <411>-orientated specimens is caused by different Schmid factors of the glide systems relative to the specimen axis. These result in a ratio of relaxation strengths: $\Delta_{<110>}:\Delta_{<411>}\approx1.5$ (see eq.3).

Table 1. Results on the α-relaxation

Speci-men	Defor-mation	H_α eV	$f_{o\alpha}$ $10^9 s^{-1}$	Δ 10^{-3}	L nm	N $10^6 cm^{-2}$
Ge0.2	0.2%	$-^{1)}$	$0.14^{2)}$	11.7	730	10.6
Ge0.5	0.5%	1.096	0.17	12.8	670	13.8
Ge1.0	1.0%	1.102	0.30	17.2	520	32.0
Ge1.8	1.8%	1.118	1.01	16.5	280	98.4
Ge3.8	3.8%	1.092	2.03	$10.1^{3)}$	170	165
Ge4.7	4.7%	1.107	4.0	10.0	150	220
Ge6.9	6.9%	1.118	6.7	$9.75^{3)}$	100	510
Ge8.0	8.0%	1.126	15.1	$8.4^{3)}$	65	962

[1]: one measurement only
[2]: evaluated with H_α=1.108eV
[3]: <411>-orientated, multiplied with Ω=1.5

An Arrhenius plot of frequency versus inverse temperature is shown in fig. 2. The maximum temperatures are mean values of the heating and following cooling curves, which differ a bit (see fig.3). This difference is caused by annealing effects at high temperatures during the measurement and decreases for one sample from one temperature cycle to the next. This results in linear regression for each deformation, i.e. the slope H_α and the intersection $f_{o\alpha}$ in fig.2 are listed in table 1.

Obviously the α-relaxation can be described by a constant activation enthalpy H_α=1.108±0.01eV, while the frequency factor increases strongly with deformation. Besides the relaxation strength is nearly constant for $\varepsilon>0.2\%$ with a small maximum at $\varepsilon\approx2\%$.

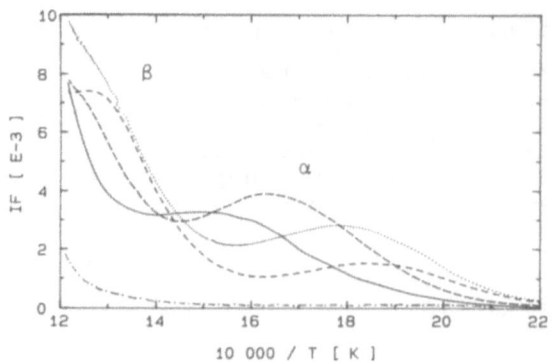

Fig. 1. Internal friction versus reciprocal temperature for
differently deformed specimens at about 0,5Hz
(constant background subtracted).
———: Ge0.2, ---:Ge1.8, ...: Ge4.7,
(<110>-oriented), ---:Ge8.0 (<411>-oriented),
-.-: undeformed.

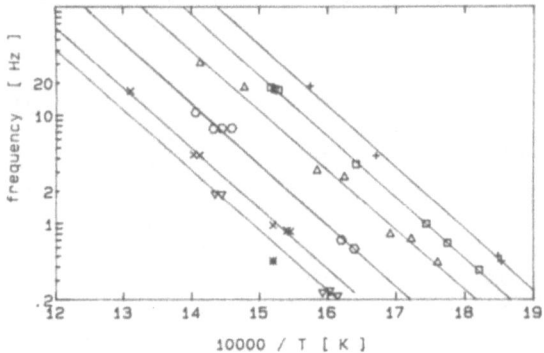

Fig. 2. Logarithm of the frequency versus reciprocal maximum
temperatures of the α-maxima for differently deformed
specimens; +: Ge8.0, □: Ge6.9, Δ: Ge3.8,   :Ge1,8,
x: Ge1.0, ∇: Ge0.5, *: Ge0.2.

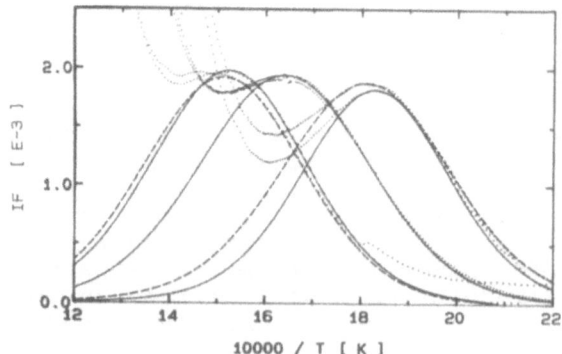

Fig. 3. Part of the internal friction versus reciprocal
temperature curves for the same specimen (Ge6.9:...)
at f=17.1, 0.35 and 3.5Hz. (constant background
subtracted). Fitted α-maxima (———, ---: heating,
cooling curves).

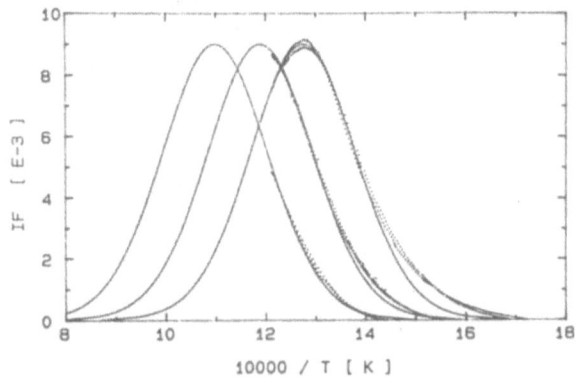

Fig. 4. Internal friction versus reciprocal temperature for
Ge6.9 at f=17.1, 0.35 and 3.5Hz (internal friction of
undeformed specimens at about the same frequency
subtracted): ... and the fitted β-peaks.

Table 2. Results on the β-relaxation

Speci-men	H_β [eV]	$f_{o\beta}$ [$10^{12}s^{-1}$]	Δ [10^{-3}]
Ge3.8	2.055	6.4	27.1
Ge6.9	2.096	11.8	37.4
Ge8.0	2.045	8.3	27.5

The internal friction values left after subtraction of
the α-maximum and of the curve of an undeformed sample at
about the same frequency are defined as the β-relaxation. It
could be fitted reasonably well for the low frequency
measurements of Ge3.8, 6.9 and 8.0 only, because only in
these cases the maximum was really passed (see fig.4). The
obtained peak was then rigidly shifted to fit the high
temperature rise of the same specimen at other frequencies.
This analysis is not as exact as that of the α-relaxation;
the Arrhenius plots in fig.5 and the results in table 2 show,
however, that the scatter is satisfactorily small.

So the β-relaxation can be described by an activation
enthalpy of H_β=2.07±0.1eV and a frequency factor of
$f_{o\beta}$=10^{13}Hz. Clear dependences on deformation cannot be
stated.

Sb-doped Germanium

Looking at fig.6 the characteristic features in the
internal friction spectra in Sb-doped Germanium in comparison
with undoped crystals can be summarized as follows:
  -) The α-maximum is broadened and flattened and shifted
     to lower temperatures. These effects increase with
     increasing doping level while the relaxation strength
     decreases. At n≈$10^{18}cm^{-3}$ Sb the α-maximum is
     suppressed completely.
  -) The α-peak anneals out at high temperatures while
     shifting to lower temperatures with doping.
  -) The β-peak moves to even higher temperatures and its
     low temperature side becomes more steep with doping.

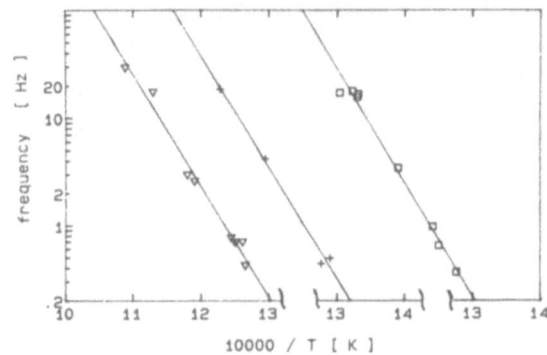

Fig. 5. Logarithm of the frequency versus reciprocal maximum
temperatures of the β-maxima for differently deformed
specimens; □: Ge8.0, +: Ge6.9, and ∇: Ge3.8

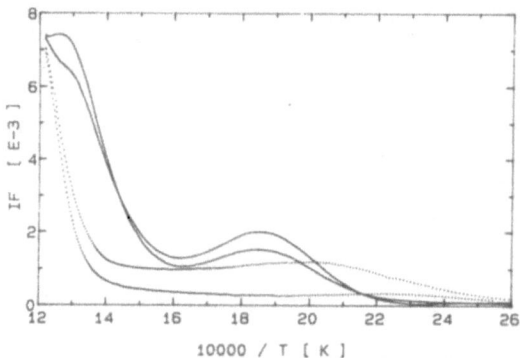

Fig. 6. Comparison of the internal friction spectra of a
Sb-doped (...: Sb3.3) and an undoped specimen (——:
Ge8.0) at f≈0.5Hz.

Table 3. Results on the α-relaxation in GeSb

Speci-men	H_α^n [eV]	$f_{o\alpha}^n$ [$10^6 s^{-1}$]	$\Delta^{1)}$ [10^{-3}]
Sb2.3	0.76±0.05	25	7.8±1
Sb3.3	0.70±0.1	10	6.6±1
Sb4.0	0.67±0.1	0.4$^{2)}$	20 +5$^{2)}$

1): virgin curves considered only
2): all measurements with large amplitude

Because of the strong annealing effect the temperature shift
with frequency was examined for directly following cooling
and heating curves only. All the measurements but the last
were not driven above 600 to 700K to avoid complete
annealing. The mean values averaged over three frequency
changes each are listed in table 3.

The α-maximum is strongly amplitude dependent in doped material: High strains of $1-2 \cdot 10^{-4}$ increase the relaxation strength up to values greater than those in intrinsic material. Furthermore they move the peak to lower temperatures, but have no influence on the peak shift with frequency. This behaviour has to be described by a lower prefactor while the activation enthalpy remains unaffected (see table 3).

DISCUSSION

Interpretation of the α and β-relaxation

The α-relaxation is the internal friction maximum at the lowest temperature which is caused by the motion of dislocations. Therefore it should be related to that step of dislocation motion with the lowest activation enthalpy. This is the motion of geometrical kinks over the secondary Peierls potential.

This case, where there are only kinks of one sign on a dislocation segment of length L and no KPF occurs was treated first by Brailsford[16]. He assumed the kinks to be "abrupt" and to behave like a one dimensional gas. Southgate and Attard[17] extended the model by including kink-kink interaction.

They got in a good approximation for the relaxation strength:

$$\Delta \approx \frac{8 \Omega G b^2}{\pi^4 E_o} \cdot N L^2 \tag{2}$$

with

$$\Omega = \cos \alpha \cdot \cos \lambda \tag{3}$$

where b is the magnitude of the Burgers vector,

$E_o = (G b^2 \beta / 4 \pi) \cdot \ln(R/r_o)$ the line energy of a dislocation, G is the shear modulus, R, r_o are the outer and inner cut-off radius of the strain field of a dislocation and β a orientation factor, α the angle of the normal of the glide plane to the tension axis and λ the angle of the stress component in the glide plane to the Burgers vector of the dislocation.

They calculated for the relaxation time:

$$\tau_\alpha = \tau_{o\alpha} \cdot \exp \frac{H_{KM}}{kT} \tag{4}$$

$$\tau_{o\alpha} = \frac{2kT}{\pi^2 E_o a^3 \nu_D} \cdot \frac{\overline{L^2}}{\sin |\overline{\Phi}|} \tag{5}$$

where $\overline{L^2}$, $|\overline{\Phi}|$ are mean values and ν_D is the Debye frequency. Similar results are obtained by Alefeld[18].

Taking the measured values τ_o, Δ and $|\overline{\Phi}|$, $\overline{L^2}$ and N can be calculated from eqs. 2 and 4 (see table 1). The evaluated shortening of the bowing out dislocation segment lengths L by a factor 10 with increasing deformation is a quite reasonable result, if L is determined by the intersection of dislocations. The observed broadening of the α-maximum can be understood by a distribution of L. Thereby the activation enthalpy H_α remains constant and can be identified with the secondary Peierls potential, i.e.:

$$H_{KM} = 1.108 \pm 0.01 eV.$$

From a theoretical point of view the description of KPF is much more complex and all the calculations carried out assume a high kink mobility[19,20]. Nevertheless, they agree with the internal friction results in fcc metals, i.e. the Bordoni relaxation. From both some characteristic features of a KPF-maximum can be given[19,21].

The relaxation time is approximately $\tau = \tau_o \cdot \exp(H_{KPF}/kT)$ where τ_o is in the order of the reciprocal Debye frequency, but may depend on N and L like τ_o $N \cdot L^n$ (0.5<n<2). For the relaxation strength a proportionality Δ $N \cdot L^m$ (1<m<2) is found. The exponents n and m both depend on special assumptions. Besides the Bordoni peaks in fcc metals are strongly broadened.

Qualitatively all these features show up in the β-relaxation found in this work. But a quantitative agreement of $H_{KPF} = H_\beta$ with the dislocation velocity data has to be proved.

Following[22] the enthalpy of a kink pair at the saddle point of formation can be described in the case of a negligible secondary Peierls potential ("elastic string model"): $H'_{KPF} = 2 \cdot H_K - 2 \cdot \sqrt{\sigma bhc}$, where H_K is the static enthalpy of a kink and $c = G \cdot b^2 \cdot h^2 \cdot \beta / (8\pi)$. The simplest possibility to take into account a non negligible secondary Peierls potential is the addition of H_{KM}, so that

$$H_{KPF} \approx 2 \cdot H_K + H_{KM} - 2 \cdot \sqrt{\sigma bhc} \tag{6}$$

This result is confirmed by a computer simulation of KPF, carried out by Jones[11].
The kink velocity described as a linear diffusion along the dislocation line is

$$v_K = v_o \cdot \exp - \frac{H_{KM}}{kT} \quad \text{with } v_o = \frac{\sigma bha^2}{2kT} \cdot \nu_D \tag{7}$$

and the KPF rate per unit dislocation length:

$$J = J_o \cdot \exp - \frac{H_{KPF}}{kT}$$

$$\frac{v_K}{b^2} = \cdot \exp \frac{-(2H_K - 2 \cdot \sqrt{\sigma bhc})}{kT} \qquad (8)$$

With these quantities the dislocation velocity for low kink mobility and in the absence of obstacles can be written[22] as:

$$v_D = 2h \cdot \sqrt{v_K \cdot J} = \frac{2h}{b} \cdot v_o \cdot \exp - \frac{H_{KPF} + H_{KM}}{2 \cdot kT} \qquad (9)$$

So the activation enthalpy of the dislocation velocity is

$$H_v = H_K + H_{KM} - \sqrt{\sigma bhc}. \qquad (10)$$

Dislocation velocity measurements at high stresses result in a generally accepted value of $H_v = 1.58 \pm 0.05 eV$ for 60° as well as 0° dislocations[2,3].
The above interpretation of the α relaxation gives
$H_\alpha = H_{KM} = 1.108 \pm 0.01 eV$ leading to

$\quad H_K = 0.52 \pm 0.06 eV$

(with $\sqrt{\sigma bhc} = 0.05 eV$, the approximate value for $\sigma = 10MPa$); this gives the kink pair formation energy
$\quad H_{KPF} = 2.05 \pm 0.1 eV$
in excellent agreement with the magnitude of
$\quad H_\beta = 2.07 \pm 0.2 eV.$

It must be emphasized that H_α and H_β can be interpreted only as the activation enthalpies for KM and KPF on a single partial dislocation, while the existence of extrinsic obstacles and correlation effects of the partial dislocations can be ruled out[23].

The doping effect

The influence of the Sb atoms on the α-relaxation could be due to two different mechanisms:

a) Dislocation segments are pinned by segregation of Sb atoms while free atoms are overcome easily.
This would explain the development of the α-peak during the temperature cycles. During the short cooling period after deformation a few Sb atoms segregate to the dislocations, shortening the free segment lengths. This results in a lowering of the relaxation strength and the relaxation time, while the distribution of segment lengths is broadened. These effects increase with annealing time and temperature and with doping concentration.

On the other hand segregation can be avoided for not too large doping concentrations by large strain amplitudes, where the dislocations are moving faster and over larger distances. However, no breakaway of the already pinned dislocations is possible at the achieved strains, confirming the high starting stresses at dislocation velocity measurements of $\sigma \approx 20 \text{MPa}^3$.

b) The activation enthalpy for kink motion H_{KM} is lowered by 0.3-0.4eV, the prefactor $f_{o\alpha}$ by a factor $\approx 10^3$. (The counteracting affect of an Sb segregation not taken into account. This would shift the β-peak to higher temperatures.) Assuming that H_K is not affected by doping, H_{KPF} should be lowered by the same amount as H_{KM} (see eq. 6) and $f_{o\beta}$ by the same factor as $f_{o\alpha}$. This consequence was checked by fitting broadened Debye peaks with corresponding $f_{o\beta}$ to the measured part of the β-relaxation. These fits resulted in $H_\beta(\text{Sb3.3})=1.82\pm0.05\text{eV}$ and $H_\beta(\text{Sb4.0})=1.74\pm0.05\text{eV}$, i.e. the β-relaxation in doped Ge is still consistent with $\Delta H_\beta = \Delta H_\alpha$.

A comparison of the effect of n-doping on the dislocation velocity and the α-relaxation (see eq. 10) indicates that the decrease of the activation enthalpy of the dislocation velocity H_v is due solely to the decrease of H_{KM}. Models concerning the doping effect are discussed in [23].

EL2 PEAKS IN GAAS

Different GaAs specimen were put into longitudinal eigen vibrations in a composite oscillator at 100kHz between 80K and 200K. With the long axis parallel to <110> and $n=3\cdot10^{14}\text{cm}^{-3}$ a peak of height 10^{-3} was observed near 130K whose exact location depended on cooling/heating rate (of the order K/min). Thermal hysteresis was minimum for faster cooling or slower heating. For <111>oriented si-bars no peaks are observed between 110 and 190K.. No peaks are found down to $5\cdot10^{-5}$ for stressing si-GaAs in <100> direction. This does not look like relaxation of a defect of <111> symmetry as proposed in various models of the EL2 center. A Kissinger type of analysis is being made for the transition between the high and low temperature states of the defect and the effect of doping as well as of lower temperatures is investigated. At higher temperatures (350K) the Mitrokhin peak[24] is observed in undoped specimens due to the ionization of EL2 centers in the pieco-charged surface regions of the specimens[25].

ACKNOWLEDGEMENTS

We would like to thank the Max-Planck-Institute für Metallforschung in Stuttgart for making available a pendulum for this work. Furthermore, thanks are due to Dr. J. Wolf, J.

Berger, W. Ulfert, Dr. U. Ziebart, Th. Albrecht and H. Waldmann for technical support during the work in Stuttgart and to H. Heymel who helped in specimen preparation. Thanks to Prof. H. Teichler for helpful discussions. The Deutsche Forschungsgemeinschaft and the Göttinger Akademie der Wissenschaften gave financial support. This is gratefully acknowledged.

REFERENCES

1) H. Alexander, in: *Dislocations in Solids*, F.R.N. Nabarro (ed.) 7: 115 (1986)
2) H. Schaumburg, Phil. Mag. 25: 1429 (1972)
3) I.E. Bondarenko, V.N. Erofeev, V.I. Nikitenko, Sov. Phys.-JETP 37: 1109 (1973)
4) P.B. Hirsch, Deformation of ceramic materials II, Mat. Sci. Res. 18: 1 (1983)
5) B.Ya. Farber, Yu.L. Iunin, V.I. Nikitenko, a) Phys. Stat. Sol.(a) 97: 469 (1986) b) personal communication to P. Haasen
6) L. Louchet, Inst. Phys. Conf. Ser. 60: 35 (1981)
7) K. Ohori, K. Sumino, Phys. Stat. Sol. (a) 9: 151 (1972)
8) K. Ohori, K. Sumino, Phys. Stat. Sol. (a) 14: 489 (1972)
9) E. Bonetti, P. Gondi, S. Valeri, Il Nuovo Cim. 33B: 103 (1976)
10) H. Veth, Thesis, Göttingen 1983
11) R. Jones, a) Phil. Mag. B 42: 213 (1980) b) J. de Phys. 44, C6: 61 (1983)
12) G. Welsch, T.E. Mitchell, R. Gibala, Phys. Stat. Sol. (a) 15: 225 (1973)
13) A.P. Gerk, W.S. Williams, J. Appl. Phys. 53: 3585 (1982)
14) H.-J. Möller, U. Jendrich, Deformation of ceramic materials II, Mat. Sci. Res. 18: 25 (1983)
15) U. Jendrich, Diploma thesis, Göttingen 1983
16) A.D. Brailsford, Phys. Rev. 122: 778 (1960)
17) P.D. Southgate, A.E. Attard, J. Appl. Phys. 34: 855 (1963)
18) G. Alefeld, J. Appl. Phys. 36: 2642 (1965)
19) A. Seeger, J. de Phys. 42, C5: 201 (1981)
20) H. Engelke, Phys. Stat. sol. 36: 231 and 245 (1969)
21) G. Fantozzi, C. Esnouf, W. Benoit, I.G. Ritchie, Progress in Mat. Sci. 27: 311 (1982)
22) J.P. Hirth, J. Lothe, *Theory of dislocations*. McGraw-Hill, New York (1968)
23) U. Jendrich, Thesis, Göttingen 1988; U. Jendrich and P. Haasen, Phys. Stat. Sol. (a) 108: 553 (1988)
24) V.I. Mitrokhin, S.I. Rembeza, U.V. Svirilov and N.P. Yaroslavtsev, S.P.-Sol. State 27: 1247 (1985)
25) D. Laszig and P. Haasen, Phys. Stat. Sol. (a) 104: K105 (1987)

INTERACTION OF IMPURITIES

WITH DISLOCATIONS IN SEMICONDUCTORS

Koji Sumino

Institute for Materials Research
Tohoku University
Sendai 980, Japan

INTRODUCTION

Interactions of impurities with dislocations give rise to a variety of interesting phenomena in semiconductors worth clarifying from both fundamental and practical viewpoints. The distribution of impurities around a dislocation is usually different from that in the absence of the dislocation. This effect has long been known as the formation of the Cottrell atmosphere in the field of metal physics. In the device production technology the technique of intrinsic or extrinsic gettering of impurities from device-active regions is now widely used. The impurity-dislocation interactions play a central role in this technique. Nevertheless, the relation between the formation of the Cottrell atmosphere and the impurity gettering is not necessarily correctly understood in many papers so far published in the field of electronic materials. Incorrect descriptions are often seen in literature. The inhomogeneity in the impurity distribution caused by dislocations results in the inhomogeneities in the electrical and optical properties within the crystal if the impurities are the agents that determine such properties of the crystal. Understanding of the nature of impurity-dislocation interaction is essential in establishing the technology to produce electronic devices of high quality.

The above is related to the effects of dislocations on the impurity distribution in the crystal. There are also many problems in the opposite sense. These are related to the effects of impurities on various kinds of activity of dislocations in semiconductors. Impurities dispersed within a crystal affect the dislocation motion through a variety of mechanisms. When impurities concernd are electrically active, for instance as donors or acceptors, they induce the shift of the Fermi level of the crystal. This may in turn affect the energy of the dislocation core or that of kinks on dislocations. The height of the energy barrier or, in other words, the activation energy for the dislocation motion is then altered, resulting in the change in the dislocation mobility. When impurities are gettered by a dislocation or the Cottrell atmosphere is developed around a dislocation, the dislocation is immobilized. The multiplication of dislocations in the crystal under a stress at elevated temperature is then suppressed very effectively. The generation of dislocations from some irregularities in the crystal as well as the mechanical strength of the crystal is affected very sensitively by both the dislocation mobility and the immobilization phenomenon. Device materials should be mechanically stable against the thermal stress during processing. No warpage or no slip of wafers is expected to take place. Thus, the strengthening mechanism of semiconductors due to impurity doping is a subject to be clarified in detail not only from the fundamental interest but also from the view of practical importance.

It is also very interesting to understand the state of impurities that are gettered at the dislocation core. Since the atomic arrangement at the dislocation core is peculiar, being much different from that in the matrix crystal, some special reaction incorporating impurities may take place there. Such reaction products may show some peculiar electrical or optical functions that are absent in the matrix region. Geometrically, dislocations in a semiconductor crystal may have dangling bonds aligned along the core. It is now widely believed that some kind of reconstruction of such dangling bonds takes place. It is interesting to know whether and how impurities can affect such reconstruction.

This paper gives a rather brief review of some aspects on impurity-dislocation interactions in semiconductors, especially those in Si and GaAs, on the basis of the works of the author's group. More detailed reviews are found in references 1 – 4.

IMPURITY DISTRIBUTION AROUND A DISLOCATION IN THERMAL EQUILIBRIUM

Individual impurity atoms, their clusters and complexes involving impurity atoms as constituents are all called here simply *impurities* for the sake of brevity. Any given impurity occupies a definite site in the crystal, and each site is occupied only one impurity. The distribution of impurities within the crystal in thermal equilibrium then obeys the Fermi-Dirac statistics.

The probability p with which an impurity occupies the site where the energy of interaction between the impurity and a dislocation is E_i is given approximately by the following equation : [5]

$$p = 1 \ / \ [1 + \ (1/C_o) \ \exp \ (- E_i \ / \ k \ T)], \qquad (1)$$

where C_o is the mean concentration of impurities in the crystal, k the Boltzmann constant, and T the temperature.

Figure 1 shows p calculated as a function of T (in a unit of °C) for a mean concentration $C_o = 10^{-6}$ for various values of E_i. The occupation probability p changes from unity to C_o within a rather narrow temperature range as the temperature increases. At temperatures around 700°C at which the diffusion of most impurity atoms is active, impurities of a concentration of 1 ppm are effectively trapped by the sites where the magnitude of the interaction energy is higher than about 1.5 eV.

The dependence of p on T is influenced also by the magnitude of C_o. Figure 2 shows the relations between p and T for $E_i = 1.5$ eV calculated for various values of C_o. The temperature range in which impurities are trapped effectively by the sites of any given value of E_i becomes wider as the impurity concentration increases.

Several kinds of mechanism are conceivable on the interaction between an impurity and a dislocation. The most important interaction at high temperature is thought to be that through the elastic strain fields of an impurity and a dislocation. The interaction energy has been calculated by the elasticity theory.[6] The magnitude of the interaction energy in this case is proportional to the inverse of the distance between the impurity and the dislocation and, thus, decreases rather rapidly as the impurity is separated from the dislocation. The maximum of the interaction energy is attained at the position of the impurity nearest to the dislocation. Usually, such position is taken to be one atomic distance apart from the center of dislocation core. For a typical impurity atom accompanying a large misfit strain in a semiconductor crystal, such as an In atom in GaAs, the maximum of the interaction energy is evaluated to be 0.67 eV.[7]

We reach the following conclusion from the above discussion : *In general, an extended impurity-rich region is never developed around a dislocation at temperatures at which the atomic diffusion is prominent as long as impurity concentrations typical in semiconductors, say 10 ppm or less, are concerned.* At first sight, this conclusion seems to conflict with the concept of the development of the Cottrell atmosphere around a dislocation which is well accepted in the theory of alloy hardening of metals. Actually there is no contradiction. The solute concentrations concerned in the discussion of alloy hardening are much higher than those concerned in the semiconductor technology, being of order of percent or more.

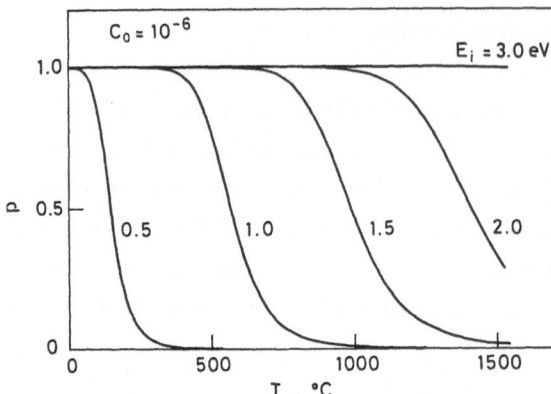

Fig. 1 The occupation probability p in thermal equilibrium of an impurity at the site with the interaction energy E_i plotted against the temperature T for an impurity concentration of 1 ppm, E_i being taken as a parameter.

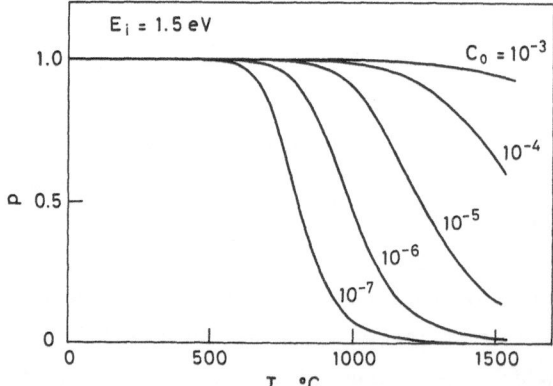

Fig. 2 The occupation probability p in thermal equilibrium of an impurity at a site with an interaction energy of 1.5 eV plotted against the temperature T, the impurity concentration being taken as a parameter.

Calculation with Eq. (1) verifies that the magnitude of p can be close to unity at high temperatures even if the interaction energy is as low as of an order of 0.1 eV for such solute concentration.

Impurity gettering in semiconductors is usually not related to the development of the Cottrell atmosphere around dislocations. Impurities are thought to be gettered by dislocations at high temperature by the following mechanisms : (1) Impurities are supersaturated and dislocations act as preferential nucleation sites of precipitates. (2) Some reaction which incorporates impurity atoms takes place at the dislocation core even if the concentration of impurities is lower than the solubility limit in the matrix region. Special reaction which is absent in the matrix region may occur at the dislocation core because of the peculiarity of atomic arrangement there. The reaction product may have a high energy of interaction with the dislocation. Occurrence of such special reaction at the dislocation core has indeed been observed for supersaturated O impurity in Si.[8,9] Impurities such as Cu, Ag, Au, Li in Si or GaAs may assume high diffusion rates at low temperatures at which the magnitude of p in Eq. (1) is close to unity. Dislocations may getter such impurities in the isolated atomic state.

KINETICS OF IMPURITY GETTERING BY DISLOCATIONS[9]

On the basis of the conclusion in a preceding section that the impurity gettering by a dislocation proceeds by means of preferential precipitation of supersaturated impurities or development of reaction products at the dislocation core, we can deduce the kinetics of impurity gettering by a dislocation rather easily with a simple assumption that the dislocation core is a perfect sink for impurity atoms that arrive there.

The change rate of the impurity concentration C at any place in the stress field of a dislocation is given by

$$\partial C / \partial t = D \, \nabla \, [\nabla C + (C / k \, T) \, \nabla E_i], \qquad (2)$$

where t is the time, D the diffusion constant of impurities and E_i the energy of interaction of an impurity atom with the dislocation. The first term on the right hand side of Eq. (2) is related to the diffusion flow of impurities that originates from the inhomogeneity in the impurity concentration while the second term the drift flow due to the force between the impurity atom and the dislocation. Taking a dislocation that is straight along the z-axis, we have the problem in the two dimensional space. Again considering the interaction through strain fields, the problem is solved in a numerical way with a high capacity computer with the initial condition that the impurity distribution is uniform at $t = 0$. A dislocation is placed at the center of a two dimensional square lattice and the cyclic boundary condition is applied on the periphery of the lattice. The dislocation density in the crystal is defined to be the inverse of the area of the lattice.

The time variation of C at any position in the lattice at a given temperature has been traced. The simulation shows that the concentration of oversized impurity atoms in the dilatation field close to the dislocation core increases slightly in a very early stage of aging. Such accumulation of impurity atoms in the dilational field soon disappears, and the impurity concentration in the region close to the dislocation core starts decreasing. With increase in the aging duration the impurity depletion region expands outwards by keeping approximately a cylinder shape of which central axis is along the dislocation line. At the same time, the number of impurity atoms absorbed at the dislocation core increases steadily. In a late stage of aging impurity atoms to be gettered are exhausted in the matrix and the number of impurity atoms gettered by the dislocation becomes saturated.

The fraction f of the number of impurity atoms gettered by a dislocation to the total number of impurity atoms in the lattice can conveniently be expressed as a function of the aging duration t by the following equation:

$$f = K\ t^{n}, \tag{3}$$

where K is a coefficient that depends on the temperature, the interaction energy and other parameters. As shown in Fig. 3, the magnitude of n in the early stage of aging ($f < 5 \times 10^{-3}$) is 0.79 irrespective of the dislocation density. Over a following wide range of f the magnitude of n depends on the dislocation density, approaching to unity as the dislocation density decreases toward zero.

The result of above simulation has been compared with experiments on gettering of O impurity in Si by dislocations. Dislocations have been introduced into Si crystals at various densities by plastic deformation followed by annealing at high temperature. The annealing eliminates defects other than dislocations that can act as preferential precipitation sites for supersaturated oxygen atoms and, at the same time, dissolves oxygen atoms that were segregated on defects at the time of deformation. Dislocations survived after annealing are rather straight and the density of which is well controlled by adjusting the amount of deformation. The crystals have been subjected to aging to cause the gettering of O impurity by dislocations. The number of gettered oxygen atoms has been determined from the decrease in the concentration of dissolved oxygen atoms that can be measured by infrared absorption. The time law of Eq. (3) has been found to hold quite well. Experimental results on the dependency of $f - t$ behaviour on the dislocation density have also been found to be in good agreement with the theoretical results.

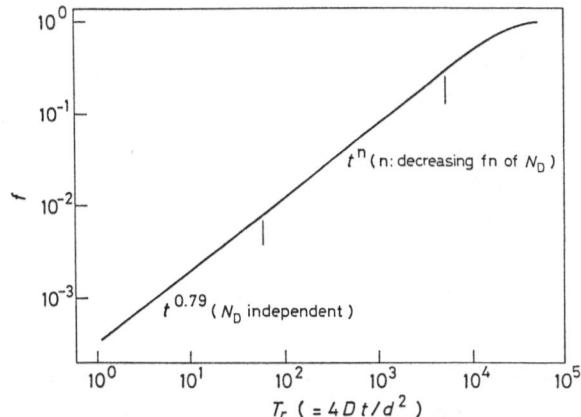

Fig. 3 The fraction f of the number of impurity atoms gettered by a dislocation to the total number of impurity atoms in the lattice calculated as a function of the reduced time T_r. D, d and t are the diffusion coefficient of the impurities, the lattice constant and the aging duration, respectively. N_D in the figure is the dislocation density. The calculation has been done for O impurity in Si.[9]

ELECTRICAL AND OPTICAL INHOMOGENEITIES CAUSED BY DISLOCATIONS

When the gettering of impurities by dislocations is incomplete within a crystal and the concerned impurities are such ones that play an important role in determining the electrical or optical properties of the crystal, the electrical or optical properties of the dislocated crystal are observed to be inhomogeneous reflecting the inhomogeneity in the distribution of impurities in the crystal.

In a Si crystal enriched with recombination centers EBIC images often show bright regions around dislocations.[10] This observation is interpreted most reasonably with an idea that some impurities acting as effective recombination centers are gettered by dislocations during a preceding heat treatment. The concentration of free electrons in an as-grown crystal of GaAs is reported to be higher in the vicinity

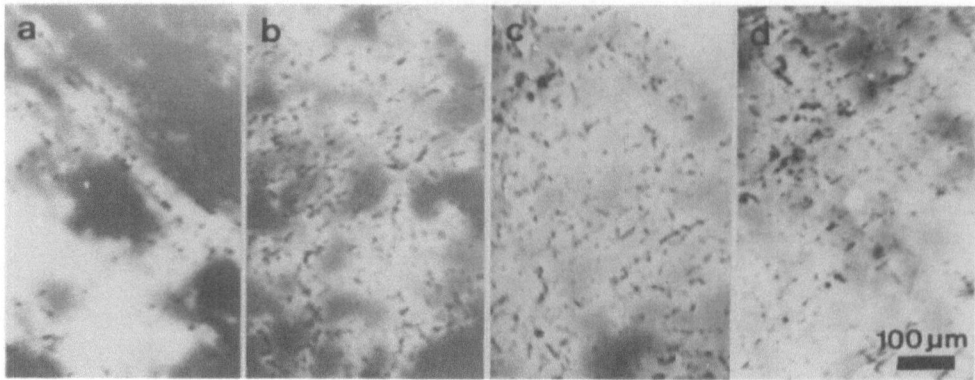

Fig. 4 Change in cathodoluminescence images of LEC-grown GaAs due to isochronal annealing. As-grown state (a), after annealing at 700 ℃ (b), at 750 ℃ (c), and at 800 ℃ (d), each for 24 h.[13]

of dislocations than in the region far from there.[11,12] This effect is often interpreted in terms of the idea that EL2 centers are enriched in spaciously extended regions around dislocations. According to the discussion in the foregoing section, it is improbable that a region enriched with EL2 centers is developed widely around a dislocation since the interaction energy between a dislocation and an EL2 center is too small to enrich EL2 centers when they are separated by more than a few atomic distances. It seems reasonable to interpret the effect to be caused by gettering of acceptor-type impurities or defects by dislocations.

The cathodoluminescence (CL) of an as-grown crystal of GaAs is usually much brighter at the regions in the vicinity of dislocations than in the regions far from dislocations as shown in a micrograph of Fig. 4 (a).[13] The bright regions are denuded zones of non-radiative recombination centers caused by gettering due to the dislocations. The overall CL intensity becomes weaker and uniform throughout the crystal if the crystal is subjected to annealing at 1050 ℃ followed by rapid cooling as shown in a micrograph of Fig. 5 (a).[13] The bright regions around dislocations are eliminated by such a treatment. Non-radiative recombination centers gettered by dislocations in the as-grown GaAs crystal are thought to be released into the matrix region by annealing at 1050 ℃. If the crystal is cooled slowly after annealing at 1050 ℃, the CL image contrast seen in Fig. 4 (a) is restored. Namely, non-radiative centers are gettered by dislocations at some intermediate temperatures.

The temperature range in which non-radiative centers gettered by dislocations at the time of crystal growth are released is determined by observing the change in the CL images of the as-grown crystal due to isochronal annealing. In the same way, the temperature range in which non-radiative centers are gettered most effectively

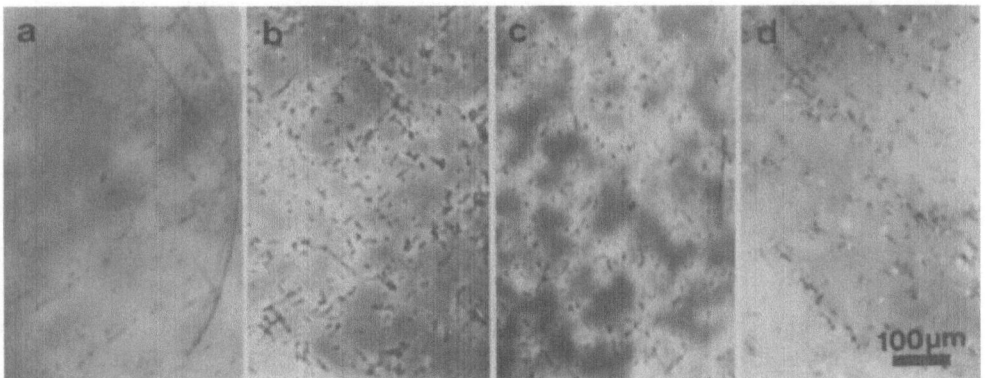

Fig. 5 Change in cathodoluminescence images of homogenized GaAs due to isochronal annealing. After annealing at 1050℃ followed by rapid cooling (a), after annealing at 700℃ (b), at 750℃ (c), and at 800℃ (d), each for 24 h.[13]

by dislocations is determined by the observation of the change in the CL images due to isochronal annealing of the crystal homogenized by the high temperature annealing followed by the rapid cooling.[13] No detectable release of the gettered centers takes place at temperatures lower than 650℃. The non-radiative centers start to be released from dislocations at 700℃. Homogenization in the CL intensity proceeds with increasing temperature as seen in a series of micrographs in Fig. 4. The centers are not completely released from dislocations even by annealing at a temperature as high as 850℃. As to the gettering process, no appreciable gettering of non-radiative centers takes place at temperature lower than 650℃. As seen in a series of micrographs in Fig. 5, the gettering takes place most efficiently at 750 ℃. Annealing at temperatures higher than 800℃ does not cause the gettering of non-radiative centers. It is interesting to note that the gettering does not take place at temperatures where releasing of gettered centers is not complete.   Existence of such thermal hysteresis in the gettering and releasing processes of the centers implies that the centers gettered by dislocations are in the state of precipitates or reaction products of which stability increases with the decrease in the temperature.

DISLOCATION IMMOBILIZATION DUE TO IMPURITY GETTERING

A fresh dislocation makes motion at a velocity which is determined by the temperature and the stress, and also by impurities dissolved in the crystal. The characteristic in the motion of dislocations in semiconductors is that the velocity depends very sensitively on the temperature and rather weakly on the stress in comparison with other kind of materials. When an originally fresh dislocation getters impurities by aging at elevated temperature, it is immobilized since the system is stabilized. An extra stress is needed to start such an aged dislocation moving. Such stress is termed the *release stress* or *unlocking stress* of the dislocation and depends on the temperature and the state of gettered impurities.

Figure 6 shows[14] how the release stress at 647°C for 60° dislocations increases with the duration of aging at 647°C in Si crystals doped with O at a concentration of 1.5×10^{17} or 7.5×10^{17} atoms/cm^3, P at a concentration of 1.2×10^{19} atoms/cm^3, or N at a concentration of 5.5×10^{15} atoms/cm^3. The release stress increases with an increase in the aging duration. The efficiency of immobilization

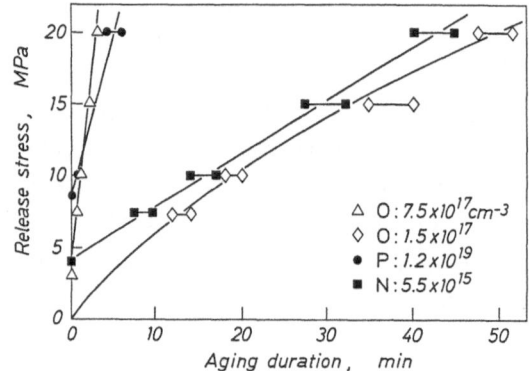

Fig. 6 Variations in the release stress at 647°C of initially fresh 60° dislocations against the duration of aging at 647°C in Si doped with O, P or N impurity the concentrations of which are given in the figure.[14]

depends on the species and the concentration of impurities. The diffusion rate of impurities is also an important factor in determining the immobilization efficiency. It is shown that N and P impurities immobilize a dislocation much more strongly than O impurity if the same number of atoms are gettered by the dislocation. Analysis of the data on the relation between the release stress and the number of gettered impurity atoms and that between the release stress and the temperature verifies that gettered impurity atoms coagulate into small particles separated by some distance along the dislocation line.[14,15]

Figure 7 shows[16] how the release stress at 350°C for α dislocations increases with the duration of aging at 350, 450 and 550°C in GaAs doped with In impurity at a concentration of 2×10^{20} atoms/cm^3 or doped with Si impurity at a concentration of 4×10^{18} atoms/cm^3. Doping with In at such a high concentration has been known to be effective to reduce the density of grown-in dislocations in GaAs grown by the liquid encapsulated Czochralski technique.[17-20] The increase in the release stress in the GaAs doped with Si is small when the temperature of aging is lower than 450°C in comparison to the case of the In-doped GaAs. Immobilization of dislocations becomes stronger in the Si-doped GaAs than in the In-doped GaAs after aging at 550°C for long durations. It is known that Si impurity immobilizes dislocations more strongly than In impurity once it is gettered by the dislocations. However, a lower diffusivity of Si impurity makes the immobilization effect remarkable only at high temperatures.

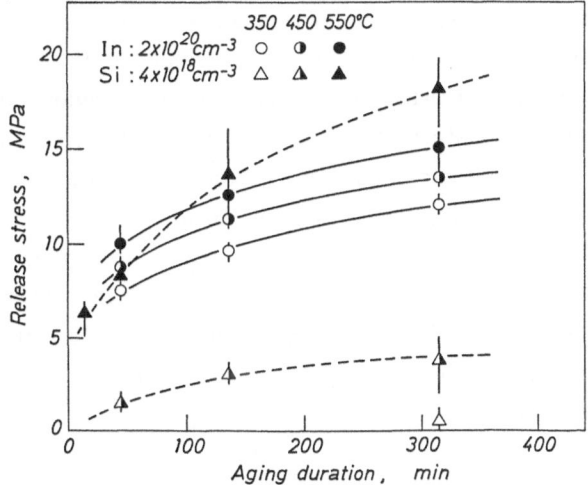

Fig. 7 Variations in the release stress for initially fresh α dislocations at 350°C against the duration of aging at 350, 450 and 550°C in GaAs doped with In or Si impurity of which concentrations are given in the figure.[16]

It is very interesting to note that the efficiency of immobilizing dislocations of any given species of impurities in GaAs depends, in general, very sensitively on the type of dislocations on which they are gettered.[16,21] For a given aging treatment In and Te impurities show much stronger immobilizing effects on α dislocations than on β dislocations, while Al and Zn impurities show stronger effects on β dislocations than on α dislocations. On the other hand, Si impurity shows almost the same immobilization effects on α and β dislocations. These points are discussed in the next section.

IMPURITY EFFECT ON DISLOCATION GENERATION

The exact mechanism of dislocation generation in a crystal has not yet been fully understood in spite of the fact that it is one of the most important problems to be solved from the practical point of view. Dislocations are generated heterogeneously when the crystal is stressed. Usually, they are observed to propagate from the surface region into the bulk crystal. It is thought that any crystal has some irregularities on the surface that act as preferential generation centers of dislocations under stress. Introducing a definite type of flaws on the crystal surface that act as effective generation centers of dislocations, we are able to study how various kinds of impurities affect the generation of dislocations. Such a flaw on the surface of a semiconductor crystal accompanies a region highly disturbed in the structure around it. An *in situ* X-ray topographic study[22] has verified that such disturbed region is converted to a dislocated microregion when the crystal is brought to a high temperature. If the crystal is under stress, some dislocations with the maximum Schmid factor come out of the microregion and penetrate into the matrix crystal and expand on a macroscopic scale. At this stage one recognizes that dislocations are generated. Dislocations are easily generated from a flaw in an undoped crystal even under extremely low stresses at high temperature. However, in a crystal doped with a certain kind of impurities there is a critical stress below which no dislocation generation takes place even though the flaw is converted to a dislocated microregion.

Figure 8 shows how the critical stress depends on the temperature in the case that surface flaws are made by a Knoop indenter at room temperature on Si crystals grown by different techniques.[22] The indented crystals were heated quickly to the temperature of the abscissa under stress. In the case of high purity Si grown by the floating-zone technique, the critical stresses are measured to be lower than 1 MPa which is the minimum stress applicable with an equipment used over the whole temperature range investigated. On the other hand, in Czochralski-grown Si fairly high magnitudes of the critical stress are measured. The critical stress increases with an increase in the temperature. O impurity dissolved in a Si crystal is known not to affect the velocity of a dislocation in motion. The mobility of a dislocation increases with an increase in the temperature. Thus, the critical stress for dislocation generation in Fig. 8 is not related to the resistance to the dislocation motion due to O atoms dissolved in the crystal. It is concluded that the critical stress is related to the release stress of dislocations in the microregion around the flaw that are immobilized during heating of the crystal due to the gettering of O impurity.

A variety of suppressing effects of impurities on dislocation generation has been observed in GaAs.[16,21,23] Figure 9 shows the critical stresses for generation of α dislocations above and β dislocations below from a scratch as a function of the temperature in GaAs doped with isovalent impurities of In of three different concentrations or Al. The critical stresses are almost the same as the detection limit of the stress of the used equipment for both α and β dislocations in undoped GaAs in the temperature range $300 - 700\,°C$. The critical stress for α dislocations in GaAs doped with In at concentrations higher than about 1×10^{20} atoms/cm^3 is very high

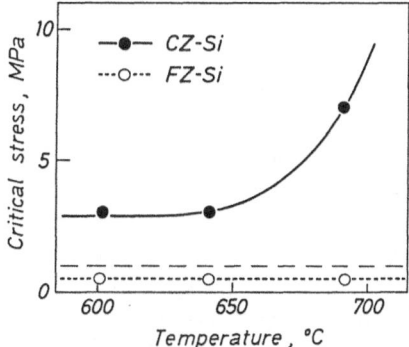

Fig. 8   Variations of the critical stress for dislocation generation from a Knoop indentation against the temperature in floating-zone Si (FZ-Si) and Czochralski-grown Si (CZ-Si).[22]

over the whole temperature range. It increases with an increase in the In concentration and decreases with an increase in the temperature. The critical stress for α dislocations in GaAs doped with In at a concentration of 3×10^{19} atoms / cm^3 or Al at a concentration of 3×10^{18} atoms/cm^3 is almost negligible at temperatures lower than $450\,°C$ and starts to increase with the temperature from $500\,°C$. As to the generation of β dislocations, Al impurity at a concentration as low as 3×10^{18} atoms/cm^3 shows a very remarkable effect similar to that In impurity of a concentration higher than 10^{20} atoms/cm^3 does on α dislocations. The effect of In impurity on the generation of β dislocations is much weaker even at concentrations higher than 10^{20} atoms/cm^3 except the temperature range higher than about $700\,°C$.

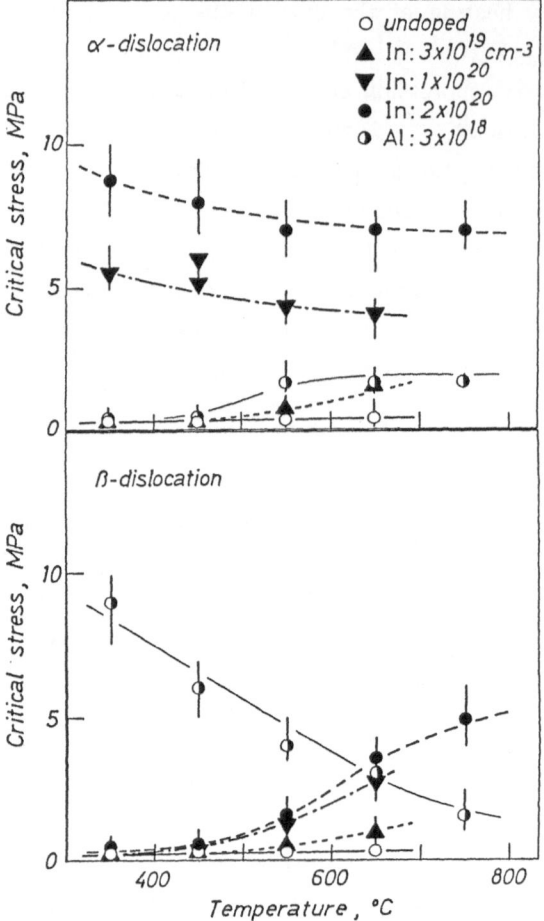

Fig. 9 Critical stresses for generation of α dislocations above
and β dislocations below from a scratch as functions
of the temperature in GaAs doped with isovalent impurity of In or Al of which concentrations are
given in the figure.[16,21] Data for undoped GaAs are
also shown.

In impurity and Al impurity are both group Ⅲ elements and have been
shown not to affect the velocities of both α and β dislocations in motion.[21] Hence,
the difference in the behaviour of the critical stress for generation between In-doped
GaAs and Al-doped GaAs seen in Fig. 9 can never be interpreted in terms of the

effects of the impurities on the dislocation mobility. It is well correlated to the
immobilizing behaviour of the two kinds of impurities mentioned in the preceding

section. Thus, as in the case of Czochralski-grown Si, the critical stress for dislocation generation is ascribed to the immobilization of dislocations due to impurity gettering during the heating of the crystal. In and Al impurities both occupy the Ga site in GaAs in the dissolved state. An α dislocation has a row of As atoms at the core. Thus, an In or Al atom can occupy the site nearest to such As atom row where the energy of interaction with the dislocation is the highest. This picture gives a reasonable explanation for why α dislocations are immobilized by In impurity in preference to β dislocations. Apparently, this kind of explanation does not apply to the case of Al-doped GaAs. In this case the concentration of Al impurity which gives rise to the effect is far lower than the concentration of In impurity concerned. Further, the misfit of an individual Al atom is much smaller than that of an individual In atom in GaAs. It is thought that some special chemical reaction which incorporates Al takes place at the core of a β dislocation at rather low temperatures but not at the core of an α dislocation.

Figure 10 shows the effects of electrically active impurities on the generation of α dislocations above and β dislocations below in GaAs.[16,21,23] Si and Te which are both donor-type impurities show quite different effects on the generation behaviour of α dislocations. Te impurity at a concentration of 6×10^{18} atoms$/$cm^3 has a strong effect on the critical stress for generation of α dislocations which is similar to that of In impurity of 2×10^{20} atoms$/$cm^3. Effect on the generation of β dislocations is also similar to that of In impurity. Si impurity has almost the same effects on the generation of α and β dislocations which are not significant at low temperatures. It is, thus, found that selective immobilization of α dislocations by Te impurity is not related to its electrical activity as donor. Zn impurity acting as acceptor affects the generation of β dislocations strongly but not that of α dislocations. Te impurity occupies the As site while Zn impurity the Ga site, both being the second nearest sites of the core atoms of the dislocations which they immobilize preferentially. Thus, we may again conclude that the reaction incorporating the impurities at the dislocation core plays an essential role in immobilizing dislocations.

Figure 11 shows[21] the critical stress for the generation of screw dislocations as a function of the temperature in GaAs doped with In, Zn or Si together with that in undoped GaAs. It is seen that screw dislocations are strongly immobilized by In and Zn impurities which are effective in immobilizing α and β dislocations, respectively. Effect of Si impurity on screw dislocations is similar to those on α and β dislocations. An α dislocation consists of a 90° partial and a 30° partial both of α type and a β dislocation the two partials of β type. On the other hand, a screw dislocation consists of a 30° partial of α type and a 30° partial of β

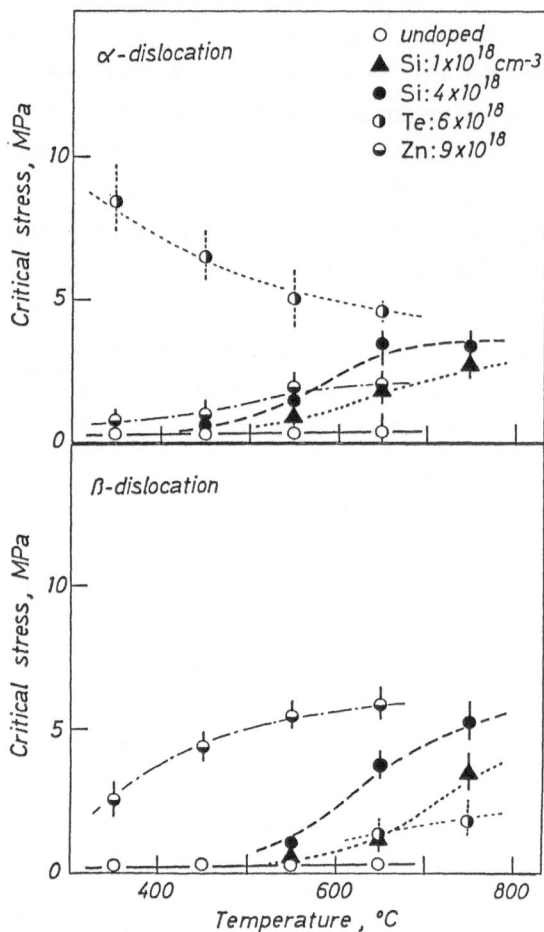

Fig. 10 Critical stresses for generation of α dislocations above and β dislocations below from a scratch as functions of the temperature in GaAs doped with electrically active impurities of Si, Te or Zn of concentrations are given in the figure.[21]

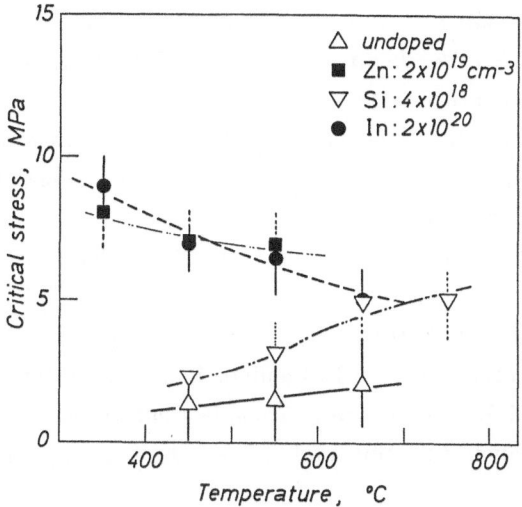

Fig. 11 Critical stress for generation of screw dislocations from a scratch as a function of the temperature in GaAs doped with In, Si or Zn impurity of which concentrations are given in the figure. Data for undoped GaAs are also shown.[21]

type. This leads to the conclusion that 30° partial dislocations play an important role in the immobilization of dislocations by impurity gettering.

Suppression of the dislocation generation by impurity gettering is now successfully utilized in the technique to grow large crystals of GaAs nearly free from dislocations.

The author is grateful to I. Yonenaga and T. Sekiguchi of his research group for their cooperation in the works, on which this review is based.

REFERENCES

1. K. Sumino, *Defects and Properties of Semiconductors : Defect Engineering*, J. Chikawa, K.Sumino and K.Wada, ed., KTK Scientific Publishers, Tokyo (1987) p.227.
2. K. Sumino, Oyo Buturi, **56**, 860 (1987).
3. K. Sumino, *Proc. 1st International Autumn School on Gettering and Defect Engineering in the Semiconductor Technology*, H. Richter, ed., Academy of Sciences of the GDR, Frankfurt (Oder) (1985) p.41.

4. K. Sumino, *Proc. 2nd International Autumn School on Gettering and Defect Engineering in the Semiconductor Technology*, H, Richter, ed., Academy of Sciences of the GDR, Frankfurt (Oder) (1987) p.218.

5. K. Sumino, *Defects in Semiconductors II : Materials Research Society Symposia Proceedings* 14, S. Mahajan and J. W. Corbett, ed., North-Holland, New York／Amsterdam／Oxford (1983) p.409.

6. J. P. Hirth and J. Lothe, *Theory of Dislocations*, John Wiley & Sons, New York (1982) p.497.

7. H. Ehrenreich and J. P. Hirth, Appl. Phys. Lett. **47**, 668 (1985).

8. M. Koguchi, I. Yonenaga and K. Sumino, Jpn. J. Appl. Phys. **21**, L411 (1982).

9. I. Yonenaga and K. Sumino, *Proc. Yamada Conf. IX on Dislocations in Solids*, H. Suzuki, T.Ninomiya, K.Sumino and S.Takeuchi, ed., Univ. Tokyo Press, Tokyo (1985) p.385.

10. Y. Miyamura and K. Sumino, to be published.

11. S. Miyazawa, T. Honda, Y. Ishii and S. Ishida, Appl. Phys. Lett. **44**, 410 (1984).

12. T. Takebe, S. Murai, K. Tada and S. Akai, Inst. Phys. Conf. Ser. No.79, 283 (1985).

13. T. Sekiguchi and K. Sumino, Jpn. J. Appl. Phys. **26**, L179 (1987).

14. K. Sumino and M. Imai, Philos. Mag. **A47**, 753 (1983).

15. M. Sato and K. Sumino, *Proc. Yamada Conf. IX on Dislocations in Solids*, H. Suzuki, T. Ninomiya, K. Sumino and S. Takeuchi, ed., Univ. Tokyo Press, Tokyo (1985) p.391.

16. I. Yonenaga and K. Sumino, J. Appl. Phys. **62**, 1212 (1987).

17. M. G. Mil'vidskii, V. B. Osvenskii and S. S. Shifrin, J. Cryst. Growth **52**, 396 (1981).

18. G.Jacob, M. Duseaux, J. P. Farges, M. M. B. van den Boom and P. J. Roksnoer, J. Cryst. Growth **61**, 417 (1983).

19. H. Kohda, K. Yamada, H. Nakanishi, T. Kobayashi, J. Osaka and K. Hoshikawa, J. Cryst. Growth **71**, 813 (1985).

20. T. Ibuka, Y. Seta, M. Tanamura, F. Orito, T. Okano, F. Hyuga and J. Osaka, *Semi-Insulating III-V Materials*, Hakone, 1986, H. Kukimoto and S. Miyazawa, ed., Ohm and North-Holland, Tokyo and Amsterdam (1986) p.77.

21. I. Yonenaga and K. Sumino, J. Appl. Phys. **64**, No. 12 (1988) in press.

22. K. Sumino and H. Harada, Philos. Mag. **A44**, 1319 (1981).

23. I. Yonenaga, K. Sumino and K. Yamada, Appl. Phys. Lett. **48**, 326 (1986).

GETTERING MECHANISMS IN SILICON

W. Schröeter and R. Kuehnapfel

IV. Physikalisches Institut and Sonder-
forschungsbereich 126 Goettingen/Clausthal,
Bunsenstr. 11-15, D-34 Goettingen, West Germany

1 INTRODUCTION

Gettering in silicon is a well established procedure by which metallic impurities are concentrated within a predetermined part of the specimen. For example, in phosphorous diffusion gettering (PDG) the gettering part is a very thin silicon layer at the Si/PSG-interface (PSG:amorphous phosphorsilicate glass [1]). This layer becomes (1) highly doped by the in-diffusion of phosphorous up to the maximum solubility (920°C: $3 \cdot 10^{20}$ cm^{-3}), and (2) supersaturated with silicon self – interstitials by injection from the advancing Si/PSG-interface (see fig. 1). In its initial state the specimen has metallic impurities more or less homogenously distributed on lattice or interstitial sites, in pairs with the dopant or in particles. PDG changes this distribution into a highly inhomogenous one with the impurities concentrated within a layer of a few thousand Angstrom width.

Because of its technological importance gettering has been well characterized with respect to optimum operation. On the other hand , the fundamental aspects of gettering have been investigated in some detail only recently and are still not clear. It is the aim of this paper to briefly summarize concept and experimental evidence of those gettering processes which are well understood.

In a second part we describe in some detail what has been found experimentally about silicide formation during PDG. The main experimental and conceptual features about the mechanisms, which bring about gettering of metallic impurities in silicon, are briefly summarized in this paper. Silicide formation during phosphorous diffusion gettering is discussed in some detail and shown to be a non-equilibrium phenomenon. A tentative model, which explains the main experimental observations, is developed on the basis of a local coupling between the current of self interstitials and the metallic impurities.

Finally we propose a dynamic gettering process which may lead to the formation of silicides in the neighbourhood of the Si/PSG-interface.

2 Basic mechanism of gettering

From the experimental results and their interpretation three different mechanisms of gettering have been considered.

(1) Gettering due to the Fermi level shift regards the variation of point defect equilibria and impurity solubilities with the Fermi level shift. During PDG phosphorous indiffusion causes a shift of the Fermi level E_F up into the conduction band within a thin layer at the Si/PSG-interface (see fig. 1). Any point defect, that introduces an acceptor level E_A into the band gap, increases its solubility with E_F moving up.

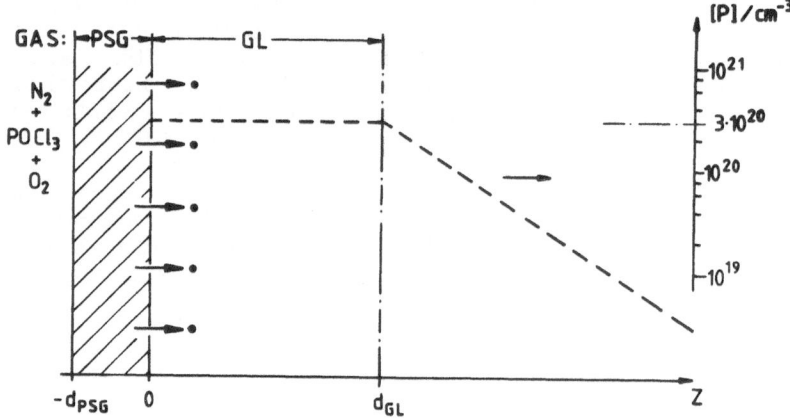

Figure 1. Characteristics of PDG: a gas mixture of N_2 as carrier gas, saturated with $POCl_3$ at 24°C, and O_2 flows past the specimen surface, kept at 920°C, with constant flow rate (1580 ml/min) and leads to (1) a PSG-layer growth into the silicon and (2) to phosphorous-doping of the near interfacial layer of silicon. The processes (1) and (2) define PDG. The gettering layer (GL) is the layer, in which the electrically active phosphorous concentration is at the solubility limit.

The concentration ratio of the negatively charged and the uncharged impurity species is determined by the gain in electronic energy according to

$$[M^{(-)}]/[M^{(0)}] = \exp((E_F - E_A)/k \cdot T)$$

The total impurity solubility is $[M]_{eq}=[M^{(-)}]_{eq}+[M^{(0)}]_{eq}$. Its variation is largest if the charged species is already dominant in the intrinsic material. For silicon at 920^0C, which is a frequently used temperature for PDG, the impurity solubility may increase by up to a factor of 10^2, for a double acceptor the maximum increase would be by 10^4.

(2)   Gettering by relaxation occurs by the preferred nucleation of impurity precipitation at extended defects, which are generated during the gettering procedure.

Many of the metallic impurities - like Fe,Co,Ni and Cu - dissolve predominantly as interstitials in silicon and have extremely large diffusivities (920^0C: 10^{-6} - 10^{-4} cm^2/s) with weak temperature dependence (activation enthalpy: 0.4 - 0.7 eV) [3,4]. Others - like Au and Pt with predominant substitutional solubility - are kicked into interstitial sites by reaction with self interstitials ($Au_s+Si_i \rightleftharpoons Au_i$) during gettering and become then highly diffusive [5]. Referring to these properties the question has been raised, which part of gettering has to be ascribed to relaxation into substitutional sites, into pairs or precipitates at preferential sites during cooling. There is good experimental evidence [6,7], that precipitation at stacking faults during cooling is one basic mechanism of intrinsic gettering, by which a small volume (denuded zone) is purified due to gettering action of a larger volume.Stacking faults offer an easy volume adaptation for metal silicides (see section 4).

On the other hand, during PDG metallic impurities, initially distributed over all the specimen, concentrate within a thin layer (<1 μm) at the Si/PSG-interface. It has been observed that cobalt [8] and iron [9] diffuse out during PDG with a smaller or about the same diffusivity, with which they diffuse in, respectively. This means that outdiffusion combined with precipitation during cooling is not an essential mechanism of PDG and will not be considered further in this work. We mention that gettering by relaxation is also responsible for haze formation, which is applied in semiconductor technology as a procedure to check the presence of metallic impurities in a wafer[10].

(3)   Dynamic gettering results from the interaction of metallic impurities with intrinsic point defects, whose equilibria have been disturbed by the gettering procedure.

The Si/PSG-interface penetrates into silicon, thereby generating interstitials, whose role for PDG has been first noticed in the bulk of the specimen. Lescronier et. al.[11] have found that the diffusion of gold to the gettering layer is by orders of magnitude faster than in normal diffusion experiments and comparable to that of interstitial impurities. These findings have been explained by the action of a surplus of self interstitials on the kick-out reaction of gold from the substitutional to the interstitial site[12-14]. This interpretation has been confirmed also for platinum by Falster [14]. It has been also shown that self interstitials in supersaturation might destabilize precipitates in the bulk [15]. Gettering needs a directed force on and a sufficient mobility of the impurity. Therefore the above mentioned results establish a very important role of self interstitials, which transform nearly immobile impurity species into highly mobile one.

3 Phosphorous diffusion gettering

Proceeding to PDG we describe the experimental results and the models derived from these to explain PDG. Most of the data, that are available now, concern the impurity distribution within the highly phosphorous doped near-interfacial layer GL.

3.1 Fermi level effect

The Fermi level effect has been investigated in some detail by Chou and Gibbons [16], Tseng et. al.[17], and Lescronier and coworkers [11]. The last group applied neutron activation analysis (NAA), deep level transient spectroscopy (DLTS), Rutherford backscattering (RBS) and channelling spectroscopy to characterize the gold distribution as a function of the phosphorous concentration and gettering temperature. To adjust the P-concentration, they span P-doped silica-films on the wafers. The injection rate of self interstitials is presumably rather weak in their experiments, so that this is not PDG in the sense of the definition given in the introduction.

In short, they observed that the gold profile follows the phosphorous profile, that accumulation of gold on a substitutional site begins as soon as the Fermi level is raised above $E_c-0.15$ eV, and that this process is reversible. From NAA and RBS the maximum gold concentration near to the interface was about $6 \cdot 10^{17}$ cm^{-3} at 900^{0}C, which is a factor of about 200 above the solubility in intrinsic silicon ($3 \cdot 10^{15}$ cm^{-3}). This factor can be accounted for by the Fermi level effect, if pairing between gold and phosphorous or an reaction of gold with E-centres is also considered[11]

3.2 Silicide formation within the gettering layer

New results on Pt, which dissolves mainly substitutionally,and on Ni, Fe and Co, which occupy interstitial sites in intrinsic silicon, made it very clear, that the model of Lescronier et. al., although correct in describing the Fermi level effect as a part of PDG, does not comprise the total phenomenon.

Ourmazd and Schroeter [18], using high resolution electron microscopy (HREM), showed that $NiSi_2$-particle epitaxially grow at the Si/PSG-interface, especially in the vicinity of SiP-particles. SiP-particle growth, occuring under well defined conditions of phosphorous and oxygen activity [19], is accompanied by emission of silicon self interstitials, about 20% of which are deposited on the Si/PSG-interface in close vicinity of the SiP-particles [20]. In the presence of nickel the self interstitial deposition apparently is replaced or accompanied by $NiSi_2$-growth[18]. Also $FeSi_2$

particle growth has been observed by HREM, again under PDG with SiP-particle growth[21].

Falster[14], using SIMS (secondary ion mass spectroscopy) and RBS has investigated PDG of platinum, which is expected to behave like gold. He found, that its concentration profile from the Si/PSG-interface into the silicon bulk does not follow the phosphorous profile and that platinum is gettered on a non-substitutional site.

These results give clear evidence that silicide formation contributes to PDG.

3.3 Role of SiP-particle growth

Another question - raised by the results of the HREM-studies - concerns the role of SiP-particle growth for silicide formation during PDG. Making use of the carefully established relations between gas mixture and P-doping, PSG-growth and SiP-growth by Negrini et. al.[19], Kuehnapfel et. al.[8] have compared cobalt concentration profiles in the gettering zone. The profiles after PDG with and without SiP-particle growth were found to be comparable, although different in some details. The differences have been revealed by Moessbauer spectroscopy, which allows to determine the local concentrations of all gettered species of cobalt separately with a depth resolution of about 30 nm. In general two gettered species are found, one represented by a single line, the other by a quadrupole doublet. The concentrations of the two species are comparable, when no SiP-particle grow, the quadrupole doublet dominates, when SiP-particle grow during PDG. In the last case there are two additional lines within the first 35nm to the interface, which represent about 30% of the total gettered amount. It has been also found that the species, which is associated with the single line, is metastable at 600^0C and transforms into the species, which is represented by the quadrupole doublet[22].

The quadrupole doublet originates in platelets consisting of two $\{111\}$-$CoSi_2$-planes (see section 4). The species represented by the single line has not been unambiguously identified till now. The same line has been observed in highly phosphorous-doped (10^{20} P-atoms/cm^3) Czochralski silicon after in-diffusion of cobalt at 700^0C and quenching[24]. It has been tentatively ascribed to cobalt complexes containing oxygen-related defects.

3.4 Coupling within the gettering layer

Two further experimental investigations have led to a plausible mechanism for the coupling of cobalt to self interstitials within the gettering layer and to a preliminary model of PDG by silicide formation.

The first of these investigations[25] considered the question how the solubility and diffusivity of cobalt is changed by phosphorous doping and which species are responsible for this change. In silicon doped with 10^{20} P-atoms/cm^3 at 700^0C negatively charged substitutional cobalt

Co$_s$ and its pair with phosphorous Co$_s$P have been found to be the dominant cobalt species [9]. Their existence leads to an increase of the cobalt solubility by several orders of magnitude in highly P-doped silicon (the cobalt solubility at 700^{0}C increases from 4 10^{11} cm^{-3} in intrinsic silicon to 2 10^{15} cm^{-3} in silicon doped with 1 10^{20} P-atoms/cm^3) and also brings about a coupling of cobalt to self interstitials within the gettering layer.

The second experimental investigation[26] has shown that two processes have to cooperate to achieve effective gettering: (1) high phosphorous doping by in-diffusion and (2) PSG-growth. Process (1) enhances the solubility of substitutional cobalt, not affecting the interstitial one[25]. Process (2) injects self interstitials, which via the kick-out reaction reduce the concentration of substitutional and enhances that of the interstitial species (see fig. 2). Each of these processes alone achieves only moderate gettering (see table 1).

Table 1.   Total gettered amount of cobalt for various boundary conditions $[Co]_{getter} = \int_o^d dz\ [Co]$
gettering temperature: 920^{0}C
gettering time: 108 min

boundary condition	$[Co]_{getter}(10^{12}cm^{-2})$		
only P-diffusion	1.7[1)		
only oxide growth	1.4[1)		
P-diffusion and PSG-growth	15.0[2)	1) d=10^3 nm	2) d=d$_{GL}$

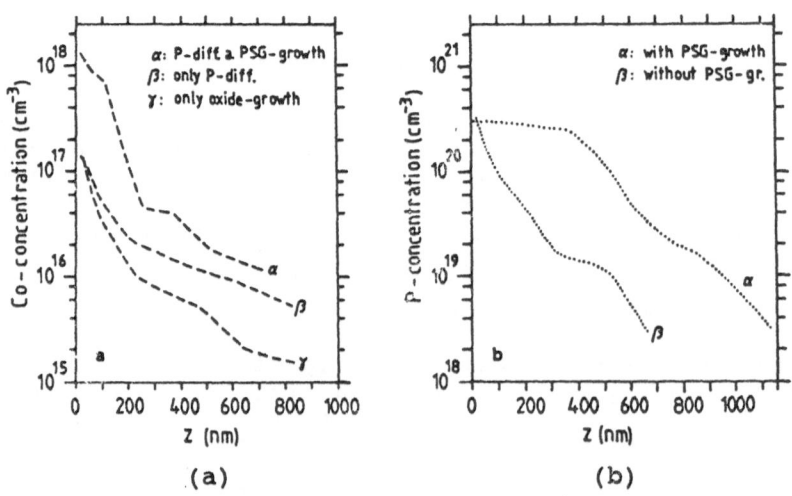

(a)   (b)

Figure 2.   Concentration profiles of cobalt(a), and of phosphorous(b) for the boundary conditions as described in the inserts, all other parameters kept constant, T = 920^{0}C, t = 108 min.

The cooperation of both processes also builds up a plateau with a constant electrically active phosphorous concentration of $3 \cdot 10^{20}$ cm^{-3}. The length of this plateau defines the width of the gettering GL layer and is given in table 2.

Table 2.   Length of the plateau with a constant electrically active phosphorous concentration of $3 \cdot 10^{20}$ cm^{-3} (d_{GL}) for different times, T=920^{0}C.

t (min)	d_{GL} (nm)
6	125
14	200
108	380

3.5 Gettering by silicide formation: a tentative model

The model, which we propose to explain gettering by silicide formation, will be described with reference to fig.3 [26].

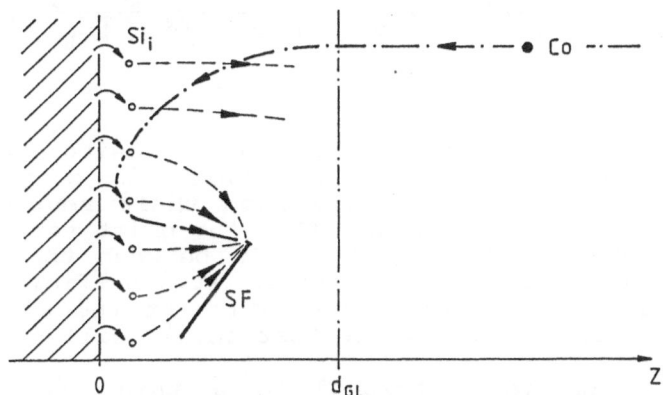

Figure 3.   Proposed mechanism for the gettering by silicide formation: the advancing Si/PSG-interface injects self interstitials, which interact with substitutional cobalt within the gettering layer. If a stacking fault (SF) is in a position near to the interface, a large self interstitial current gives rise to a cobalt drift towards the stacking fault and leads to a supersaturation of cobalt and cobalt silicide formation there.

At the Si/PSG-interface, which advances during PDG into the silicon, self interstitials are generated. As a consequence a current of self interstitials is directed from the interface to some sinks in the bulk - like stacking faults or dislocations. Please note, that during PDG with SiP-particle growth self interstitials are also generated at the advancing SiP/Si-interface. A part of these interstitials has been observed to condensate at the Si/PSG-interface and to form epitaxial silicon[20] or, if nickel was present in the specimen, epitaxial NiSi$_2$[18].

The local current density $J_{si,i}$ increases with decreasing distance between source and sink. We propose that configurations with large local currents of self interstitials can lead to silicide formation. We use a rather simplified calculation to show that the coupling of self interstitials to substitutional cobalt leads to a cobalt drift current, whose magnitude increases with increasing $j_{si,i}$. We assume that the current of self interstitials between source and sink is one-dimensional and also consider the cobalt current as one-dimensional. Actually the long range transport of cobalt from the bulk to the gettering layer and within is not one-dimensional. We further consider only two cobalt species, Co_i and Co_s, and assume local equilibrium with respect to the kick-out reaction, i.e. $[Co_s] [Si_i]= K [Co_i]$. A more realistic treatment of this problem would not change our qualitative conclusions. The current of cobalt atoms is given by:

$$J_{Co} = - D_i \cdot \frac{\partial [Co_i]}{\partial z} - D_s \cdot \frac{\partial [Co_s]}{\partial z}$$

Taking into account the kick-out reaction, one easily obtains the following relation:

$$J_{Co} = - \frac{K \cdot D_s + [Si_i] \cdot D_i}{K + [Si_i]} \frac{\partial [Co]}{\partial z} + \frac{k}{(k + [Si_i])^2} \frac{D_i - D_s}{D_{Si,i}} J_{Si,i} \cdot [Co]$$

The first term represent a cobalt diffusion current, the second a cobalt drift current, which is coupled to the current of self interstitials from their source to their sinks. It has been shown[26] that the drift term gives rise to a cobalt concentration gradient between the source and sink of the self interstitials. For sufficiently large drifts the local cobalt concentration at the sink will exceed the cobalt solubility so that cobalt will precipitate.

This mechanism is restricted to a thin layer near the interface, since large concentration gradients of self interstitials can build up only there. Near to the self interstitial source the cobalt drift will generate a cobalt undersaturation, which initiates long-range cobalt diffusion from the bulk (see fig.3).

One signature of the gettering strength caused by this mechanism are the large concentration gradients [8], which are measured near to the Si/PSG-interface, a second one is the shape metastability of the $CoSi_2$-precipitates[23,26].

4 Metastable silicide formation from supersaturation

Silicide particles, that are metastable with respect to their shape or their composition, have been also detected after diffusion and quenching from high temperatures. They have been explained as a result of large supersaturation. Actually the identification of the Moessbauer quadrupole doublet, representing one of the two dominant cobalt species after PDG, comes from HREM-studies of Co-[23] and Ni-doped[28] silicon, brought into supersaturation by quenching. Because of large solution enthalpies (2.1 and 1.5 eV for Co and Ni in Si, respectively[3]) unusually large driving potentials of the order of 0.5 to 1 eV/impurity atom are obtained in these systems by quenching.

$CoSi_2$ and $NiSi_2$, both crystallizing in the CaF_2-structure with a misfit with respect to silicon below 1%, grow as platelets consisting of two {111}-planes, and - as has been demonstrated for $NiSi_2$-platelets[28] - bordered by dislocations with Burgers vector $\underline{b} = 1/4 \cdot <111>$[22]. It has been argued, that diffusion of interstitial nickel to the dislocation and its incorporation into the $NiSi_2$-phase within the dislocation core is the largest-rate process to lower supersaturation (see Bene's hypothesis[29]). The fact, that the coordination of the metallic atoms in the platelets is reduced, accounts for the quadrupole doublet in the Moessbauer spectrum. At moderate annealing, so that long-range diffusion of cobalt is not possible, the $CoSi_2$-platelets transform into more compact particles[23].

At present it has to be regarded as an open question, whether for gold in silicon gettering by silicide formation can dominate over gettering due to the Fermi level shift. The phase diagram of Au/Si shows only a eutectic and no gold silicides. However, it has been shown recently, that gold silicides precipitate at stacking faults from a supersaturated solution[30]. The variation of the stacking fault diameter during gold silicide growth indicates, that the silicide particles adapt their volume by exchanging silicon interstitials with the stacking fault.

5 Gettering and the defect chemistry at the Si/SiO_2-interface

An important factor for modern device technology is the reliability of the gate oxide. Metallic impurities gettered at the stacking fault act as nucleation centers for oxide decomposition at the Si/SiO_2-interface, which finally, since SiO is volatile, leads to void formation within the interface during inert ambient anneals[31,32]. SiO-formation is associated with enhanced hole trapping in the oxide, a reduction of the effective oxide thickness and low-field breakdown.

Acknowledgement

The authors are grateful to A. Ourmazd, D. Gilles, M. Seibt and J. Utzig for valuable discussions C.A. Warwick, K. Ahlborn, T. Tuetken and M. Schrader for help in the preparation of the manuscript.

References

[1] Solmi S., Alotti G., Nobili D., and Negrini P., J. Electrochem. Soc. 123, 654 (76)

[2] Shockley W., and Moll J., Phys. Rev. 119, 1480, (60)

[3] Weber, E. R., Appl. Phys. A30, 1 (83)

[4] Gilles D., and Utzig J., Defects in Semiconductors, Budapest 1988, to be published

[5] Goesele U., Frank W., and Seeger A., Appl. Phys. 23, 361 (80)

[6] Ueda O., Nauka K., Lagowski J., and Gatos H. C. in: Defects in Semiconductors 1986, ed. by H. J. von Barderleben, Material Science Forum 10 - 12, 145 (86)

[7] Graff K., Hefner H. A., and Hennerici W., J. Electrochem. Soc. 135, 952 (88)

[8] Kuehnapfel R., Schroeter W., and Gilles D. in: Defects in Semicondcutors 1986, ed. by H. J. von Bardeleben, Material Science Forum 10 - 12, 151 (86)

[9] Gilles D., Thesis Goettingen 1987, to be published

[10] Seibt M., and Graff K., J. Appl. Phys. 63, 4444 (88)

[11] Lescronier D., Paugham J., Pelous G., Richou F., and Salvi M., J. Appl. Phys. 52, 5090 (81)

[12] Bronner G. B., and Plummer J. D., MRS Proc. 36, 49 (85), and J. Appl. Phys. 61, 5286 (87)

[13] Ourmazd A., MRS Proc. 59, 331 (86)

[14] Falster R., Appl. Phys. Lett. 46, 737 (85)

[15] Polignano M. L., and Gerofolini G. F., J. Appl. Phys. 64, 869 (88)

[16] Chou S. L., abd Gibbons J. F., J. Appl. Phys. 46, 1197 (75)

[17] Tseng W. F., Koji T., Mayer J. W., and Seidel T. E., J. Appl. Phys. 33, 442 (78)

[18] Ourmazd A., and Schroeter W., Appl. Phys.. Lett. 45, 781 (84)

[19] Negrini P., Nobili D., and Solmi S., J. Electrochem. Soc. 122, 1254 (75)

[20] Bourret A., and Schroeter W., Ultramicrsocopy 14, 97 (84)

[21] Ourmazd A., and Schroeter W., MRS Proc. 36, 25 (85)

[22] Shaikh A. A., Schroeter W., and Bergholz W., J. Appl. Phys. 58, 2519 (85)

[23] Utzig J., J. Appl. Phys. 64, 3629 (88)

[24] Gilles D., and Schroeter W., to be published

[25] Gilles D., and Schroeter W. in: Defects in Semiconductors 1986, ed. by H. J. von Bardeleben, Material Science Forum 10 - 12, 169 (86)

[26] Kuehnapfel R., and Schroeter W., to be published

[27] Martin G., Phil. Mag. A38, 131 (78)

[28] Seibt M., and Schroeter W., Phil. Mag. B in print

[29] Bene, A., J. Appl. Phys. 61, 1826 (87)

[30] Baumann F., and Schroeter W., Phil. Mag. Lett. 57, 75 (88)

[31] Liehr M., Bronner G. B., and Lewis J. E., Appl. Phys. Lett. 52, 1892 (88)

[32] Rubloff G. W., Hoffmann K., Liehr M., and Young D. R., Phys. Rev. Lett. 58, 2379 (87)

EFFECT OF IMPURITY SEGREGATION ON THE ELECTRICAL

PROPERTIES OF GRAIN BOUNDARIES IN POLYCRYSTALLINE SILICON

S. Pizzini, F. Borsani, A. Sandrinelli, and D. Narducci

Dipartimento di Chimica Fisica ed Elettrochimica, Via Golgi 19
20133 Milano

F. Allegretti

INFN, Sezione di Roma, Piazza A.Moro 2, 00185 Roma

INTRODUCTION

Solute segregation at (or near) grain boundaries (GB) of polycrystalline solids has been widely observed and it is recognized[1] to be driven by an electrical field (in the case of charged species) or by an elastic-strain field at inherently distorted GB regions, in close similarity with the case of dislocations, where elastic strain is known to produce the so called "Cottrell atmosphere" or a solute impurity cloud[2] .

The set up of a chemical potential gradient could be an additional cause of segregation, if impurities undergo chemical reactions at residual dangling bonds, with the formation of stable chemical compounds. Although the direct contribution of chemical potential gradients towards segregation may be considered very modest if GB are at least partially reconstructed (as an example, a density of 10^4 dislocations cm^{-2} demands only a few 4×10^{11} impurities cm^{-3} to saturate the 100% of dangling bonds), nevertheless it can activate the set up of local strain fields capable to drive the segregation further.

One of the expected consequences of the segregation of impurities at GB is a modification of their electrical activity, as the result either of a sensible decrease of the density of trap states (which should induce the "passivation" of GB) or of a sensible increase of the local density of electrically active impurities, which should turn out to the electrical degradation of GB.

Both these effects are well known in practice, but often experienced very empirically, without a sound knowledge of the physical phenomena underlying them. Furthermore, both the majority and minority carrier properties are influenced by impurity segregation, as any modification of the trap states density induces as well a modification of the height of the potential barrier associated to the onset of a space charge region at GB and responsible of the thermally activated mobility of carriers in polycrystalline semiconductors (see Theoretical Section).

Albeit the idea that the electrical activity of GB in polycrystalline semiconductors is dominated by impurity segregation is far being new[3-6] and well supported primarily by an impressive series of experiments carried out by Kazmerski[7-11], also at an atomic scale resolution[12], still the entire matter is not satisfactorily settled down, both phenomenologically and theoretically.

Aim of this paper is to contribute to an improvement of this picture, with the support of some new results and with a particular emphasis to the segregation of non metallic impurities (boron, oxygen and carbon), whose influence is generally under-evaluated with respect to that of metallic ones. Polycrystalline silicon is chosen as a model material, in virtue of the relative ease to obtain it in very reproducible conditions and of the extended literature available on related subjects.

THEORETICAL BACKGROUNDS

On the base of the presently available models, a GB is described as a double Schottky junction[13-16] (or, alternatively, as a SIS junction, when an insulating oxide is segregated on it). Consequently, each GB is associated to an electrical potential barrier whose height E_B and width w_B depend on the distribution of the different interface states of energy E_t within the gap, on their density N_t and on the density N_D or N_A of shallow acceptors and donors.

Such interface states may be generated:
a) by dislocations introduced in the material as a consequence of the release of the elastic strain present in regions of crystallographic mismatch between adjacent grains (intrinsic dislocation states)
b) by impurities segregated in the Cottrell atmosphere of the dislocations (extrinsic dislocations states) or/and
c) by unsaturated bonds (intrinsic GB trap states)
d) by impurities segregated at GB (extrinsic GB trap states)
e) by point defects (vacancies and selfinterstitials) generated at internal surfaces as a consequence of local elastic strain conditions.

In the single, monovalent trap state approximation, the width of the space charge region w_B is given by the following equation[15]

$$w_B = \frac{N_t}{2N_D} \left[1 + 0.5 \exp\left(\frac{E_t + E_B - E_F}{kT} \right) \right]^{-1}$$

where E_F is the intrinsic Fermi level at the surface and k is the Boltzmann constant.

As it is shown in Fig. 1, w_B ranges from 10^{-6} to 10^{-4} cm for N_t ranging from 10^{11} to 10^{13} cm^{-2}. Under the same approximation, the barrier height is given by

$$E_B = \frac{e^2 w_B^2 N_B}{2\varepsilon\varepsilon_0}$$

where e is the absolute value of the electronic charge.

For entirely depleted grains ($w_B > L$, L being the grain size), E_B may be calculated to be

$$E_B = \frac{e^2 L^2 N_D}{8\varepsilon\varepsilon_0}$$

For partially depleted grains, instead

$$E_B = E_F - E_t + kT \ln 2 \left[\frac{eN_t}{(8\varepsilon\varepsilon_0 N_D E_B)^{1/2}} - 1 \right]$$

which, when solved iteratively, shows (see Fig. 2) that E_B is a very sensitive function of both the trap density N_t and of N_D or N_A. Depending on the width of the space charge region, the barrier transit could occur via thermionic emission or quantum-mechanical tunneling. So, charge transfer processes in polycrystalline materials must be described with flux equations where the carrier transit probability is given by

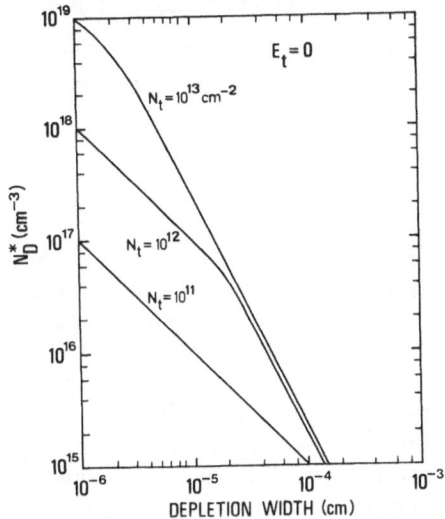

Fig. 1. Influence of the impurity concentration and of the trap density on the depletion width for the case of traps located midgap (From ref. 15)

$$\Omega = \Omega_0 \exp(-\alpha w_B) \exp\left(-\frac{E_B}{kT}\right)$$

where the first exponential accounts for quantum-mechanical tunneling. The corresponding conductivity equation looks therefore like

$$\sigma = \left(\frac{e^2 L n_0 v_c}{kT} \right) \Omega$$

where n_0 is the electron concentration in the neutral region and v_c is the collection velocity although in most of the case the second exponential in Ω dominates the transit probability.

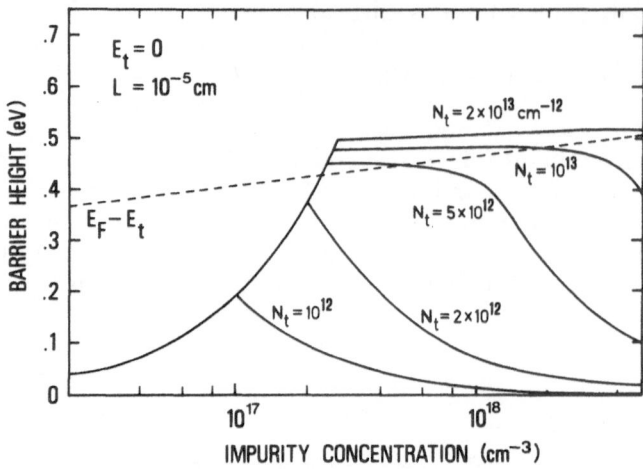

Fig. 2. Influence of the dopant concentration on the barrier height. Reference is made to a grain size of 0.1 micrometer (from ref.15)

Such model has been demonstrated to hold for the electrical conductivity of polycrystalline silicon grown by CVD techniques, but often fails in more complex systems, like the ternary or multinary compound semiconductors. In this case, especially if the amount of disorder and elastic stress within the GB region largely increases as a consequence of the segregation of non metallic impurities having atomic volumes different from that of the atoms of the host lattice, eventually accompanied by the formation of new phases, we expect the onset of a different conduction regime (polaronic conduction) like in disordered or amorphous semiconductors.

It is here very instructive to mention that for these systems a formally equivalent equation is known to hold[17]:

$$\sigma = \nu_{hop} \frac{e^2}{RkT} x(1-x)\exp(-2\alpha R)\exp\left(-\frac{W}{kT}\right)$$

where R is the average distance between the couple of filled vs empty trap sites involved in a hopping event, x is the atomic fraction of filled traps, T is the absolute temperature, ν_{hop} is the hopping frequency and the activation energy W is the sum of two terms $W = W_H + W_D/2$, the first being the polaron hopping energy and the second the effective average energy difference between the states involved in the hop.

However, while in the former case E_B is supposed to be independent of the temperature, in the second one both W_H and W_D are shown to be temperature dependent for $T < \Theta/2$, where Θ is the Debye temperature. Therefore, if localization occurs in the GB region as a consequence of structural and compositional disorder, we expect a transition from a T-independent to a T-dependent activation energy for $T = 325$ K, as the Debye temperature for single crystal silicon is 645 K[18].

Trapping, generation and recombination processes also depend on the density of trap sites N_t and on their energy E_t in the case of the Shockley-Read-Hall mechanism. In the case of recombination at a single set of surface traps, the recombination rate could be written:

$$R_{SRH}^{surf} = \frac{(p_s n_s - n_i^2)\sigma_p\sigma_n v_{therm}N_t}{\left[\sigma_n\left(n + N_c \exp\dfrac{E_t - E_c}{kT}\right) + \sigma_p\left(p + N_v \exp -\dfrac{E_t - E_v}{kT}\right)\right]}$$

where σ_p, σ_n are the capture cross sections for holes and electrons, n_i is the intrinsic concentration of carriers, N_c, N_v are the effective densities of states relative to the conduction and valence band, respectively and v_p and $v_n(= v_{therm})$ are the thermal velocities of electrons and holes.

In the case of Auger recombination, instead, the recombination probability involves a three bodies event and is given by en equation of the type $\Omega = np^2$ or $\Omega = pn^2$, for holes or electrons as the minority carriers, respectively. It is known to occur, therefore, only in heavily doped bulk silicon[19] and we may suspect to occur within the impurity cloud of a GB or of any other extended defect, in the case of strong dopants accumulation.

Segregation of metallic impurities at GB invariably has the consequence to influence the recombination rate of GB. Activation of GB could result either as a consequence of the generation of additional trap states (if impurities trapped at GB by elastic interaction forces or by chemical forces behave as deep levels) or as a consequence of the interaction between metallic and non metallic impurities (oxygen, carbon, dopants) if the resulting impurity cluster behaves as a deep level.

Deactivation of GB could instead occur when to the metallic-non metallic impurity complex pertains a level far from midgap or even out of the gap, as in the case of strong chemical bonding or when segregation of dopants could induce a collapse of the potential barrier (see Fig. 2).

It must be anticipated however that segregation of dopants could be totally inactive on the GB properties if clustering of segregated dopant atoms occurs[20].

EXPERIMENTAL EVIDENCE OF IMPURITY INTERACTIONS AT GB

Albeit quite easily forecastable from first principles, the fact that a substantial interaction occurs between impurities segregated at GB is now also experimentally supported by a number of direct evidences, mostly obtained from the analysis of the valence Auger transitions lineshapes and energies.

As the first example, in chronological order, the comparison of the $Si - L_{2,3}VV$ transition in undoped polycrystalline silicon with that observed in correspondence to iron rich precipitates in heavily contaminated polycrystalline silicon succeeded in indicating a predominant silicide bonding for segregated iron[21] (see Fig. 3).

Further examples are reported in a long series of papers of Kazmerski[10-12]. Very significant with respect to the aims of this present discussion is the case of hydrogen interaction with GB on which boron or oxygen were already segregated[12]. In the case of boron rich GB, the analisys of the Si LVV and B KLL transitions indicates that hydrogen segregates with the silicide configuration (Si-H bond) without any apparent interaction with boron (see Fig. 4). In the case of oxygen rich GB, instead, hydrogen interacts primarily with segregated oxygen and takes the hydroxyl configuration. Apparently, the energetics of the interaction between impurities is dictated here by chemical forces ranging within the impurity cloud, as in the case of the interaction of hydrogen with GB on wich aluminium was segregated.

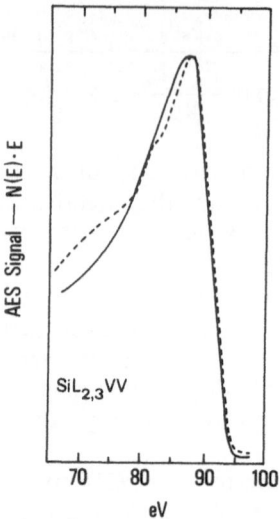

Fig. 3. AES spectra for undoped polycrystalline silicon and for iron doped (dotted line) polycrystalline silicon.

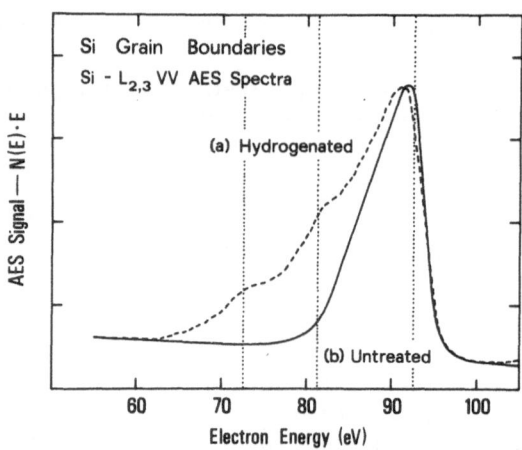

Fig. 4. AES spectra for untreated and hydrogen treated GB in polycrystalline silicon, indicating the onset of Si-H bonding.

Also here, in fact, and independently of the extent of the Al-Si interaction (which may be easily driven by heating), hydrogen takes an Al-H configuration in oxygen poor grain boundaries (ca. $10^{17} cm^{-3}$) while the hydroxyl configuration predominates in oxygen rich GB (ca. $10^{19} cm^{-3}$ from SIMS measurements).

We have however some evidences that in addition to chemical reaction occurring within the GB region, chemical equilibria between impurities in the volume of the grains and elastic deformation fields set up within the GB region as the consequence of atomic volumes misfits have a noteworthy influence on the configuration of the impurity cloud.

As a significant example of the role of chemical equilibria occurring in the volume of the grains, we quote the segregation of oxygen and carbon which has been recently demonstrated[22] to be driven by the reaction of C-O pairs formation in solution.

As the thermodynamics of C-O pairs formation was unknown (while the stability of the C-O pairs was already proposed by Newmann[23] on the base of the fine structure details of the 9 μm band in the presence of both oxygen and carbon) we carried out a systematic investigation on the dependence of the C-O pairs concentration on the carbon and oxygen concentration, using FTIR spectroscopy for the detection of the adsorptions relative to these species.

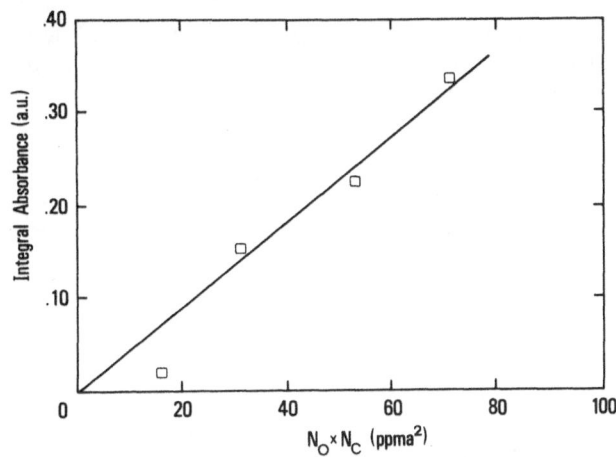

Fig. 5. Concentration dependence (from the absorbance of the 1104 cm^{-1} IR band) of C-O pairs on carbon and oxygen concentration.

The results are reported in Fig. 5 as the dependence of the absorbance of the C-O band at 1104 cm^{-1} (measured at 70 K) on the product of the oxygen and carbon concentration. They show not only that a chemical equilibrium is well established at T = 723 K, being the concentration of C-O pairs

$$N_{C-O} = KN_O N_C$$

proportional to the product of oxygen and carbon concentration, but also that the formation of thermal donors (which incidentally are known to be stable in the same temperature interval) is substantially inhibited (see Figs. 6 and 7) by the presence of C-O pairs.

Consequent to the set up of this equilibrium in solution are the features of carbon and oxygen at GB segregation[24]. In fact, in oxygen rich or in carbon rich samples the sole segregation of oxygen or carbon occurs (see Fig. 8) on most of the GB, which is accompanied by a strong enhancement of their recombination activity (see next Section).

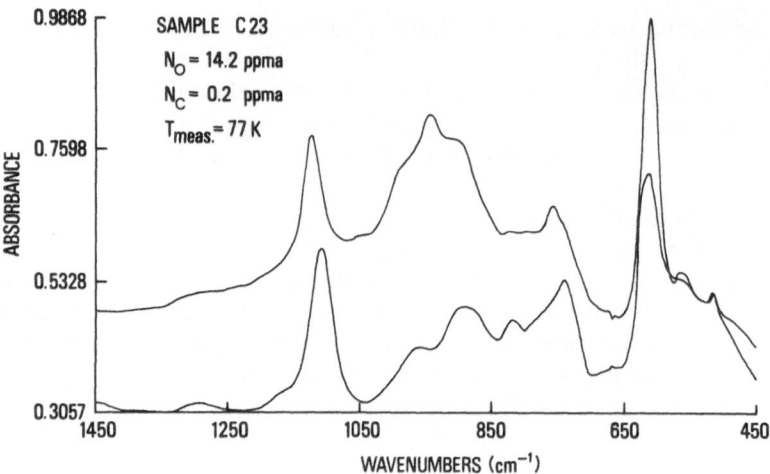

Fig. 6. FTIR spectrum for a carbon poor, single crystalline sample. a) before heat treatment, b) after heat treatment at 725 K in order to induce thermal donor generation

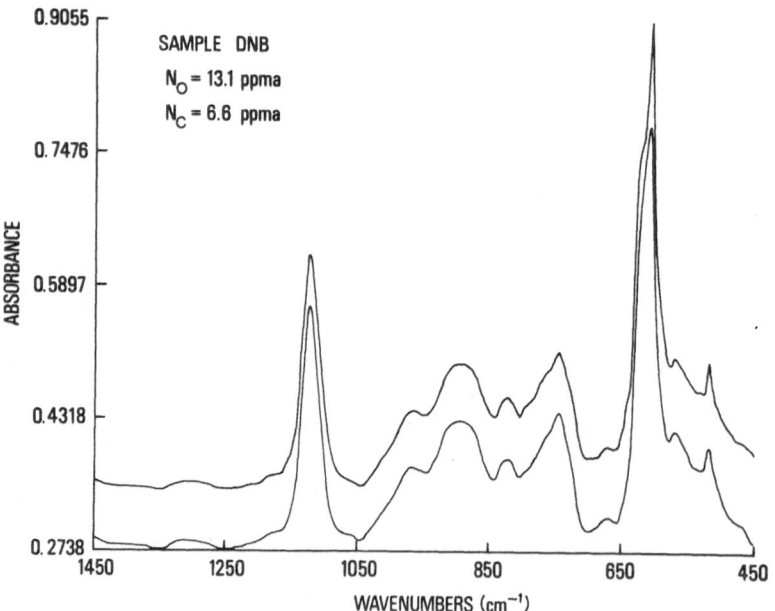

Fig. 7. FTIR spectrum for a sample containing both carbon and oxygen. a) before heat treatment b) after heat treatment at 725 K.

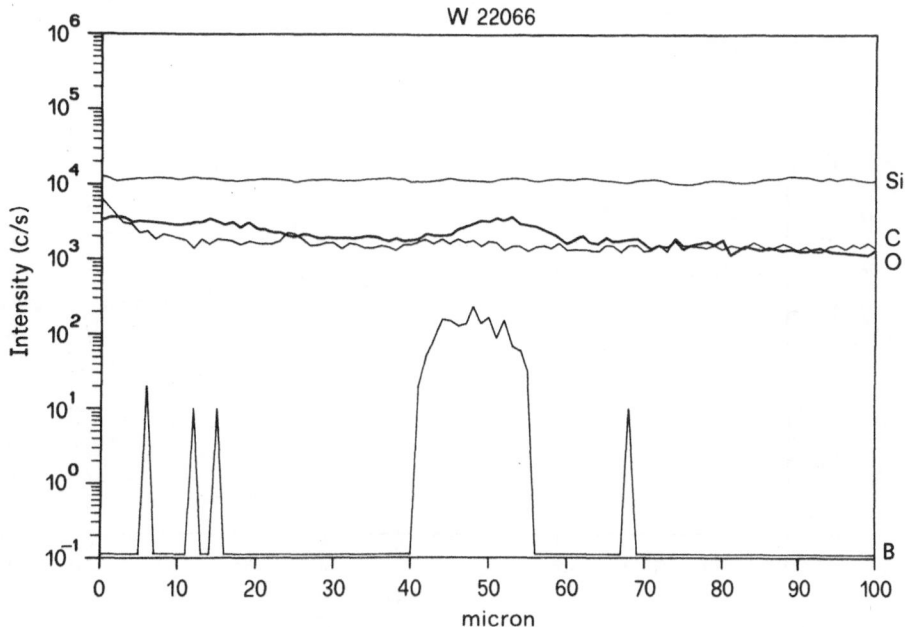

Fig. 8. SIMS lateral scanning profile for a GB of a carbon rich
(N_O = 1.4 ppma, N_C = 7.3 ppma) polycrystalline silic
on sample.

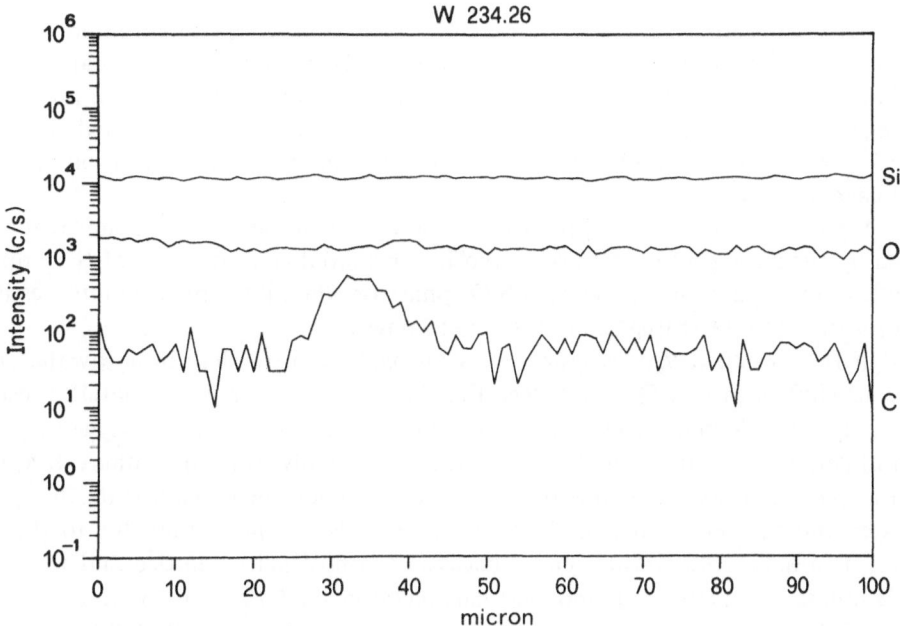

Fig. 9. Carbon segregation on a strongly recombining GB of a sample with
equal concentration of oxygen and carbon (N_O = 7.7 ppma, N_C = 7.1
ppma).

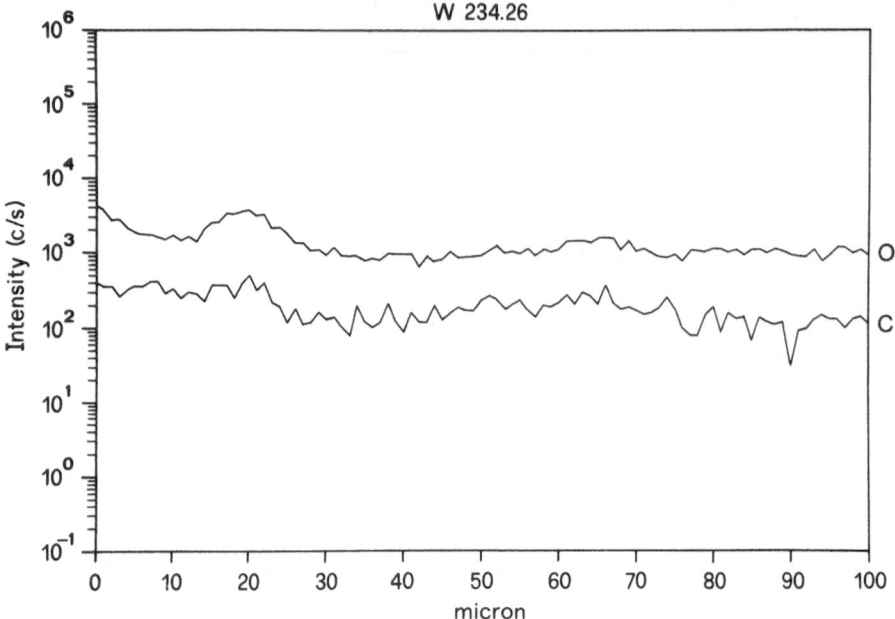

Fig. 10. SIMS lateral scanning profile for a slightly recombining GB in a sample with equal concentration of oxygen and carbon (N_O = 7.7 ppma, N_C = 7.1 ppma).

In equiconcentrated samples ($N_O - N_C = 0$), instead, not only a minority of GB shows enhanced electrical activity, which is invariably associated to carbon segregation (see Fig. 9), but on most of the slightly recombining GB carbon and oxygen segregation occurs simultaneously (see Fig. 10). It could be concluded, therefore, that it is the pairing equilibrium in solution which governs the segregation of oxygen and carbon at GB.

A good example of a competition between elastic fields and chemical forces is given by the strong differences observed in the spatial configuration of the impurity cloud, when oxygen segregates as a SiO_2 phase or when it simply accumulates in the GB region, without segregation of a second phase.

In fact, when oxygen segregates as an oxide phase - detectable by the simultaneous shift of Si and O signals (see Fig. 11) - carbon segregates spatially resolved from oxygen[25]. When, instead, no oxide phase segregation occurs, oxygen and carbon undergo segregation together and undistinguishably from each other. It appears that a spatially resolved segregation of impurities could be explained on the base of the demanding volume accommodation law[26] while assuming that due to the large molar or atomic volume differences between the host silicon lattice and the silicon oxide phase ($V_{SiO_2}/V_{Si} = 2$) and the carbon atoms ($V_C/V_{Si} = 0.65$) a compressive elastic field is set up at the Si/silicon oxide interface and a tensile field is set up in correspondence of the region of strong carbon accumulation, which are however mutually compensated within the GB region.

The case of spatially unresolved segregation (which mostly occurs in equiconcentrated solutions) could be instead explained by assuming that C-O pairs segregate at GB. In both cases, local mechanical equilibrium conditions are set up, albeit the range of local equilibrium is on atomic scale in the case of simultaneous C-O segregation, which is driven by chemical forces, while is on macroscopic scale (few micrometers) in the other cases.

Incidentally, but in good agreement with these findings, we found that also the dislocation generation process is driven by the excess carbon in solution (see Fig. 12), this result being a direct support of the role of the oxygen-carbon compensation on the relieve of the local elastic stress[24].

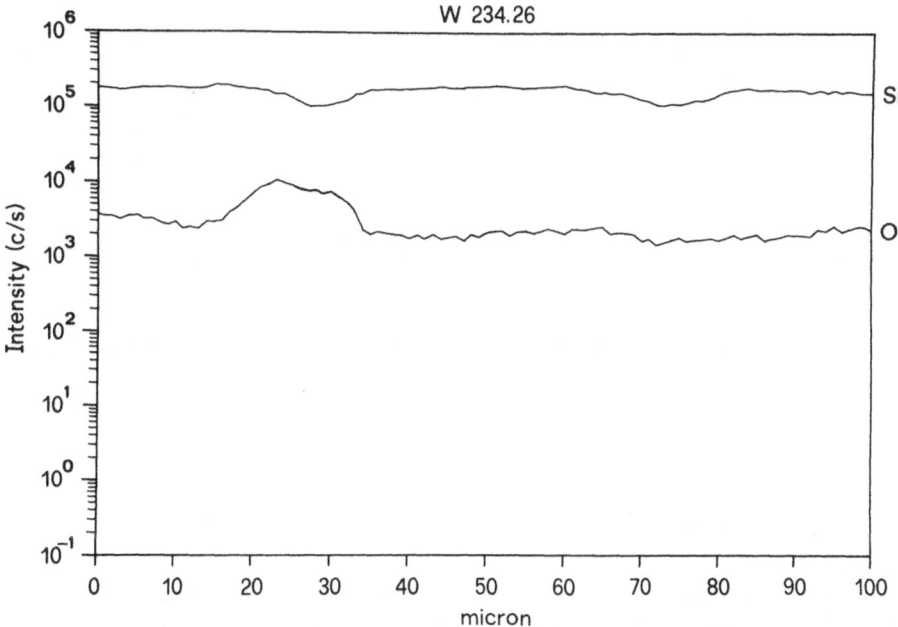

Fig. 11. Silicon oxide precipitation on a slightly recombining GB of a equiconcentrated sample.

As the last example of segregation dominated by a chemical equilibrium, we report the still puzzling case of boron segregation, which accompanies the carbon segregation and which is signaled by a hundredfold increase of the boron background at GB in the SIMS lateral profile reported in Fig. 8. As we know that B-C complexes are reported to be stable[27], with an ionization energy of 37 meV, slightly lower than that of the unpaired acceptor (49 meV), we would assume that pairing dominates the segregation features also in this case.

However, if we look at the results of our previous IR microscopy measurements[28] which showed that only a 25 % increase of the free carriers density is observed in correspondence with the GB region and we recall the already referred experiments of Kazmerski[11-12] of hydrogen segregation at boron rich GB, which did not reveal any appreciable interaction of hydrogen with boron, we may conclude that boron segregates already as an electrically inactive species, possibly as B-B pairs[20]. In the other case we shall expect B-H association, which leaves the acceptor

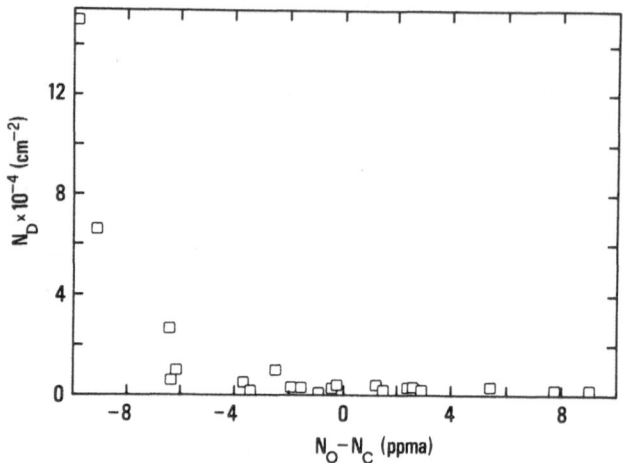

Fig. 12. Dependence of the dislocation density on the excess oxygen or carbon in solution.

atom threefold coordinated with its remaining Si neighbours and, therefore, electrically inactive[29].

It turns out conclusively from all these experimental evidences that:
a) the physical models for carrier transport and recombination relying on pure Schottky barriers or Auger recombination at GB, as described in the theoretical Section, are of little heuristic use in the present circumstances;
b) unless a very accurate and complete analysis of the segregation features at GB is preliminarily carried out, correlations between chemical and electrical properties are meaningless, as most of the correlations proposed in the past.

EFFECT OF IMPURITIES SEGREGATION ON THE ELECTRICAL PROPERTIES OF GB IN POLYCRYSTALLINE SILICON

So far, only indirect evidences of electrical effects coming from impurity segregation are reported in literature. In general, in fact, the enhancement of the electrical activity of GB and other extended defects (negligible in the as grown state) which is observed after annealing at high temperature is attributed to impurity segregation without any specific experimental proof that segregation occurred.

The fact that rapid thermal annealing (RTA) induces homogeneous recombination, while normal annealing induces marked inhomogeneities in recombination behaviour at GB is taken, for example, as a good proof that segregation occurred[30].

The direct proof that the enhancement of the electrical activity observed after annealing in the 850 − 1150 °C range (which, incidentally, is the range of temperatures used for p-n junction formation) is associated to impurity segregation has been recently reported by Pizzini et al[24].

These experiments were made on p-type polycrystalline silicon samples, which contained different amounts of oxygen and carbon and which were of EG quality, with metallic impurities only detectable with neutron activation analysis in the sub ppb level.

After some preliminary tests, and in order to have a one to one correspondence between electrical activity and segregation behaviour, the samples were carefully mapped at the Scanning Electron microscope first, in the secondary electrons and in the EBIC configurations.

Once a number of strongly recombining GB was localized, individual GB were systematically analyzed on a local scale by LBIC first (using a 780 nm laser radiation) and then by SIMS.

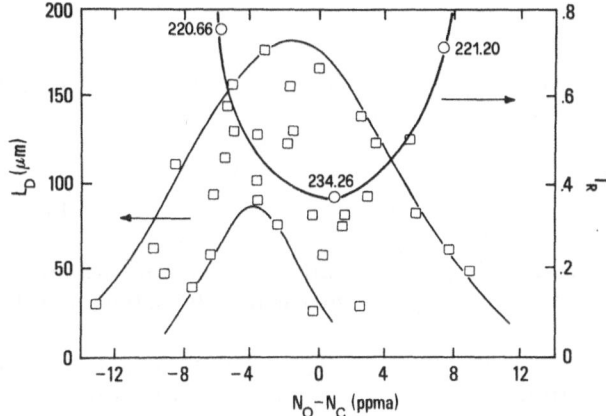

Fig. 13. Dependence of the diffusion length L_D and of the electrically active GB linear density ratio T_R on carbon and oxygen content. The numbers used to label the T_R values indicate the sample code.

The results are displayed in Table 1 and in Fig. 13 in terms of:
a) the equivalent width F of a statistically sizeable number of GB, where F is the area of the EBIC profile[31]

$$F = \frac{1}{I_o} \int_{-\infty}^{+\infty} I^\bullet(x)dx$$

which could be approximated by the product $(I^\bullet/I_o)w_{0.5}$ where $I^\bullet/I_o$ is the maximum EBIC contrast and $w_{0.5}$ is the half width of the EBIC profile.

b) the ratio $T_R = L_{GB}^*/L_{GB}$ of the linear density of strongly recombining GB L_{GB}^* over the linear GB density L_{GB} for each specific sample considered. This last parameter has been introduced[24] in order to identify a specific sample through its effective average recombination activity.

c) the average diffusion length L_D of a specific sample, as measured by the SPV technique[31].

It appears first from the results reported in Fig. 11 and in Table 1 that a correlation exists between segregation features and electrical activity. In fact, the maximum in L_D observed in Fig. 12 corresponds to a minimum in T_R and indicates that in equiconcentrated samples $N_O = N_C$ the majority of GB, which have been shown in the former Section to be associated to unresolved C + O segregation and then to an atomic scale equilibrium conditions, are electrically inactive. Here, conditions of local stress relief within the GB region prevent metallic impurities segregation to occur and therefore electrical GB activation, being oxygen and carbon by themselves electrically inactive.

Table 1. Comparison of the Equivalent Strengths of Various GB and TB

Sample Nr.	N_O	N_C	T_R	I/I_0	$W_{0.5}$	F
GB/220.66[a]	1.4	7.3	.74	.34	39.5	13.4
TB/220.66[a]				.30	11.3	3.4
GB/234.26[b]	7.7	7.1	.37	.29	52_l-37_r	15.3_l-10.9_r
TB/234.26[b]				.083	92	7.6
GB/221.20[c]	13.7	7.0	.70	.27	53	14.1

[a]Carbon (and boron segregation at strongly recombining GB. Twin boundaries (TB) also electrically active, but less than GB.

[b]Carbon segregation at strongly recombining GB: asymmetry effects from LBIC profile analysis. Twin Boundaries (TB) show strong electrical activity.

[c]No significant segregation of oxygen on strongly recombining GB. On a similar sample ($N_O = 12.5$, $N_C = 7.1$ ppma) IR microscopy shows excess free carriers and low mobility effects.

In carbon rich and in oxygen rich samples, instead, where large values of T_R indicate that the majority of GB are electrically active the set up of local elastic deformation fields, associated to excess carbon or oxygen segregation, favours the electrical activation of the extended defects, although the very reason of this activation is apparently still unclear, at least in the case of carbon segregation.

In fact, although it has been already shown[32] that in carbon rich solutions the carbon segregation is associated to an excess of dislocation density (see Fig. 13), the electrical activity of GB is totally uncorrelated from their being dislocation decorated[33].

Moreover, the bright halo effects - that in the case of oxygen rich GB may be attributed to denuded zone formation and impurity gettering at a silicon oxide phase segregated[34] at the GB - are totally suppressed in the case of carbon segregation, in good agreement with previously reported results on the effect of high carbon content on the stability of the denuded zone at the surface of single crystal silicon[35]

Eventually, also Auger recombination effects associated to the contemporary boron and carbon segregation may be considered not to be significant, as apparently most of boron is clustered to an inactive B-B pairs. It turns out that the elastic deformation field associated to the sole carbon segregation must be the source of the recombination activity, via point defects generation, these latter behaving as deep levels.

It turns out on the base of these experimental evidences that the enhancement of the electrical activity of GB is associated either to metallic impurity segregation at carbon or oxygen aggregates or precipitates or, eventually, to point defects (vacancies and interstitials) generated in correspondence to strongly stressed regions of the crystal.

CONCLUSIONS

It has been shown that without considering a synergy between chemical and mechanical forces, neither the segregation features at GB nor the electrical activity of GB could be appropriately described.

It has also been shown that the segregation phenomena at GB are much more complicated than previously considered and that the simple models used in the literature for discussing the electrical properties of microcrystalline silicon are only tenable as coarse and naive approximations to the the physics of the processes involved.

We hope that a quantitative discussion of all this matter could result consequent to an extensive experimental and theoretical work now in progress in our group, concerned with the energetics of the impurity interaction and of the charge transfer processes at extended defects in silicon.

ACKNOWLEDGEMENTS

We wish to thank EniChem for having entirely supported our work in the frame of a three years Research agreement.

LITERATURE

1. M. F. Yan, R. M. Cannon, H. K. Bowen, Space charge contributions to solute segregation near grain boundaries, in: "Grain boundaries in semiconductors", G. E. Pike, C. H. Seager, H. J. Leamy, eds., North Holland, New York (1982)
2. R. W. Cahn, "Physical Metallurgy", North Holland, Amsterdam (1970).
3. M. O. Cower, T. O. Sedwick, Chemical vapour deposited polycrystalline silicon, J. Electrochem. Soc. 119:1565 (1972)

4. A. L. Fripp, L. H. Slack, Resistivity doped polycrystalline silicon J. Electrochem. Soc. 120:146 (1973)
5. M. M. Mandurah, K. C. Saraswat, C. R. Helms, Dopant segregation in polycrystalline silicon, J. Appl. Phys. 51:5755 (1980)
6. H. J. Queisser, Electrical properties of dislocations and boundaries in semiconductors, in: "Defects in Semiconductors II", S. Manharajan, J. W. Corbett, ed., North Holland, New York (1983)
7. L. L. Kazmerski, P. E. Russel, Chemical and electrical characterization of polycrystalline semiconductors, J. Phys. (Paris) 431-172 (1982)
8. L. L. Kazmerski, P. E. Russel, P. J. Ireland, C. H. Herrington, J. R. Dick, R. J. Matson, K. M. Jones, Grain boundaries in silicon solar cells, J. Vac. Sci. Techn. A2:1120 (1984)
9. L. L. Kazmerski, Silicon grain boundaries, correlated chemical and electrooptical characterization, Proc. 17th IEEE Photovoltaic Specialist Conference (1984)
10. L. L. Kazmerski, Polycrystalline silicon: impurity incorporation and passivation, Proc. 6th E. C. Photovoltaic Solar Energy Conf., D. Reidel, ed., Doordrecht (1985)
11. L. L. Kazmerski, Scanning tunneling microscope and complementary microchemical investigations of hydrogen and shallow acceptors at silicon grain boundaries, Proc. 8th E. C. Photovoltaic Solar Energy Conf., D. Reidel, ed., Doordrecht (1988)
12. L. L. Kazmerski, Atomic level imaging and microanalysis of GB in polycrystalline semiconductors, Proc. Symp. Polycrystalline Semiconductors (Polyse), Springer Verlag, Berlin (in press)
13. C. H. Seager, T. G. Castner, Zero bias resistance of GB in neutron doped polycrystalline silicon, J. Appl. Phys. 49:3879 (1978)
14. J. X. W. Seto, The electrical properties of polycrystalline silicon films, J. Appl. Phys. 46:5247 (1975)
15. G. Baccarani, B. Ricco', G. Spadini, Transport properties of polycrystalline silicon films, J. Appl. Phys. 49:5568 (1978)
16. N. F. Mott, E. A. Davis, "Electronic Processes in non crystalline solids" Clarendon Press, Oxford (1971)
17. C. Kittel, H. Kroemer, "Thermal Physics" W. H. Freeman and Co. S. Francisco (1980)
18. Y. S. Kim, C. I. Drowley, C. Hu, A new method of measuring diffusion length and surface recombination velocity, Proc. 14th IEEE Photovoltaic Specialist Conference (1980)
19. L. Passari, E. Susi, Recombination mechanism and doping density in silicon, J.Appl. Phys. 54:3935 (1983)
20. G. F. Ccrofolini, L. Meda "Physical chemistry of silicon" Springer Verlag, Berlin (1989)
21. S. Pizzini, L. Braicovich, L. Calliari, M. Gasparini, C. Mari, F. Redaelli, M. Sancrotti, Segregation of impurities at GB and other compositional inhomogeneities in cast silicon ingots, Proc. 4th E. C. Photovoltaic Solar Energy Conf., D. Reidel, ed., Doordrecht (1982)
22. P. Cagnoni, Interaction between impurities and extended defects in polycrystalline silicon (in italian), Thesis University of Milan, Dept. Physics (1987)

23. R. C. Newman, "Infrared studies of crystal defects", Taylor & Francis, London (1973)
24. S. Pizzini, F. Borsani, A. Sandrinelli, D. Narducci, M. Anderle, R. Canteri, On the influence of the Cottrell atmosphere on the recombination losses at GB in polycrystalline silicon, Proc. Symp. Polycrystalline Semiconductors (Polyse), Springer Verlag, Berlin (in press)
25. S. Pizzini, P. Cagnoni, A. Sandrinelli, M. Anderle, R. Canteri, Grain boundary segregation of oxygen and carbon in polycrystalline silicon, Appl. Phys. Lett. 51:676 (1987)
26. T. Y. Tan, Exigent volume of precipitation and formation of oxygen precipitates in silicon, in: "Oxygen, carbon, hydrogen and nitrogen in crystalline silicon", MRS Symposia Proceedings, Vol. 59, Materials Research Society, Pittsburgh (1986)
27. R. C. Newman, Carbon in crystalline silicon, ibidem
28. A. Borghesi, M. Geddo, G. Guizzetti, S. Pizzini, D. Narducci, A. Sandrinelli, A. Zachman, IR microcharacterization of GB in polycrystalline silicon, Solid St. Comm. (1989) (in press)
29. S. J. Pearton, Hydrogen in crystalline silicon, in: "Oxygen, carbon, hydrogen and nitrogen in crystalline silicon", MRS Symposia Proceedings, Vol. 59, Materials Research Society, Pittsburgh (1986)
30. A. Barhadi, H. Amzil, J. C. Muller, P. Siffert, Thermal activation and hydrogen passivation of grain boundaries, Proc. Symp. Polycrystalline Semiconductors (Polyse), Springer Verlag, Berlin (in press)
31. G. Donolato, Theory of beam induced current characterization of GB in polycrystalline solar cells, J. Appl. Phys. 54:1314 (1983)
32. S. Pizzini, A. Sandrinelli, M. Beghi, D. Narducci, F. Allegretti, S. Torchio, G. Fabbri, G. P. Ottaviani, F. Demartin, A. Fusi, Influence of extended defects and native impurities on the electrical properties of polycrystalline silicon, J. Electrochem. Soc. 135:155 (1988)
33. F. Borsani, Segregation phenomena and electrical activity of GB in silicon (in Italian), Thesis University of Milan, Dept. Physics (1988)
34. S. Martinuzzi, Activation and Passivation of recombination activity of GB in polycrystalline semiconductors, Proc. Symp. Polycrystalline Semiconductors (Polyse), Springer Verlag (in press)
35. A. Poggi, E. Susi, Effect of high carbon content on denuded zone stability in intrinsic gettering processes, Proc. 2nd GADEST Conference, Garzau (DDR) 119 (1987)

THE EXTENDED NATURE OF POINTLIKE DEFECTS IN SILICON

G. F. Cerofolini

EniChem
Via Medici del Vascello 26
20138 Milano MI, Italy

ABSTRACT

The problem of the localized or extended nature of pointlike defects in silicon is considered. Evidence from different areas (solid solubility of substitutional impurities, mobility in Mott - Anderson Si:As metal, and supershallow levels in p-type silicon) suggests that each atomic perturbation of the crystal periodicity is associated with a cloud of displaced atoms involving about 10^3 lattice sites.

INTRODUCTION

Having recently written a book on physical chemistry of, in and on silicon [1], I have considered some of the most obvious queries which could be raised when considering such a topic, *viz.*:

   (i) which is the atomic configuration of point defects? (e.g., is the self-interstitial quasi free or does it have a dumb-bell configuration?)
  (ii) has each defect only one configuration or are several configurations possible?
 (iii) which is the electronic structure?
  (iv) which charge states are associated with each defect and where are they located in the gap?
   (v) can the defect be actually considered pointlike (i.e., do the remaining atoms remain on their lattice location) or does the deformation extend to long range?
  (vi) does an entropic or enthalpic barrier exist for Frenkel-pair recombination?
 (vii) which are the defect diffusivities in relation to their charge states?
(viii) is the surface an effective generation-recombination centre for point defects?
  (ix) to which extent does this generation-recombination rate depend on surface conditions (free, oxidated, nitridated, etc.)?

Most of them, however, remained unanswered. That these questions are truly puzzling queries is confirmed, for instance, by the following statements, taken from recent literature:

"at room temperature [...] vacancies and interstials are mobile in silicon" [2]

"the diffusivity of vacancies is much higher than that of self-interstitials" [3]

"once formed, vacancies are fairly immobile" [4].

In this lecture I shall focus my attention only over one particular problem, i.e., query (v). Of course, one could have direct evidence for the localized or extended nature of a pointlike defect x if he knew its formation entropy, i.e., its equilibrium concentration $N_x(T)$ at the temperature T. However, to the best of my knowledge, no direct measurement has ever been reported; $N_x(T)$ is usually inferred from other measurements, such as anomalous diffusivity, formation of stacking faults, etc.

The absence of direct measurements has allowed several speculations. For instance, in an early paper Van Vechten [5] suggested that the vacancy is a true pointlike defect formed by a void with the equilibrium (*octahedral*) shape. More recently, however, the same Author postulated that the vacancy may be an extended defect resulting from a particular configuration of an amorphous cluster [6] — indeed, assuming that the vacancy formation enthalpy is approximately 5 eV [7] and the energy excess of the amorphous phase is 0.13 eV/atom [8], the vacancy and an amorphous cluster with about 40 atoms should be energetically equiprobable, and the second could arrange in a configuration proper of the extended vacancy.

Direct measurement of the deformation field originated by vacancies or self-interstitials is possible by x-ray diffraction in the multiple-crystal arrangement [9]. This technique gives evidence for a strain in the crystal, thus suggesting long-range effects. However, the absence of information about the absolute concentration of defects and their aggregation state makes it difficult to quantify the range of the relaxation field around the defect. Multiple-crystal x-ray diffraction data give evidence, however, that this field can adequately be parametrized by a strain (no disorder, only deformation) superimposed on a static disorder (characterized by a mean square displacement around a lattice site). The relative weight of one parameter compared with the other is essentially dependent on the nature of the defect. On another side, techniques such as the aligned Rutherford backscattering spectrometry (RBS) are able to determine the concentration of self-interstitials (provided that it is high enough) and, used in combination with the multiple-crystal x-ray diffraction, should be able to characterize the displacement field. This combination was used for the characterization of the Si:H system obtained by ion implantation, but the the sensitivity of the RBS technique was found to be inadequate for an exact quantitation of self-interstitials. Not only did this study show a low yield for Frenkel pair production [10], but also proved the existence of non-linear phenomena which are easily explained by assuming that a relevant number of atoms is displaced around each defect [11]. This number was found to be of the order of 10^3.

In the following I shall consider two situations giving evidence that the replacement of an atom by another, i.e. the insertion in the crystal of a substitutional impurity, produces a perturbation which extends to a number of host atoms of the order of 10^3. In turn, this fact suggests that point defects are endowed with a relaxation field involving the same number of atoms.

SOLID SOLUBILITY

The solid solubility n_{sol} of substitutional impurities can be written as follows

$$n_{sol} = n_{Si} \exp\left(-\frac{\Delta H}{k_B T}\right) \exp\left(\frac{\Delta S}{k_B}\right) \quad ,$$

where n_{Si} is the atomic density of silicon ($n_{Si} = 5 \times 10^{22}$ cm^{-3}), k_B is the Boltzmann constant, ΔH and ΔS are the solution enthalpy and entropy, respectively. In silicon (and in germanium too) the solid solubility is mainly governed by entropic factors, provided that the difference of tetrahedral radii of the guest atom and the host crystal atom (r and r_{Si}, respectively) is sufficiently small (say, < 20 %). Indeed, in this case one finds $n_{sol} \approx 10^{20}$ cm^{-3}

approximately independent of T in the interval 900 − 1100 °C (see Fig. 1), quite irrespective of the impurity [12]. A similar fact is observed in germanium in the temperature range 500 − 800 °C [13]. The large entropy associated with the dissolved impurity (of about −6 k_B, due to the above value of n_{sol}) in turn suggests that a relevant number of dregrees of freedom is involved in the replacement of a silicon atom by the impurity. In other words, one can interpret this entropy in terms of disorderd cloud around each impurity.

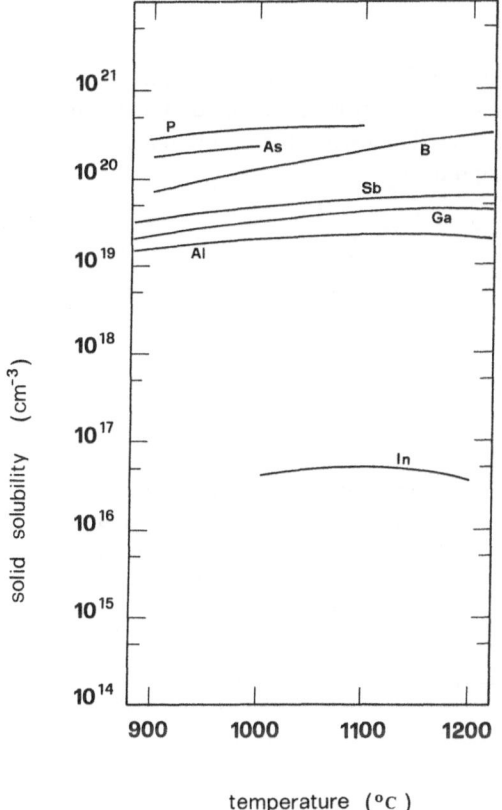

Fig. 1. Solid solubility of group III and V dopants in silicon. The solubility is approximately constant with T in the range 900 − 1200 °C and, except for indium, is in the range $2 \times 10^{19} - 2 \times 10^{20}$ cm^{-3}.

Naive Model

A far simplified model is the following:
In the low concentration limit, one can look at the dissolved impurities, each surrounded by its own displacement cloud, as independent of one another. This picture starts to lose meaning when these clouds begin to be superimposed. Any further attempt to insert a new atom into the crystal requires not only the formation of a displacement cloud but the modification of an already displaced cloud (i.e., two displacing fields compete with one another).
It is not implausible to assume that this fact limits the solubility of an impurity in the crystal. If N_{dis} is the number of atoms forming the disordered cloud, the above hypothesis gives

$$N_{dis} n_{sol} \approx n_{Si} \ , \tag{1}$$

i.e.,

$$N_{dis} \simeq 2.5 \times 10^2 - 2.5 \times 10^3 \ ,$$

because of the typical values of solid solubility, 2×10^{19} cm$^{-3} < n_{sol} < 2 \times 10^{20}$ cm^{-3}. A similar model was assumed to explain the Na$^+$/K$^+$ ratio in biosystems [14].

Strain Entropy

The following, more accurate, analysis is taken from ref. [1].
Considering all entropic contributions to the solid solubility, one obtains that this quantity is limited by a contribution per impurity atom ΔS_{str} which behaves as the difference of tetrahedral radii $r - r_{Si}$ between impurity and silicon [12]:

$$\Delta S_{str} \propto (r - r_{Si})$$

at least for $r > r_{Si}$ (see Fig. 2). The behaviour of ΔS_{str} vs. $r - r_{Si}$ clearly shows that this contribution is associated with a lattice deformation. To a first approximation, the deformed lattice can be seen as a lattice with an increased atomic density but with the same force constants (because of the negligible enthalpic contribution).

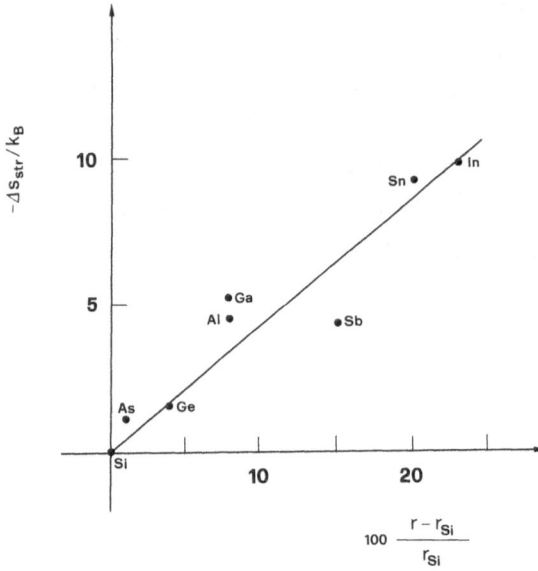

Fig. 2. Correlation between strain entropy $-\Delta S_{str}$ and tetrahedral radius r of substitutional impurities in silicon. The considered temperature is 1273 K. Indium, which was excluded in the naive estimate, is described by this correlation.

I shall assume that a zone with a different atomic density is associated with a local Debye temperature Θ. In this view, the perturbed region must be thought of as formed by a relative large number N_{dis} of displaced atoms in the proximity of the guest impurity. If the local Debye temperature is varied from Θ to $\Theta' = \Theta + \Delta\Theta$, the vibrational entropy per displaced atom ΔS_{vib} is varied by an amount [15]

$$\frac{\Delta S_{vib}}{k_B} = 3 \left[\frac{\Theta'}{2T} \coth \frac{\Theta'}{2T} - \ln \left(2 \sinh \frac{\Theta'}{2T} \right) \right]$$

$$-3 \left[\frac{\Theta}{2T} \coth \frac{\Theta}{2T} - \ln \left(2 \sinh \frac{\Theta}{2T} \right) \right]$$

$$\simeq -3 \left[\frac{\Theta}{2T} \frac{1}{\sinh(\Theta/2T)} \right]^2 \frac{\Delta\Theta}{\Theta} . \tag{2}$$

The basic idea is the following: the strain entropy per impurity atom is due to the change of vibration entropy in the displacement cloud, $\Delta S_{str} = \Delta S_{vib} N_{dis}$. Together with weak equality (2), this hypothesis gives

$$\frac{\Delta S_{str}}{k_B} \simeq -3 \left[\frac{\Theta}{2T} \frac{1}{\sinh(\Theta/2T)} \right]^2 \frac{\Delta\Theta}{\Theta} N_{dis} \quad . \tag{3}$$

The Debye temperature Θ is 640 K for silicon and 290 K for germanium, while the ranges of temperature of interest are $1100 - 1400$ K for silicon and $800 - 1100$ K for germanium; therefore in the ranges of interest $\Theta/2T \simeq 0.3$. For $\Theta/2T < 1$ the quantity in square parentheses is very close to 1 ($x \to 0 \Rightarrow x/\sinh x \sim 1$); $-\Delta S_{str}/k_B$ is in the range $0 - 10$, so that eq. (3) shows that even modest changes of Debye temperature are sufficient to allow for very high strain entropy provided that N_{dis} is sufficiently high.

The change of Θ in the displacement cloud can be evaluated working in the Debye model. In fact, the cloud of displaced matrix atoms can be described, in the first approximation, as the unperturbed crystal at a somewhat different density n where the force constants are not significantly changed. The influence of n on Θ is given by

$$\Theta = \left(\frac{9}{4\pi} \right)^{\frac{1}{3}} \frac{\hbar\pi}{k_B} \left(\frac{1}{c_l^3} + \frac{1}{c_t^3} \right)^{-\frac{1}{3}} n^{\frac{1}{3}} \quad , \tag{4}$$

where $\hbar$ is the reduced Planck constant, and c_l and c_t are the longitudinal and transversal sound velocities [15]. These quantities can be assumed to be constant. Indeed, if the displacement cloud extends over a few atomic distances, only high frequency modes are modified and the sound velocities (related to the dispersion relationships in the zero-frequency limit) will therefore remain unchanged.

Eq. (4) states therefore that a local increase of density, from n to $n + \Delta n$, produces therein an increase of Debye temperature from Θ to $\Theta + \Delta\Theta$, where

$$\frac{\Delta\Theta}{\Theta} = \frac{1}{3} \frac{\Delta n}{n} \quad . \tag{5}$$

The evaluation of N_{dis} is a very difficult task because one does not know how the mismatch $r - r_{Si}$ (a kind of stress) is relaxed in the vicinity of the impurity. However, the following relationship is probably not far from reality:

$$\begin{aligned} N_{dis} &\approx \frac{4}{3}\pi[\kappa(r - r_{Si})/\epsilon]^3 n \\ &= 36\pi[\kappa(r - r_{Si})]^3 (n/\Delta n)^3 n \end{aligned} \tag{6}$$

where ϵ ($\epsilon = \frac{1}{3}\Delta n/n$) is the average strain in the displacement cloud and κ is the fraction of $(r - r_{Si})$ which is not absorbed by the bond itself; according to Baldereschi and Hopfield analysis [16] I shall assume $\kappa = 0.4$. Inserting eqs. (5) and (6) into eq. (3) one has

$$\begin{aligned} \frac{\Delta S_{str}}{k_B} &\simeq -36\pi \left[\frac{\Theta}{2T} \frac{1}{\sinh(\Theta/2T)} \right]^2 \left(\frac{n}{\Delta n} \right)^2 [\kappa(r - r_{Si})]^3 n \\ &\simeq -36\pi \left(\frac{n}{\Delta n} \right)^2 [\kappa(r - r_{Si})]^3 n \quad . \end{aligned} \tag{7}$$

Since experimentally one has $\Delta S_{str} \propto (r - r_{Si})$, eq. (7) gives $(r - r_{Si}) \propto \Delta n/n \propto \epsilon$, i.e. the average strain ϵ varies in proportion to the stress $(r - r_{Si})$ — the deformation remains in the elastic range.

Inserting the numerical values for silicon at 1000 °C one has

$$N_{dis} \approx 5 \times 10^2 \quad ,$$

$\Delta\Theta$ of the order of 1 K, and the values of ϵ listed in Table 1. This estimate of N_{dis} agrees with the naive estimate as given by eq. (1).

Table 1. Strain produced by a few substitutional impurities in silicon (T = 1273 K)

element	$(r - r_{Si})$ 10^{-10} cm	$\Delta S_{str}/k_B$	ϵ
Al	9	4.5	2.5×10^{-3}
Ga	9	5.2	2.3×10^{-3}
In	27	9.7	8.9×10^{-3}
As	1	1.1	0.2×10^{-3}
Sb	19	4.3	7.9×10^{-3}

These results confirm that $\epsilon \propto (r - r_{Si})$, i.e that the deformation is elastic. Once the elastic limit is exceeded, an enthalpy term (a kind of 'deformation energy') is required for the description of solid solubility. This limit should occur at $(r - r_{Si})/r_{Si} \simeq 0.25$ because most systems characterized by a large enthalpy term (e.g., Bi, Au and Ag in Si, and Pb in Ge) satisfy the condition $(r - r_{Si})/r_{Si} > 0.25$.

CARRIER MOBILITY IN HEAVILY DOPED Si:As

When a semiconductor crystal is doped at concentration n^D exceeding a value n^D_*, it undergoes an insulator-metal transition. The critical value n^D_* is given by the Mott condition [17]

$$(n^D_*)^{\frac{1}{3}} a^* \simeq 0.25 \quad , \tag{8}$$

where a^* is the shallow dopant radius. Regarding the dopant as a hydrogenlike atom, the effective mass approximation gives

$$a^* = \epsilon \frac{m_e}{m^*} a^H \quad ,$$

where a^H is the Bohr radius (a^H = 0.53 Å), ϵ is the semiconductor dielectric constant (in silicon ϵ = 11.7), m_e is the electron mass, and m^* is the renormalized carrier effective mass [1]. This quantity is related to the dopant ionization energy E^*_{ion} by

$$E^*_{ion} = \frac{1}{\epsilon^2} \frac{m^*}{m_e} E^H_{ion}$$

For group V donors in silicon (for which $E^*_{ion} \simeq 50$ meV, that gives $m^* \simeq 0.5 m_e$ and $a^* \simeq 10$ Å) the Mott criterion (8) predicts therefore that the metal-insulator transition occurs at $n^D_* \simeq 10^{19}$ cm^{-3}.

In a variable-temperature resistivity and Hall mobility study of the Si:As system, Cerofolini et al. [18] found that the impurity band is gradually formed as n^D increases and that the metallic properties become fully apparent at $n^D > n^D_* \simeq 7 \times 10^{18}$ cm^{-3} (this result is consistent with Newman and Holcomb determination, $n^D_* = 7.8 \times 10^{18}$ cm^{-3}) [19]. In this concentration regime, the doped semiconductor can be described as a highly impure Bloch metal and the mobility is a smoothly decreasing function of T. The experimental curves of Hall mobility μ_{Hall} (supposedly equal to the mobility μ) vs. n^D are shown in Fig. 3. The isothermal curve $\mu(n^D)$ at low T (say, below 50 K where phonons are thought of not to limit carrier free motion) shows that at $n^D \simeq 4 \times 10^{18}$ cm^{-3} something exceptional happens: the addition of one more dopant increases rather than to decrease the mobility. This fact can be interpreted by admitting that as n^D exceeds 4×10^{18} cm^{-3} a superlattice is formed, the superlattice being gradually more and more completed as n^D increases.

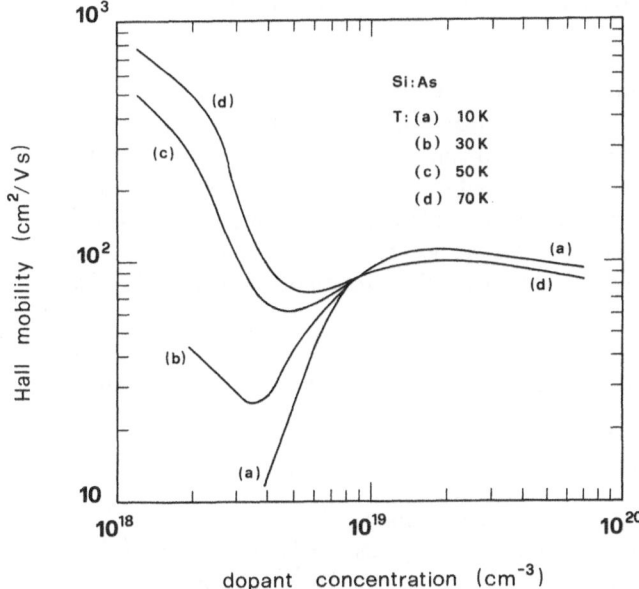

Fig. 3. Isothermal curves of Hall mobility μ_{Hall} vs. dopant concentration n^D in Si:As. For $n^D > 7 \times 10^{18}$ cm^{-3} μ_{Hall} is a deacreasing function of T, thus suggesting that the metal-insulator transition takes place at this concentration.

In the following I shall propose a semiquantitative explanation of these curves in the following hypotheses:
1) the carriers are described by the elementary theory of Fermi gas;
2) electron conduction takes place in the impurity band;
3) the actual impurity band is described in terms of an imperfect superlattice; and
4) a displacement field involving approximately 10^3 atoms surrounds each dopant.

Degenerate Fermi Gas

Assume first that most of dopants are ionized (this fact is true in the metallic region $n^D > n^D_*$ or in its vicinity at T sufficiently high). Provided that the impurity-band effective mass does not differ greatly from the electron mass, $m^* \approx m_e$, the Fermi temperature T_F,

$$T_F = \left(\frac{\hbar^2 \pi^2}{2 m^* k_B} \right) \left(\frac{3}{\pi} n^D \right)^{\frac{2}{3}} ,$$

has a lower limit of about 500 K for 5×10^{18} cm$^{-3} < n^D < 10^{20}$ cm^{-3}. Therefore, in the analysis of experimental data the electron gas can be assumed to be completely degenerate in the range $10 - 100$ K. For a completely degenerate electron gas the mobility is given by [20]

$$\mu = \frac{e\tau}{m^*} = \frac{e\lambda}{p_F} \tag{9}$$

where $-e$ is the electron charge, τ is the momentum relaxation time, λ is the carrier mean free path, and

$$p_F = \hbar (3 \pi^2 n^D)^{\frac{1}{3}} \tag{10}$$

is the Fermi momentum. Combining (9) and (10) one obtains that the mobility does not contain the details of the impurity band (described by the effective mass m^*).

Mean Free Path

The calculation of the mobility μ requires therefore the knowledge of the mean free path. To calculate λ one can first try to evaluate the concentration of superlattice defects and then how does the deformation field around each impurity affect the scattering.

Superlattice Defect Statistics

The basic idea to deal with a disordered superlattice is the following:
Let ν_0 be the site concentration in the ideal superlattice. When dopant atoms are substitutionally inserted into the crystal, they can hold superlattice or off-site positions. The basic hypothesis is that off-site dopants as well as void superlattice sites behave as scattering centres and possibly induce in the crystal a number of additional scattering centres. The increase of occupied superlattice sites $\nu_0 d\theta$ (where θ is the occupied fraction) varies with dopant concentration n^D, the constant of proportionality being the fraction of empty sites $(1 - \theta)$:

$$\nu_0 d\theta = (1 - \theta) d n^D \quad . \tag{11}$$

The solution of this equation gives $\theta = 1 - \exp(-n^D/\nu_0)$; the density ν_{sl} of superlattice defects is therefore given by

$$\nu_{sl} = n^D - \nu_0\theta + \nu_0(1 - \theta) = n^D + \nu_0 - 2\nu_0\theta \quad .$$

This density has a minimum at the dopant concentration n^D_m for which $d\nu_{sl}/d n^D = 0$, i.e. $1 - 2\nu_0(d\theta/d n^D) = 0$, which inserted into eq. (11) gives $\theta = \frac{1}{2}$, i.e. $n^D = \nu_0 \ln 2$. The average distance l_{sl} between superlattice defects is therefore given by

$$l_{sl} = \left(\frac{1}{\nu_0}\right)^{\frac{1}{3}} \left\{ 1 + \frac{n^D}{\nu_0} - 2 \left[1 - \exp\left(-\frac{n^D}{\nu_0}\right) \right] \right\}^{-\frac{1}{3}} \quad . \tag{12}$$

The study of the function $1 + x - 2[1 - \exp(-x)]$ shows, of course, a maximum at $x = \ln 2$ and a slow variation in a wide range of x. However, the distance l_{sl} cannot *a priori* be considered as the mean free path because each defect is surrounded by its displacement cloud; each atom of the cloud, being a perturbation to the finite translational invariance of the crystal, is indeed be expected to be a scattering centre. In the low concentration limit the concentration of scattering centres ν_s is given by

$$\nu_s = N_{dis}\nu_{sl}$$

and the mean distance between scattering centres l_s is accordingly

$$l_s = N_{dis}^{-\frac{1}{3}} l_{sl} \quad . \tag{13}$$

When however n^D increases and eventually exceeds n^D_* these additional scattering centres are embedded in the band above the mobility edge and disappear. At a concentration around n^D_* there is hence a sudden change in the concentration of scattering centres:

$$\nu_s = \begin{cases} N_{dis}\nu_{sl} & \text{for} \quad n^D \ll n^D_* \\ \nu_{sl} & \text{for} \quad n^D \gg n^D_* \end{cases} \tag{14}$$

Provided that eq. (14) is valid in the neighborhood of n^D_*, the left-to-right limit ratio

$$\frac{\nu_s(n^D)_{n^D = n^D_{*-}}}{\nu_s(n^D)_{n^D = n^D_{*+}}} = N_{dis}$$

is therefore a measure of the number of atoms in the displacement cloud.

Mobility

The behaviour $\mu = \mu(n^D)$, showing a sudden increase of μ with n^D just around n^D_*, suggests that this mechanism is active; assuming that the mean free path λ is given by the distance between scattering centres ($\lambda = l_s$), eq. (14) gives

$$\frac{\mu(n^D)_{n^D=n^D_*-}}{\mu(n^D)_{n^D=n^D_*+}} = N_{dis}^{-\frac{1}{3}} \quad . \tag{15}$$

Considering the isothermal curve at 10 K and assuming $n^D_* \pm 0.4\, n^D_*$ as representative of $n^D_{*\,\pm}$, one has $\mu(n^D_{*\,-}) \simeq 10$ cm^2/V s and $\mu(n^D_{*\,+}) \simeq 100$ cm^2/V s, so that eq. (15) gives

$$N_{dis} \approx 10^3 \quad .$$

It is worthwhile noticing that this estimate does not depend on the hypothesis that the Hall factor γ_{Hall} ($\gamma_{Hall} = \mu_{Hall}/\mu$) is unity; any other value of the Hall factor leads to the same estimate of N_{dis} provided that γ_{Hall} does not vary in the interval 4×10^{18} cm$^{-3} < n^D < 1 \times 10^{19}$ cm^{-3}.

Of course, a model of mobility must explain not only the *ratio* $\mu(n^D_{*\,+})/\mu(n^D_{*\,-})$, but also the *values* $\mu(n^D_{*\,+})$ and $\mu(n^D_{*\,-})$. To verify to which extent the model is able to explain the observed mobilities, consider first the low temperature behaviour (say, around 10 K). In the region $n^D > n^D_*$ the mobility first increases slowly with n^D up to a value $n^D_m \simeq 2 \times 10^{19}$ cm^{-3} and hence decreases slowly; the whole behaviour can be described by eqs. (9) and (10), the mean free path $\lambda = l_s$ being given by eqs. (12) and (13) with $\nu_0 = n^D_*$. At $n^D \simeq n^D_m$ the value of μ is approximately 100 cm^2/V s in agreement with the experimental findings.

It is now possible to consider the behaviour at higher T:
In the <u>metallic region</u> ($n^D > n^D_*$) μ decreases slowly with T in agreement with the additional scattering by phonons.
In the <u>semiconductor region</u> ($n^D < \frac{1}{2}n^D_*$) μ increases with T in agreement with a carrier Rutherford scattering.
In the <u>intermediate region</u> ($\frac{1}{2}n^D_* < n^D < n^D_*$) the impurity band is already formed and it dominates conduction at low T and the mean free path is given by eq. (13); as T increases, however, the conduction band is gradually populated and the mobility becomes that proper of, Rutherford scattering on ionized impurities, with mean free path of the order of $(n^D_*)^{-\frac{1}{3}} \simeq 10^{-6}$ cm, i.e., remarkably higher than that typical of scattering in impurity band. Of course, the details of the increase of μ with n^D are lost because eq. (14) does not describe them.

The small difference between the values of N_{dis} obtained from solid solubility and mobility data is only seeming and can be explained by observing that an atom can be regarded as 'displaced' when its average displacement from lattice position exceeds the root mean square displacement due to thermal vibrations. Since solid solubility data are referred to temperatures in the range 1100 − 1400 K, while mobility data come from experiments in the range 10 − 100 K, the small discrepancy actually upholds the proposed interpretation.

CONCLUSIONS

Several facts concerning acceptors are not explained by the standard theory of shallow dopants [1]. A detailed list is discussed in ref. [21] which also gives references to the original literature; here I quote only: the acceptor inactivation by hydrogen (the inactivation being effective only for dopants in the ground state), the existence of supershallow levels in Si:In, and equilibrium and transport properties in Si:In.

My proposal to overcome the difficulties of the standard theory of shallow dopants was formalized in a description (the deep dopant description) where the ground state electronic configuration of the acceptor is assumed to be sp^2 [21,22]. This description explains 'super-shallow' levels observed with thermal techniques (with activation energy, $\simeq 18$ meV, about one order of magnitude lower than the ionization energy, 160 meV, measured by optical techniques at low temperature) only by admitting that the activation energy is actually a free energy, given by the difference of the optical energy minus an entropic contribution. The strain entropy considered in the previous section, of approximately $5 - 10\,k_B$ at temperature around $1200 - 1300$ K, is able to reduce an enthalpy by an amount sufficient to explain the observed supershallow level at 300 K; for $T \rightarrow 0$ K the thermal ionization energy tends to the optical one because $T\Delta S_{str}$ vanishes.

However, can a displacement cloud of approximately 10^3 atoms be considered as a small thermodynamic system?

"The energy of the universe is constant. The entropy of the universe tends to a maximum".

These statements are taken from the 1865 paper of Clausius on entropy [23]. Today people are more prudent, though nobody doubts of the validity of the first and second laws of thermodynamics for limited, however large, systems.

More problematic is the smallest size for which a system can be described by thermodynamic quantities. It is well known that the properties of small systems gradually change and tend those of bulk systems as the size increases; however, the typical size at which the properties of the bulk phase are gained depends strongly on the considered property. For instance, the ionization energy of mercury clusters Hg_k coincides with the atomic one for $k = 5 - 10$ and reaches the value of the metallic one for $k = 60 - 70$; the interatomic distance in copper particles reaches the bulk value for particle diameter of approximately 10 Å; and the melting temperature of gold particles approaches asymptotically the bulk value (the difference becoming lower than 10 %) for diameters larger than $50 - 100$ Å (these data are taken from the short review in ref. [24]).

None of these pieces of evidence is useful to understand which is the minimum size large enough
1) to deserve a statistical description, and
2) to make this description accurate.
The successful description of amorphization during ion implantation as an essentially thermal phenomenon by myself and my coworkers [25] suggests that a diameter of approximately 30 Å (involving 10^4 particles) is large enough for a thermodynamic description; if the considerations of this work are correct, the minimum number of particles can be reduced by one order of magnitude (10^3); in a cloud with this size there are roughly as many atoms on the border as inside — one can hardly imagine that smaller systems can be described by bulk thermodynamic quantities.

ACKNOWLEDGEMENTS

I wish to thank Dr. L. Meda (SGS-Thomson Microelectronics) for helpful comments.

REFERENCES

1. G.F. Cerofolini and L. Meda, *Physical Chemistry of, in and on Silicon*, Springer-Verlag, Berlin (1988)
2. M. Lannoo and J. Bourgouin, *Point Defects in Semiconductors I*, Springer-Verlag, Berlin (1981)
3. W. Zulehner and D. Huber, Crystals **8**, 1 (1982)
4. G. Das, Mat. Res. Soc. Symp. Proc **14**, 87 (1983)
5. J. Van Vechten, Phys. Rev. B **10**, 1482 (1974)
6. J. Van Vechten, Proc. 13th Intl. Conf. Defects in Semiconductors, The Metallurgical Society, Warrendale PA (1985) p. 293
7. R. Car, P.J. Kelly, S. Oshiyama and S.T. Pantelides, Phys. Rev. Lett. **52**, 1854 (1984)
8. E.P. Donovan, F. Spaepen, D. Turnbull, J.M. Poate and D.C. Jacobson, Appl. Phys. Lett. **42**, 698 (1983)
9. M. Servidori, Nucl. Instrum. Meth. B **19/20**, 443 (1987)
10. L. Meda, G. F. Cerofolini, R. Dierckx, G. Mercurio, M. Servidori, F. Cembali, M. Anderle, R. Canteri, G. Ottaviani, C. Claeys and J. Vanhellemont, Nucl. Instrum. Meth. B, in press
11. G.F. Cerofolini, L. Meda, C. Volpones, R. Dierckx, G. Mercurio, M. Anderle, R. Canteri, F. Cembali, R. Fabbri and M. Servidori, Nucl. Instrum. Meth. B, in press
12. P. Cappelletti, G.F. Cerofolini and G.U. Pignatel, Phil. Mag. A **46**, 863 (1982)
13. P. Cappelletti, G.F. Cerofolini and G.U. Pignatel, Phil. Mag. A **47**, 623 (1983)
14. G.F. Cerofolini, Adv. Coll. Interface Sci. **19**, 103 (1983)
15. A.A. Maradudin, E.W. Montroll and G.H. Weiss, *The Theory of Lattice Dynamics in the Harmonic Approximation*, Pergamon Press, New York, NY (1963)
16. A. Baldereschi and J.J. Hopfield, Phys. Rev. Lett. **28**, 171 (1972)
17. N.F. Mott, Intl. Rev. Phys. Chem. **4**, 1 (1985)
18. G.F. Cerofolini, L. Meda, E. Mazzega, M. Michelini and G. Ottaviani, unpublished
19. P.F. Newman and D.F. Holcomb, Phys. Rev. B **28**, 638 (1983)
20. J.M. Ziman, *Electrons and Phonons*, Clarendon Press, Oxford (1963)
21. G.F. Cerofolini and R. Bez, J. Appl. Phys. **61**, 1435 (1987)
22. G.F. Cerofolini, Phil. Mag. B **47**, 393 (1983)
23. R. Clausius, Ann. d. Phys. **125**, 353 (1865)
24. T. Stace, Nature **331**, 116 (1988)
25. G.F. Cerofolini, L. Meda and C. Volpones, J. Appl. Phys. **63**, 4911 (1988)

STRUCTURAL AND CHEMICAL CHARACTERIZATION OF SEMICONDUCTOR INTERFACES BY HIGH RESOLUTION TRANSMISSION ELECTRON MICROSCOPY

A. Ourmazd

AT&T Bell Laboratories
Holmdel, NJ 07733
U.S.A.

INTRODUCTION

Modern growth techniques allow the fabrication of multi-layered systems, with electronic properties that can be turned to the particular application in mind. These systems necessarily contain many interfaces, whose structure can profoundly influence their electronic characteristics. Since the individual layers are sometimes only a few lattice parameters thick, a full characterization of these interfaces requires atomic resolution both spatially and chemically.

High Resolution Transmission Electron Microscopy (HRTEM) can produce compelling images, which ofter appear to present a direct two-dimensional projection of the atomic structure of the sample. There is little doubt that such images have helped elucidate the structures of a variety of important systems. But it must also be admitted that the temptation towards a simplistic interpretation of lattice images has, on occasion, led to unjustifiable conclusions. It is the purpose of this paper to present a brief description of the fundamentals of lattice image formation so as to promote a critical appreciation of the strengths and weaknesses of HRTEM. At the same time, I will attempt to highlight some of the recent developments, which now allow the simultaneous structural and chemical characterization of semiconductor interfaces on an atomic scale.

FUNDAMENTALS OF LATTICE IMAGE FORMATION

On the simplest possible level, a transmission electron microscope provides a highly magnified image by illuminating a sample with a parallel beam of electrons, which are focused onto a screen after passing through the sample. The inadequacy of this purely optical picture of the TEM becomes immediately apparent when it is recalled that, although the electron wavelength at the usually employed energies of a few hundred keV approaches 10^{-2} Å, the resolution of the most modern HRTEMs has not yet reached 1Å. The severe aberrations of electromagnetic lenses which are responsible for this loss of resolution can seriously complicate the way in which a lattice image is related to the sample structure, even when the relevant information lies within the "resolution" of the TEM. In addition, electrons interact with matter more strongly than photons and are, in general, scattered in ways not characteristic of photon/solid interactions. In order to understand the basic physics of high resolution image formation, two separate processes, electron/sample and electron/lens interactions, must be addressed.

Fig. 1 illustrates the simplest way electron/sample interaction can be viewed. The part of the electron wavefront that passes through a region of high atomic potential experiences a refractive index higher than the part passing through a region of low potential. The sample thus acts as a diffraction grating, and in exact analogy with optics, the emerging electrons give rise to a diffraction pattern. On this picture, an electron undergoes a single scattering event, and a diffracted beam suffers a phase change of $\pi/2$ with respect to the undeviated beam. Under these so-called kinematical conditions, the diffraction pattern is simply the Fourier transform of the sample structure. Fig. 2, however, shows the way in which the phases and amplitudes of the diffracted electron beams change with sample thickness for InP, a typical semiconductor. The "kinematical" phase relationship is rapidly destroyed with increasing thickness and does not hold even for the smallest achievable thicknesses (~ 30 Å). Under these so-called dynamical conditions, the diffraction pattern is no longer related to the sample structure through a simple Fourier transformation. The kinematical phase relationship, however, returns with increasing sample thickness and periodically thereafter, thus manifesting "pendellosung" oscillations [1].

The electromagnetic lens brings the diffracted beams to interference on the screen, thus forming a lattice image. A lattice image, therefore, is first and foremost an electron interferogram. In bringing the beams to interference, a perfect lens with no aberrations would simply carry out the inverse Fourier transformation necessary to form an image of the sample projected potential. However, this is in general not the case with an imperfect lens. The performance of such a lens can be characterized by its Contrast Transfer Function (CTF), which describes how, at a given lens defocus, the phase and amplitude of the transmitted information are changed as a function of the spatial frequency of the information. At the so-called optimum or Scherzer defocus, $(1.5\lambda C_s)^{1/2}$, the CTF consists of a passband followed by damped oscillations (Fig. 3). The relative phases of the spatial frequencies lying within the first zero of the CTF are not changed, while frequencies beyond the first zero can undergo relative phase changes as well as amplitude attenuation. Consequently, within the first zero of the CTF (also known as the point-to-point resolution of the microscope) information is transmitted without significant modification, while outside the first zero the lens aberrations can decisively affect the transmitted information.

Due to the dynamical nature of electron/sample interaction, as well as the aberrations of the lens, a lattice image is, in general, not a simple representation of the sample structure. However, if the sample thickness is chosen such that the kinematical phase relationship between the beams holds, and only the information lying within the first zero of the CTF is allowed to contribute to an image formed at

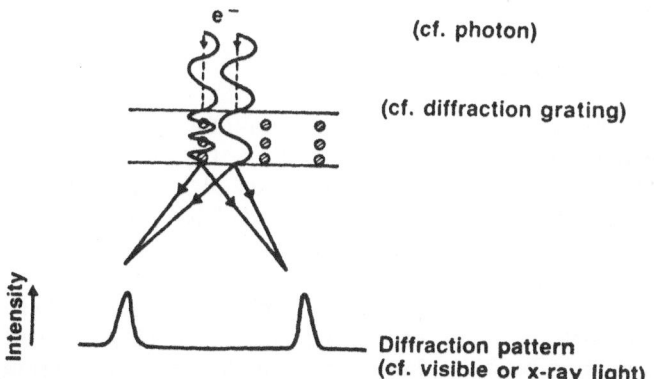

Fig. 1   Simplest representation of electron/sample interaction, giving rise to electron diffraction. In this "kinematic" approximation, the analogy with photons is exact.

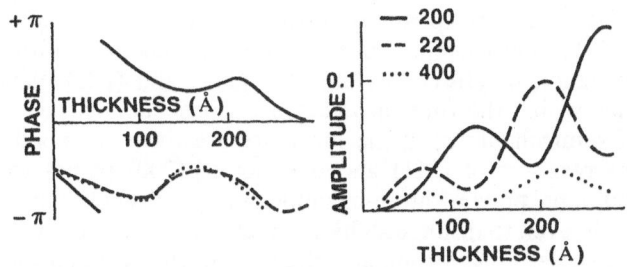

Fig. 2   Variation and phase and amplitudes of beam with thickness for InP. The electron beam is incident along <100>.

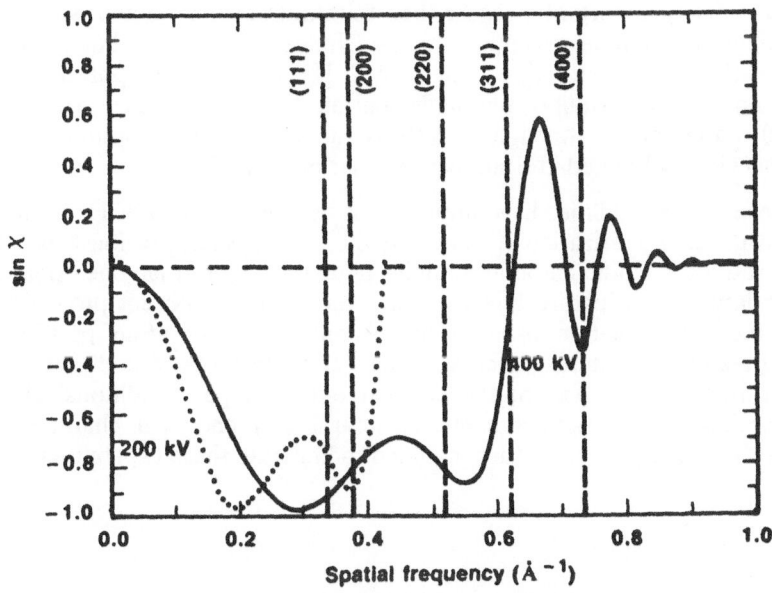

Fig. 3   Contrast Transfer Function at 200 kV (dotted) and 400 kV (solid). The vertical lines represent the various planar spacing of Si.

optimum defocus, the resulting image is a faithful representation of the sample structure to within the point-to-point resolution of the microscope. Such an image is known as a *structure* (rather than a lattice) image and, if individual atomic columns are resolved, the term *atomic* structure image is used. The hallmark of a structure image is that atomic columns appear black and the tunnels white on a positive print. Structure images, while being particularly simple to interpret, are sensitive only to the overall geometry of the structure. Images obtained under other conditions, for example at other defocus values or "non-kinematical" thicknesses, can be sensitive to small changes in atom positions, but are not simple representations of the structure. Accurate and reliable structure determination, therefore, requires the fitting of a series of experimental images, where a parameter such as defocus or sample thickness is varied, with a corresponding series of computer simulations. In practice, the structure image is used to deduce the overall structure, which is then refined by fitting the other members of the series.

Although it is in principle possible to deduce information lying well outside the first zero of the CTF, in practice it is difficult to do so reliably. Until the emergence of the latest generation of HRTEMs, only the (111) and (200) planar spacings of semiconductors lay within the first zero (Fig. 3). Consequently, the vast majority of lattice images of semiconductors in the literature are those obtained in the <110> projection, where two sets of (111) and one set of (200) planes can be resolved. Under these conditions, pairs of atom-columns appear as single dark or white blobs and it is not possible even to distinguish between atoms and tunnels without additional information (see Fig. 9a). Fig. 3 compares the CTF of the best 200 kV HRTEM with that of the latest 400 kV instruments, which have recently become available. Most significant is the fact that the (220) reflections of all semiconductors lie comfortably within the first zero of the CTF, and a number of other reflections are within its reasonable proximity. The (220) reflections, which in *Si* correspond to a spacing of 1.9Å, are critically important, because their faithful transmission allows the imaging of individual atomic columns in semiconductors. Indeed, this feature has been utilized to obtain the first atomic structure images of *Si* [2], in the two important orientations <100> and <111>. Additionally, individual atomic columns have been resolved in the <110> orientation, although at defocus values larger than optimum, and lattice images have been obtained in a number of other orientations [2].

These new capabilities have implications beyond the field of HRTEM, because they allow the characterization of semiconductors in ways not previously possible. The ability to resolve individual atomic columns, to do so in more than one orientation, and to obtain lattice or structure images of a given area in different projections makes possible the determination of three-dimensional maps of atom positions around individual features of interest, such as defects and interfaces. Somewhat surprisingly, the aberrations of the lens and dynamical effects impart additional flexibility to HRTEM, allowing the simultaneous structural and chemical characterization of materials. The examples below are intended to illustrate these capabilities.

THE STRUCTURE OF THE Si/SiO_2 INTERFACE

The Si/SiO_2 interface is a corner-stone of semiconductor technology. This, together with its intrinsic scientific interest, has stimulated sustained activity for over thirty years. It is thus now generally, but not universally, accepted that the SiO_2 is structurally amorphous and chemically stoichiometric at distances in excess of 10 Å from the interface. No such general consensus exists so far as the interfacial structure itself is concerned. Briefly, the various proposals can be divided into three general categories. a) The c-$Si \rightarrow a$-SiO_2 transition is proposed to occur via a stable, bulk phase of c-SiO_2. Due to the structural similarity between Si and cristobalite, this oxide represents the most frequently proposed crystalline interfacial phase. However, it has a lattice parameter 40% larger than Si, and it is difficult to achieve a commensurate Si/cristobalite interface. b) A metastable, sub-stoichiometric oxide is thought to affect the c-$Si \rightarrow a$-SiO_2 transition. It is then necessary to postulate a metastable phase of remarkable stability, able to exist at the relatively high temperatures employed in Si oxidation. It is possible that the special conditions present at the interface may stabilize such a phase. c) It has been shown that the c-$Si \rightarrow a$-SiO_2 could occur abruptly, with no intervening crystalline or sub-stoichiometric phase at the interface [3].

HRTEM has naturally been employed to investigate the structure of this interface, and in particular, detect the presence of any crystalline oxide [4]. The resultant lattice images show an abrupt transition from c-Si to "amorphous" material. The important question regards the interpretation of such images. An examination of the $Si(100)$ surface shows that although the surface is structurally four-fold symmetric, the dangling bonds reduce this to two-fold symmetry. An epitaxial oxide would necessarily be tied to the dangling bonds, which rotate through 90° on crossing a Si surface step consisting of an odd number of atomic layers. This implies that, unless an epitaxial oxide were fourfold symmetric, it would be polycrystalline with a grain size determined by the spacing between the steps on the Si surface. Lattice images of the Si/SiO_2 interface are obtained from cross-sectional samples about 100Å thick. Thus, for an interfacial oxide phase of less than four-fold symmetry, the image would consist of the superposition of the images of a number of oxide grains rotated by 90° with respect to each other due to the presence of Si surface steps. Consequently, unless the grain size is comparable to, or larger than, the sample thickness, the detection of a crystalline oxide phase of less than four-fold symmetry is most unlikely. All that can be surmised from the usual lattice images of the Si/SiO_2 interface is that, if a crystalline interfacial oxide is present, it is less than four-fold symmetric.

This discussion also provides guidelines for the further investigation of this important system. The interfacial structure can only be uniquely determined by HRTEM if the spacing between Si surface steps is much larger than the sample thickness (~ 100 Å). Modern MBE growth of Si on Si can produce samples of sufficient perfection [5]. Fig. 4 shows lattice images of the (001) Si/SiO_2 [110] and [110] projections for an MBE grown, oxidized Si sample. The presence of an interfacial oxide layer is immediately apparent, while the difference between the two lattice images shows the oxide to be indeed less than four-fold symmetric. Since we have been able to obtain lattice images of this phase in three projections and diffraction patterns in four projections, we have determined a three-dimensional map of atom positions at the Si/SiO_2 interface (Fig. 5) [5]. Modeling of lattice images indicates the interfacial oxide to be tridymite, a stable, bulk phase of SiO_2. This crystalline phase forms a strained, commensurate, epitaxial oxide layer on the Si substrate, and is thus characterized by a critical layer thickness, beyond which strain relaxation must occur. In this way the small thickness of the crystalline oxide epitaxial oxide layer on the Si substrate (~5 Å) and the production of amorphous SiO_2 can be understood as simple consequences of strain relaxation [5].

SIMULTANEOUS RESOLUTION AND IDENTIFICATION OF INDIVIDUAL ATOM-COLUMNS IN COMPOUND SEMICONDUCTORS

The ability to utilize the (220), and, in other orientations, the (400) reflections, allows the location of individual atomic columns, and thus a complete structural characterization of elemental semiconductors. In the case of compound semiconductors, however, a full structural characterization requires not only the location of individual atomic columns, but also their identification. In GaAs, for example, it is necessary to determine whether a particular atomic column belongs to the Ga or the As sublattice. In other words, a complete structural characterization of compound semiconductors needs both structural and chemical information.

The (220) reflections, necessary for the resolution of the atomic columns, are relatively insensitive to chemical changes. (200) reflections, on the other hand, are kinematically forbidden in elemental semiconductors, because the contributions from the two fcc sublattices of the diamond lattice cancel out at the (200) position in reciprocal space. These reflections are consequently extremely sensitive to the chemical nature of the occupants of the two sublattices. Thus, in order both to resolve and identify the individual atomic columns, the (220) and (200) reflections must make comparable contributions to the lattice image. In principle, this can be realized in the $<100>$ projection, where four (220) and four (200) beams can be used to form a lattice image, with the advantage that all the contributing beams lie within the first zero of the (CTF). Fig. 2, however, shows the amplitudes and phases of these reflections for InP as a function of sample thickness. In regions where the amplitudes of the (220) and (200) reflections are comparable, their phase relationship is far from kinematical, and a lattice image formed in such a region would not be a simple representation of the sample projected potential. On the other hand, in the kinematical region, the (220) amplitude is dominant and identification of the different chemical species is not possible. This dilemma can be resolved by recognizing that, since the (220) beams lie close to the first zero of the CTF, their contribution to the image can be finely controlled by adjusting the lens defocus or microscope accelerating voltage (Fig. 6). Thus, one can choose a sample thickness where the kinematical phase relationship holds, and attenuate the (220) amplitudes to a level comparable with those of the (200) beams. Fig. 7 shows an image of InP obtained by this method, where the lens defocus is used to "tune" the (200) and (220) contributions. Because the defocus deviation from optimum is small, the image is a simple representation of the sample projected potential and the individual atomic columns are directly resolved and identified [6]. As in all structure images, the atomic columns appear black. We have used the same approach directly to resolve and

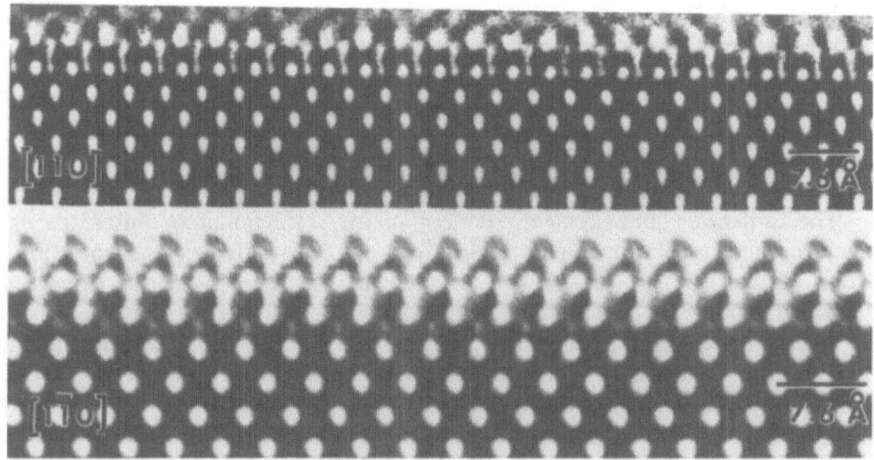

Fig. 4   Lattice images of (100)Si/SiO_2 interface in two orthogonal <110> projections. The Si sample was prepared by MBE growth of a 2000 Å Si layer on a Si substrate. Note the flatness and sharpness of the interface, and the clear presence of an ordered interfacial layer.

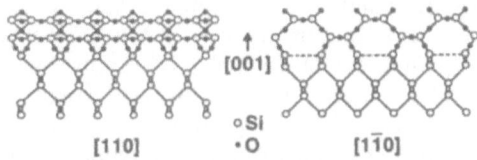

Fig. 5   Schematic representation of the Si/SiO_2 interfacial atomic configuration deduced from lattice imaging. The dashed lines indicate possible dimerisation of unsaturated bonds.

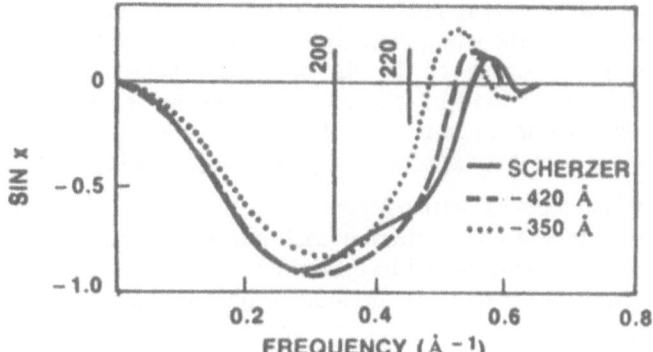

Fig. 6   Contrast Transfer Function for 400 kV HRTEM at different objective lens defoci. The vertical lines indicate the positions of (200) and (220) reflections for InP.

141

identify the individual atom-columns in GaP and GaAs [6]. This technique can therefore be used for any semiconductor and, by suitable extension, for other systems, where chemical and structural information is simultaneously required.

CHARACTERIZATION OF INTERFACES IN MULTI-LAYERED SYSTEMS

The *structure* of a perfect semiconductor/semiconductor interface is well-known; the atoms occupy zinc-blende sites on both sides of the interface. The question of the atomic configuration at such an interface is thus *chemical* and not structural in nature. The complete characterization of the interfacial configuration thus requires the development of a technique capable of yielding chemical and spatial information on an atomic scale. This can again be achieved by utilizing the chemical sensitivity of the (200) reflections. As an example, consider the InP/InGaAs system. Fig. 8 shows the way the amplitudes of the (200) and (220) beams change with thickness for these two materials. Over the thickness range 70 - 150 Å, the (200) beam is dominant in InP, while in InGaAs the (220) beam is stronger. This implies that a lattice image formed in this thickness range should, under suitable defocus conditions, exhibit a strong change in periodicity across the interface. Figs. 9a and b represent the usual <110> lattice images of the interface between InP and InGaAs, while the lattice image of Fig. 9c was obtained in the <100> orientation from the same area of the same interface, using the approach just described [7]. The InP gives rise to strong (200) fringes, while the InGaAs is represented by the fine (220) fringes at 45° to the interface. The interface, shown by the <110> image to be "abrupt" and "flat", can be seen to be rough on an atomic scale. In addition to the coarser interfacial steps, revealed as strong protrusions of the (220) fringes from InGaAs onto the (200) fringes of InP, a number of weaker (220) "spots" also protrude into the (200) InP fringes. These arise from interfacial steps, whose extent in the interface plane is a small fraction of the sample thickness (~100 Å). It appears, therefore, that this approach can sensitively detect small variations in composition at the interface. This sensitivity is due to the fact that chemical changes in the sample are translated into readily detectable periodicity changes in the lattice image, and, additionally, because "random" effects, such as surface contamination, do not give rise to definite spectral changes.

As outlined above, the application of "chemical lattice imaging" to the InP/InGaAs, establishes the inadequacy of normal structural lattice images to reveal interfacial roughness, but it does not necessarily imply that all semiconductor heterointerfaces are indeed rough. In the case of the technologically more mature GaAs/AlGaAs system, the photoluminescence (PL) linewidth, and particularly the so-called monolayer splitting of the PL lines have been interpreted in terms of atomically abrupt and smooth interfaces, where the the spacing between interfacial steps, (the island size) has been estimated at several microns. It is thus important to examine GaAs/AlGaAs interfaces of the highest optical quality.

Figure 10 is a chemical lattice image of the $GaAs/Al_{0.37}Ga_{0.63}As$ MBE sample, grown with a two minute interruption at each interface. Again the sample thickness and imaging conditions correspond to maximum chemical sensitivity, reflected in the strong change from (220) to (200) periodicity on crossing from GaAs to AlGaAs.

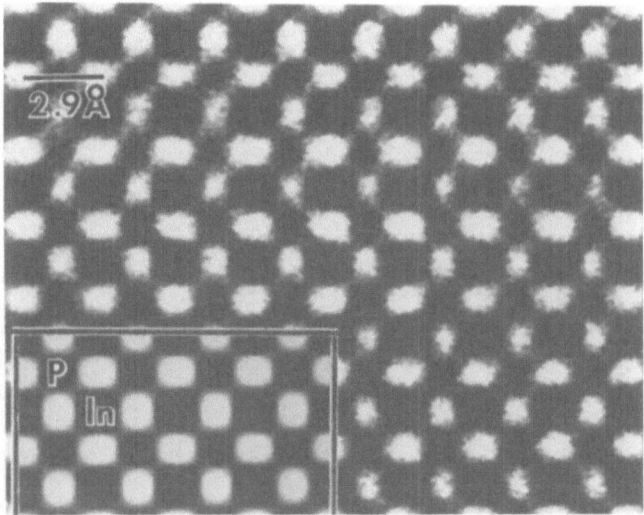

Fig. 7   Atomic structure image of InP in the <100> projection, where the
           individual atomic columns are directly resolved and identified. The
           atom-columns appear black, the tunnels white. Bottom inset is the
           simulated image. Note the heavier In atoms appear larger and darker
           than the lighter P atoms.

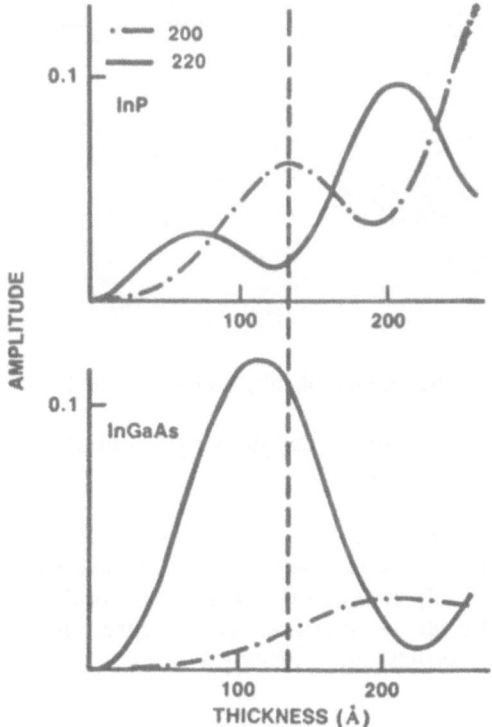

Fig. 8   Variation of amplitudes of (200) and (220) beams with thickness in InP
           and InGaAs. The vertical lines shows an ideal thickness for maximum
           change in lattice image periodicity on crossing an interface between InP
           and InGaAs.

143

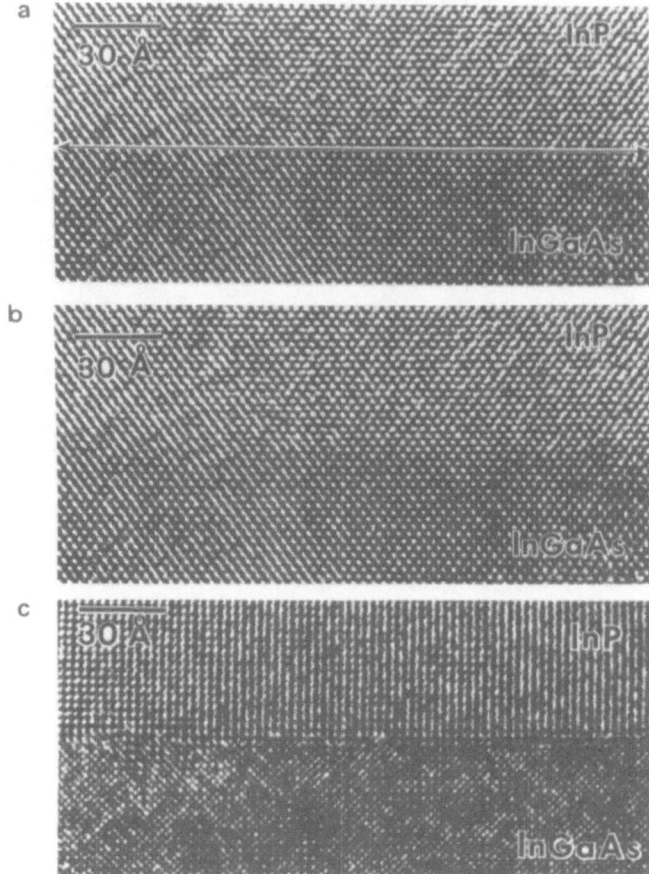

Fig. 9   (a) <110> lattice image of InP/InGaAs interface.

(b) Same image without line drawing attention to the interface, which is now difficult to recognize.

(c) <100> lattice image of the same region of the same interface, obtained as described in the text. Note the protrusion of the finer InGaAs fringer onto InP, as well as the presence of weaker (220) spots in the (200) fringes of InP at the interface. These represent interfacial roughness not revealed by normal lattice imaging.

Fig. 10   Chemical lattice image of GaAs/$Al_{0.37}Ga_{0.63}$As quantum well produced after two minutes of growth interruption at each interface. Carefull inspection reveals interfacial roughness.

Visual examination of the image directly reveals the presence of interfacial roughness. Thus, even at this qualitative level of inspection, interfaces of the highest optical quality appear rough. However, the general practice of evaluating lattice images by visual inspection is subjective and unsatisfactory. I describe below a digital pattern recognition approach, which allows the quantification of the local information content of lattice images, leading to their quantitative evaluation.

QUANTIFICATION OF LOCAL INFORMATION CONTENT OF IMAGES

The information content of a lattice image is degraded by the presence of surface contamination and damage, and electron shot noise. For esample, to distinguish GaAs from AlGaAs in the presence of such effects, it is necessary to possess perfect templates for the images of each material, and to quantify the way noise causes departures from these ideal templates. Briefly, we achieve this as follows [8]. First, we obtain an "ideal", noise-free image of the projected unit cell for each of the two materials, by averaging over many lattice image unit cells far from the interface. The resulting images act as templates for the recognition of GaAs and $Al_{0.37}Ga_{0.63}As$ (Fig. 11a). Each $2.8 \times 2.8 \, \mathring{A}^2$ template is divided into a 35×35 pixel array, and the intensity in each pixel recorded. We define a vector $\mathbf{R}^t$, whose components are the intensity values at each pixel of the template. Thus the ideal GaAs and AlGaAs unit cells are each represented by a 1225-component vector template. The change in composition from GaAs to AlGaAs can now be defined in terms of the angle θ_c between these two vectors (Fig. 11b). We represent a given unit cell of the real (ie. noisy) image also by a vector $\mathbf{R}$. For "bulk" GaAs far from the interface, the vectors $\mathbf{R}_{GaAs}$ from the individual lattice image unit cells form a distribution about $\mathbf{R}^t_{GaAs}$, with the angles between $\mathbf{R}^t_{GaAs}$ and $\mathbf{R}_{GaAs}$ representing the effect of noise. The standard deviation σ of this distribution quantifies the noise present in the image of GaAs (Fig. 11c). A similar procedure is used to quantify the noise in "bulk" AlGaAs. The noise present is such that $\mathbf{R}_{GaAs}$ and $\mathbf{R}_{AlGaAs}$ form similar normal distributions about their templates, with the template vectors $\mathbf{R}^t_{GaAs}$ and $\mathbf{R}^t_{AlGaAs}$ about 12σ apart. This means that as the Al concentration is increased from 0 to 0.37, the template vector $\mathbf{R}^t$ rotates through an angle which is 12σ. Thus, our vector representation of the lattice image allows us to distinguish between GaAs and $Al_{0.37}Ga_{0.63}As$ with total confidence.

We next consider vectors $\mathbf{R}^i$ representing the image unit cells in the vicinity of the interface. Our image simulations and control experiments show that the projection of $\mathbf{R}^i$ onto the plane defined by the template vectors rotates monotonically and linearly between the two template vectors as the Al content is changed. Thus the angular position of $\mathbf{R}^i$ with respect to the template vectors $\mathbf{R}^t_{GaAs}$ and $\mathbf{R}^t_{AlGaAs}$ yields the Al content of the $2.8 \times 2.8 \, \mathring{A}^2$ unit column. At all the interfaces we have analysed, a substantial number of vectors $\mathbf{R}^i$ fall more than 3σ away from both the template vectors $\mathbf{R}^t_{GaAs}$ and $\mathbf{R}^t_{AlGaAs}$. Such vectors represent, with an error probability of at most 3 parts in 10^3, unit cells of $Al_xGa_{1-x}As$, with $0 < x < 0.37$. The actual composition of the unit cell represented by a vector $\mathbf{R}^i$ must fall within $\sim 3\sigma$ of the value deduced from the angular position of $\mathbf{R}^i$. Thus, the approach we have outlined allows the determination of the Al content of individual unit cells $2.8 \times 2.8 \times 75 \, \mathring{A}^3$ in size. A unit cell *at* a hypothetical, atomically abrupt interface is composed of 1/4 of $Al_{0.37}Ga_{0.63}As$

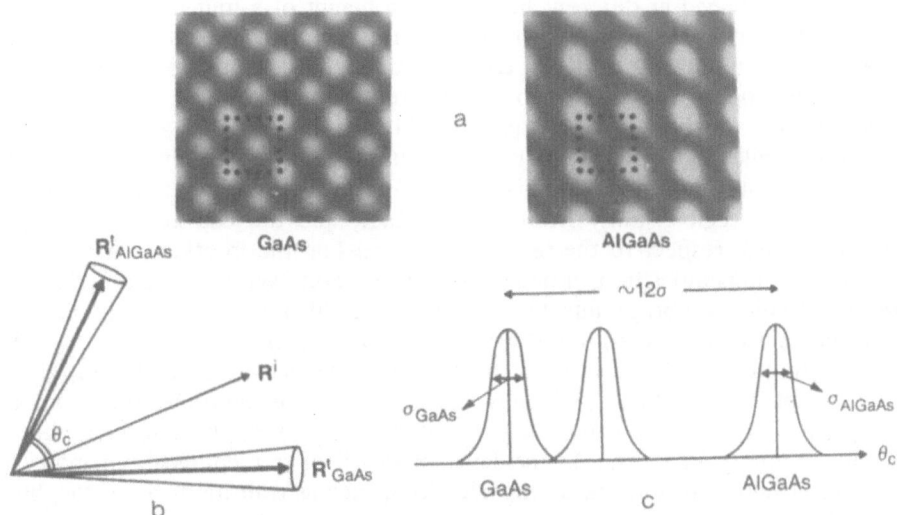

Fig. 11   a) Averaged, noise-free images of GaAs and $Al_{0.37}Ga_{0.63}As$. The unit cells used as templates for pattern recognition are the dotted 2.8Å squares. b) Schematic representations of the template vectors $\mathbf{R}^t_{GaAs}$ and $\mathbf{R}^t_{AlGaAs}$, the distribution of $\mathbf{R}_{GaAs}$ and $\mathbf{R}_{AlGaAs}$ about them, and an interfacial vector $\mathbf{R}^i$. c) Schematic representation of the distributions produced by the GaAs and $Al_{0.37}Ga_{0.63}As$ unit cells about their templates. Note that the angular position of $\mathbf{R}^i$ denotes the most likely composition only. The actual composition falls within a normal distribution about this point.

and 3/4 of GaAs or vice versa, depending on the choice of origin (Fig. 12). Thus an abrupt interface would be characterised by a transition from GaAs to $Al_{0.37}Ga_{0.63}As$ via a single unit cell, with an Al content of 1/4 x 0.37 or 3/4 x 0.37, depending on the unit cell position adopted.

QUANTITATIVE, SPATIALLY RESOLVED CHEMICAL MAPS

Fig. 13 is a representation of a chemical lattice image of the interrupted MBE-grown GaAs/AlGaAs sample analysed as described above, with the individual lattice image unit cells placed at different heights. The height of a unit cell represents the angular position of its vector **R** with respect to the template vectors. The height can thus be interpreted in terms of the Al concentration, going from zero to 0.37 from the bottom to the top of the three-dimensional representation. The blue and yellow regions indicate those unit cells falling within 3σ of $Al_{0.37}Ga_{0.63}As$ and GaAs respectively, while the other colours break up the data into different standard deviation, or composition bands. This representation allows a quantitative display of the noise in the regions away from the interface, and the change in the angular position of **R** with respect to the templates on crossing the interface. It provides a spatial map of the composition, at near-atomic resolution. With our choice of unit cell origin, an atomically abrupt interface would result in a transition from GaAs to AlGaAs via a single (green) interfacial unit cell at a height 1/4 that of a blue $Al_{0.37}Ga_{0.63}As$ unit cell (Fig. 12b). It is clear that this is not realised, and that the interface is not smooth. Fig.13, which is typical, shows that the transition from GaAs to $Al_{0.37}Ga_{0.63}As$ takes place over 2 - 4 unit cells, and that the interface contains significant atomic roughness. At the level of detail of our composition maps, the assignment of values for interfacial imperfections, such as transition width, roughness, and island size, is a matter of definition. Also, without extensive sampling caution is required in deducing quantitative values for the spacing between interfacial steps, however they are defined. Nevertheless, it is clear that significant atomic roughness at the ~50 Å lateral scale is present.

CONCLUSIONS AND FUTURE POSSIBILITIES

In this review, I have attempted to emphasize the importance of an elementary appreciation of the fundamentals of lattice image formation, because, under most circumstances, the aberrations of the microscope and the dynamical nature of electron/sample interaction can profoundly affect the recorded information. Only a systematic experimental procedure, augmented by computer simulation, allows the reliable interpretation of lattice images. Nevertheless, the recent past has shown HRTEM to be a powerful tool for the structural analysis of semiconductors. Chemical questions, on the other hand, have not, until recently, begun to be properly addressed.

The improved resolution of the latest generation of electron microscopes now makes possible the structural characterization of semiconductors and their interfaces on a scale not previously possible. The ability to obtain lattice images in several projections by tilting between them, and the possibility to resolve individual atomic columns in three important orientations, combine to allow three-dimensional mapping of atom positions around individual features of interest. Intellectually perhaps more

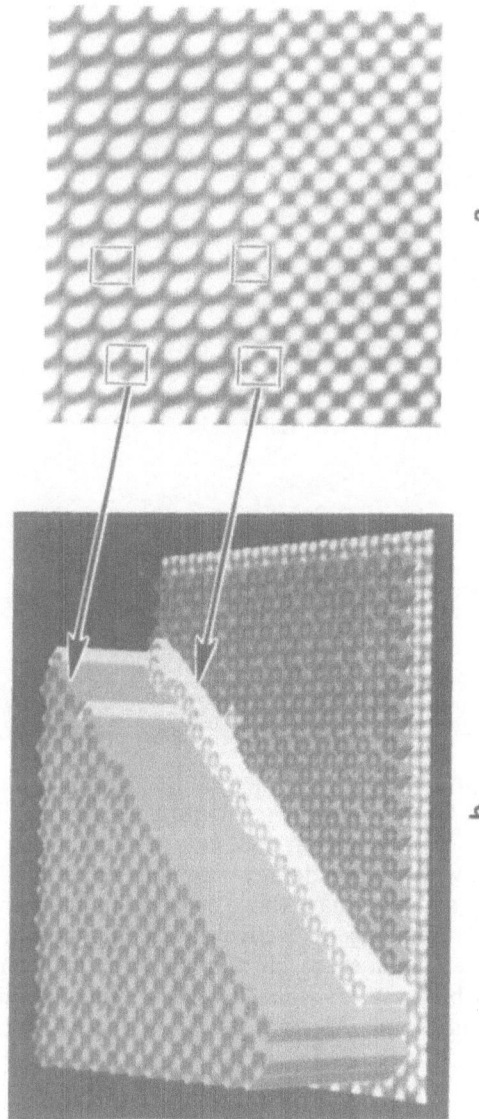

a

b

Fig. 12 a)   Computer-generated, noise free image of a perfect GaAs/Al$_{0.37}$Ga$_{0.63}$As interface. Note the two possible choices of unit cell. A unit cell *at* the interface has an Al content 1/4 or 3/4 that of bulk Al$_{0.37}$Ga$_{0.63}$As, depending on the choice of origin.

b) Three dimensional representation of the Al content of each 2.8 Å square unit cell, determined by the algorithm described in the text. Note the single row of unit cells with an Al content 1/4 that of Al$_{0.37}$Ga$_{0.63}$As, as expected.

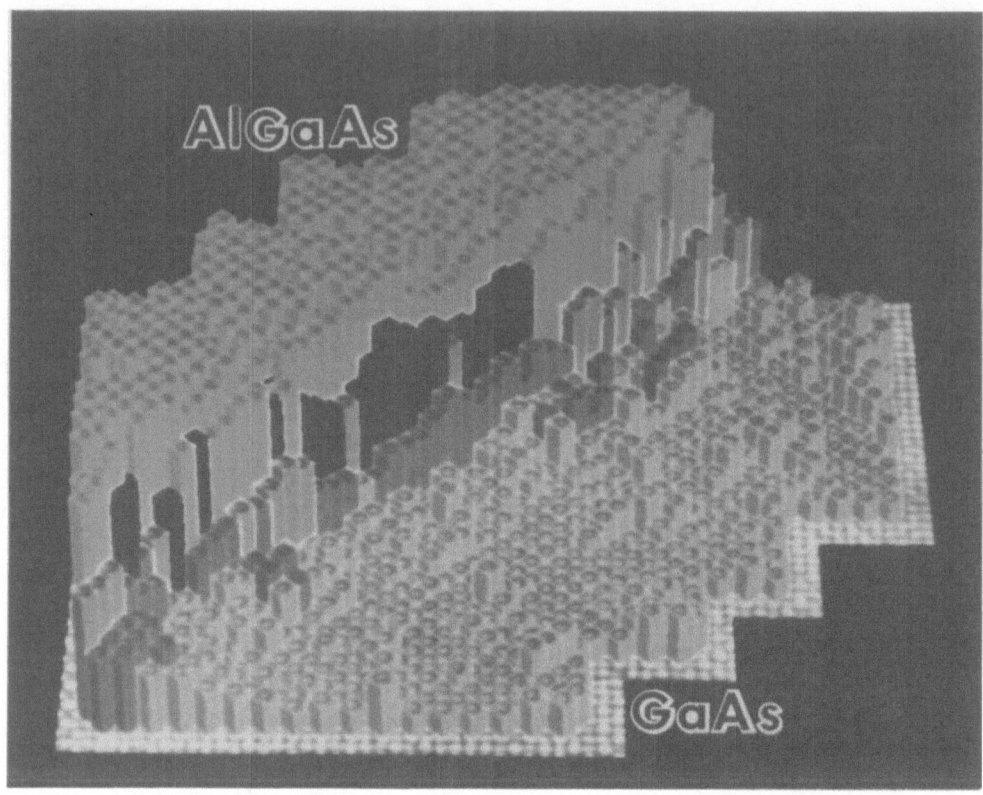

Fig. 13 Three-dimensional representation of the analysed lattice image of $Al_{0.37}Ga_{0.63}As$ grown on GaAs after a two minute interruption. The unit cells are 2.8 Å squares. The height of each cell represents the angular position of its vector **R** with respect to the template vectors, which are about 12σ apart. Yellow and blue mark those cells which fall within 3σ of GaAs and $Al_{0.37}Ga_{0.63}As$ templates respectively. Green, magenta and red represent 3σ bands centred about three, six and nine σ points from GaAs. Outside the yellow and blue regions, the Al content of each unit cell is intermediate between GaAs and $Al_{0.37}Ga_{0.63}As$, with confidence levels given by normal statistics.

appealing, however, is the new capability of simultaneously obtaining structural and chemical information on an atomic scale. Ironically, this is possible because the aberrations of the lens impart to the microscope the character of a band-pass filter, which can be tuned to enhance the chemical sensitivity of lattice imaging.

The combination of chemical lattice imaging and pattern recognition allows the quantitative determination of the composition of inhomogeneous samples at high spatial resolution. The application of such an approach to important interfacial systems, such as GaAs/AlGaAs now provokes questions, which may lead to a deeper understanding of the interrelation between the structure and properties of materials.

The same approach is also being used to investigate interfacial reactions, such as interdiffusion in the CdTe/HgCdTe system [9]. These results illustrate the importance of obtaining *local* chemical information in such inhomogeneous systems, where the proximity of the surface and the substrate appears to play a significant role.

ACKNOWLEDGEMENTS

I have benefitted from valuable discussions with J. Shah, W.Schröter, and C.A. Warwick. This work would not have been possible without the expert collaboration of D.W. Taylor, and of crystal growing colleagues J. Bevk, J.E. Cunningham, W.T. Tsang, C.W. Tu, and R.J. Fischer.

REFERENCES

1. For a detailed discussion of HRTEM, see e.g.: J. C. H. Spence, Experimental High Resolution Transmission Electron Microscopy (Oxford University Press, New York, 1980).

2. A. Ourmazd, K. Ahlborn, K. Ibeh, and T. Honda; Appl Phys Lett 47, 685 (1985).

3. S. T. Pantelides and M. Long, in *Physics of SiO$_2$ and its interfaces*, ed. S. T. Pantelides (Pergamon Press, New York, 1978), p. 339.

4. See, e.g., S. M. Goodnick, D. K. Ferry, C. W. Wilmsen, Z. Liliental, D. Fathy and O. L. Krivanek, Phys Rev B **32**, 8171 (1985).

5. A. Ourmazd, D. W. Taylor, J. A. Rentschler and J. Bevk, Phys Rev Lett, **59**, 213 (1987).

6. A. Ourmazd, J. A. Rentschler and D. W. Taylor, Phys Rev Lett, **57**, 3073 (1986).

7. A. Ourmazd, W. T. Tsang, J. A. Rentschler and D. W. Taylor, Appl Phys Lett, **50**, 1417 (1987).

8. A. Ourmazd, D.W. Taylor, J. Cunningham, and C.W. Tu, to be published.

9. Y.O. Kim, A. Ourmazd, R.D. Feldman, J.A. Rentschler, D.W. Taylor, and R.F. Austin, to be published.

INTERACTION BETWEEN POINT-DEFECTS, DISLOCATIONS AND A GRAIN BOUNDARY:

A HREM STUDY

J. Thibault-Desseaux, and J.L. Putaux

Département de Recherche Fondamentale, Service de Physique
Centre d'Etudes Nucléaires
F-85X-38041 Grenoble Cedex - France

H.O.K. Kirchner

Institut de Sciences des Matériaux
Université Paris Sud, Bat. 413
F-91405 Orsay - France

INTRODUCTION

The increasing trend towards miniaturization and large scale integration of semiconductor devices has posed the challenge of producing
materials which not only meet certain electronic specifications but also
fulfil sometimes rather particular mechanical constraints. This implies
that lattice defects created during production and materials processing
of the devices have to be known and controlled. Only a microscopic and
even nanoscopic knowledge of the defect structure allows the necessary
stringent control of the over-all properties.

High resolution electron microscopy (HREM) is a valuable means of
research which gives access to the very early stages of material degradation. This technique allows to gain crystallographic information in
real (and not reciprocal) space, in particular the imaging of atomic
columns provides a nanoscopic description of defect interaction on the
atomic scale.

This paper is an attempt to summarize new HREM results on the
interactions between point defects, dislocations and grain boundaries
mainly in silicon. Most of the observations were performed on deformed
bicrystals. The possibility of obtaining dislocation free bicrystals
allowed a well-defined description of the mechanisms occuring either
within the grains or at the boundary. The silicon bicrystals containing
a symmetrical tilt boundary $\Sigma=9$ (122) were strained in tension or compression by A.Jacques and A.George in the Ecole des Mines (Nancy-France).
Strain experiments were also carried out on a germanium monocrystal provided by X. Baillin (CEN Grenoble). The specimens were cut and thinned
<u>after deformation</u> for the HREM observations in a JEOL 200 CX microscope
(C_s=1.05 mm, 200 kV). The [011] common tilt axis is parallel to the
electron beam. The grain boundary and the created dislocations are
viewed end-on in projection suitable for HREMicroscopy.

Dislocations induced by the deformation are mainly 60° or screw dislocations. They glide in a dissociated configuration on the {111} planes and are generally aligned along the [011] directions. During their movement they react with other defects such as point defects, other dislocations and the grain boundary (GB). Nucleation of loops on the partials occurs, dipoles and locks are formed, the dislocations enter the GB and decompose into grainboundary dislocations which suffer the same effects : glide, climb and interaction with the other GB dislocations.

Each type of interaction will be considered in the grain and in the boundary respectively, therefore we will first describe the interaction between the dislocations and the grainboundary.

DISLOCATION-GRAINBOUNDARY INTERACTION

The absorption of dislocations by grainboundaries has been extensively investigated. It was established (1,2,3) that GB close to the coincidence orientation absorbes the incoming dislocations and that these dislocations decompose in the GB giving perfect GB dislocations whose Burgers vectors belong to the DSC lattice(Discrete Shift Complete lattice (4)). Recent in-situ experiments by X-rays topography and 1 MeV electron microscopy (5,6) revealed that in general the twin GB $\Sigma=9(122)$ is a strong obstacle for the passage of the dislocations whose Burgers vector is not common to both grains. However, under certain conditions dislocations can cross through the GB from one grain (7,8) into the other one. It remains unclear whether the mechanism is direct and leaves a GB residue or is indirect through a suitable recombination of GB dislocations. High resolution experiments showed (9) the decomposition of the incoming dislocations into DSC dislocations. The Bürgers vectors of the GB dislocations are determined using the method described in (10). The Burgers vector determination is based on the SF/RH convention, the line [0$\bar{1}\bar{1}$] pointing into the paper.

60° dislocations pile-ups are created at the GB (Fig.1a). Under the applied shear stress the leading partial enters the GB. The mechanism which permits the dissociated dislocation to integrate the GB is not really known. Numerous mechanisms have been evoked (11).

The central question is if the integration itself is a two dimensional process, with the participating dislocations remaining straight but dissociating or if the process is a three-dimensional one akin to cross-slip in the bulk. If three dimensional configurations like the emission of partial loops etc.. are invoked, thermal activation (which can occur only over small volumes) can help to overcome reaction barriers that could not be traversed under external or internal stresses alone. Unfortunately the activated configurations of such a process are of short duration and must remain unobservable. Only from observations of the reaction products one can take hints concerning one or the other possibility of reaction.

The final step of the integration (in compression) is shown in Fig.1b. The 60° dislocation 60°(p) at the tip of the pile-up gliding on the primary (111) plane was largely dissociated in the grain. After its integration it is decomposed into three ele◆ntary DSC dislocations (Fig.1b) following the reaction referring to grain I :

$$b_{60}^-(p) = b_{90}^1 + b_{30}^{1-} \rightarrow b_c + b_g + b_{30}^{1-}$$
$$\rightarrow a/9[\bar{1}2\bar{2}]_I + a/18[\bar{4}11]_I + a/6[\bar{1}21]_I .$$

The - sign is for the negative screw component aligned along the [011] direction. In this case the initial partials belong to the DSC lattice. It has been shown by HREM (9) that, in compression the leading partial b_{90}^1 decomposes into $b_c + b_g$ when it touches the GB. b_c is normal to the GB and to the tilt axis and sessile in the GB whereas b_g is parallel to the GB and therefore glissile in the GB : this is the reason why the b_g dislocation generally glides away like on Fig.1b . The integration of a 60° dislocation whose partials do not belong to the DSC leads to a similar decomposition in the GB. In the instance presented fig. 1c (compression), the incoming dislocation 60°(a) glides on the additional (111) planes and the residue found in the GB is an isolated b_{30}^{4-} indicated in the reaction (referring to grain I) :

$$b_{60}^-(a) = b_{90}(a) + b_{30}^-(a) \rightarrow b_{30}^{4-} + 2b_g = a/6[\bar{1}1\bar{2}]_I + 2a/18[\bar{4}11]_I .$$

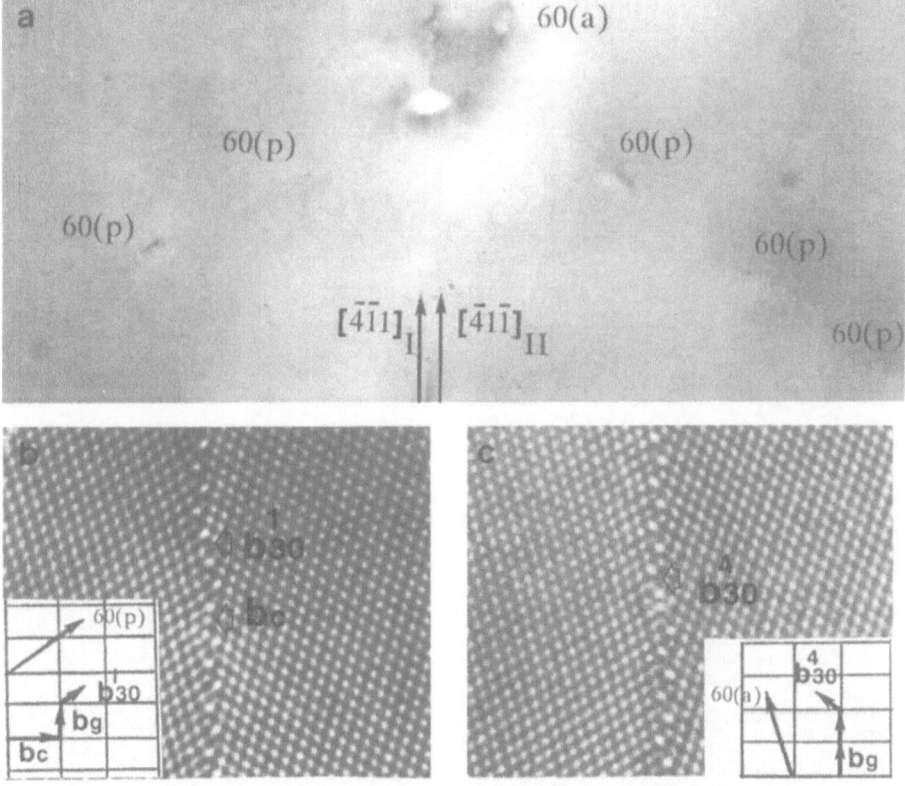

Fig.1. Interaction between dislocations and a GB. a) two 60° dislocations (60(p)) pileups on the primary planes in each grain and the tip of a pile-up on the additional planes (60(a)) are shown. b) two DSC dislocations b_c and $b\frac{1}{30}$ left by the complete integration of a 60(p) dislocation(courtesy M. Elkajbaji). c) DSC residue b_{30}^4 left by the entrance and the decomposition of a 60(a) whose initial partials did not belong to the DSC lattice.

Due to the internal stresses the glissile b_g dislocations generally disappear from the field of view. As the b_g dislocation is associated with a GB step :ho = |1/18[122]a|, the GB migrates lateraly from ho when b_g glides. Futhermore the slip of b_g makes the GB slide. It must be noticed that no residue corresponding to the entrance and the decomposition of a screw dislocation was found. This is in good agreement with the in-situ experiments(5) that showed an easy transmission of the dislocations with Burgers vector common to both grains.

DISLOCATION-POINT DEFECTS INTERACTION

In the grain

The HREM observations revealed that the 60° dislocations trapped during the deformation by dipoles formation or by the interface are subject to point defects interaction (12). This phenomenon was also confirmed in silicon by weak-beam observations(13) and is essentially a three dimensional one.

It seems that during the deformation the point defects involved are mainly interstitials. A perfect interstitial loop 1/2[110]a or

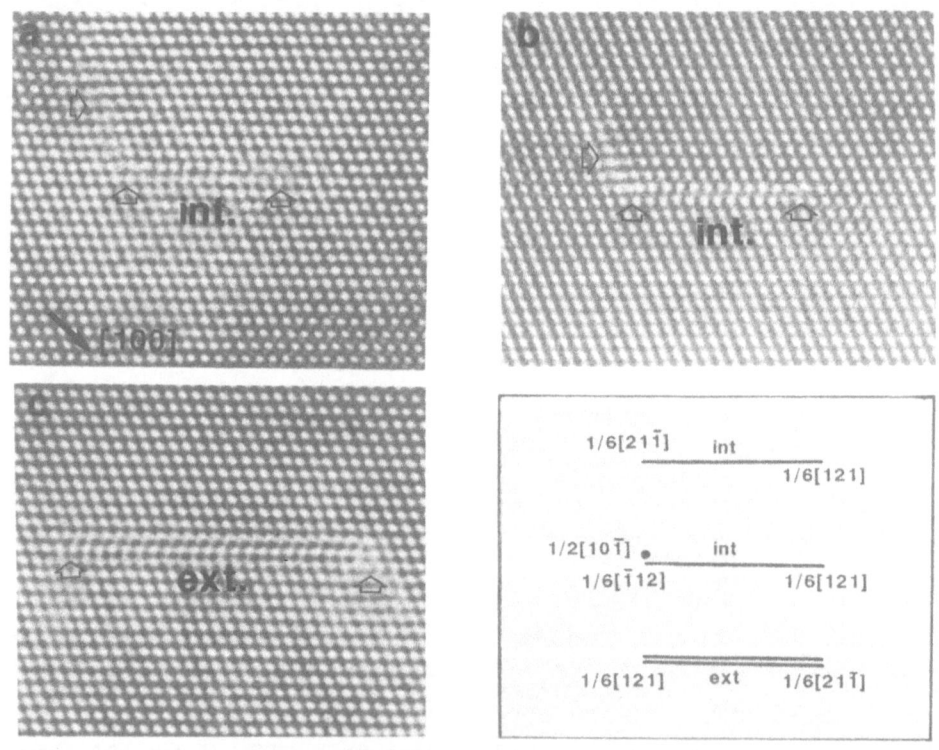

Fig.2. Interaction between point-defects and a 60°dislocation. A perfect interstitial loop is nucleated on the 90° partial of the 60° dissociated dislocation a) in silicon b) in germanium. c)60°D dissociated with an extrinsic stacking fault consequently to the redissociation(d) of the perfect loop nucleated on the plane contiguous to the initial SF plane.

1/2[10$\bar{1}$]a is nucleated on the 90° partial a/6[21$\bar{1}$] of the 60° dissociated dislocation a/2[110] (fig. 2a) aligned along [011]. This phenomenon
occurs exactly in the same way in a deformed Ge monocrystal (fig. 2b).

$$b_{60} = 1/2[110]a = 1/6[21\bar{1}]a + 1/6[121]a$$
$$\rightarrow 1/2[110]a + 1/6[\bar{1}\bar{2}\bar{1}]a + 1/6[121]a$$

or

$$\rightarrow 1/2[10\bar{1}]a + 1/6[\bar{1}12]a + 1/6[121]a$$

HREM cannot distinguish between the two cases. In the first case the
dipole can desappear whereas in the second case a screw dislocation is
left. The perfect nucleated dislocations can then dissociate and parti-
cipate again to the plastic flow.

If the nucleation occurs only on one atomic plane (2 interstitial
columns are absorbed) other complex configurations can emerge such as
the one presented on fig. 2c. If the climbed dislocation 1/2[10$\bar{1}$]a redis-
sociates on the plane contiguous to the initial intrinsic stacking fault
then an extrinsic stacking fault is formed. In spite of their energetically favourable nucleation, no Frank loop was detected. The stair-rod
involved at the edge of the two stacking faults would be 1/6[0$\bar{1}$1]a following the reaction:

a/6[211]..^{i}..a/6[121] $\rightarrow$ a/3[111]..^{e}..a/6[0$\bar{1}$1]..^{i}..a/6[121].

The two stacking faults lie on two different (111) planes. The nucleated
fault would be extrinsic because of the interstitial character of the
loop. This kind of configuration was observed only close to the boundary
(12) where the stair-rod is entirely contained in the GB itself.

The supersaturation of point defects (12) deduced from the osmotic
force needed for the perfect dislocation to climb is found to be of
about $C/C_0 = 10^6$. This could be provided by the deformation. Although the
individual three-dimensional loops nucleated cannot be identified by
HREM, the atomic scale of our observations allowed to exclude the mechanism via formation of Frank loops which later transform to complete
ones: from the very beginning complete loops are nucleated.

In the boundary

Just like bulk dislocation within the grain, also the GB dislocations moves in the interface by glide and climb. The decomposition of
the running-in dislocation (fig. 1b) is the proof of the climb of the
DSC dislocations within the GB. This phenomenon was also observed under
ion (14) or electron (15) irradiation.

Because of the knowledge of their exact core structure, the movement of the elementary DSC dislocations b_c and b_{30} can be well described
and the number of point defects columns consumed in the process can be
determined(16). As shown on Fig.3 , in order to move over one GB period
the b_c dislocation consumes twice as many interstitials as the b_{30} does.
The component normal to the GB of b_c is twice the one of b_{30}. However,
at present it is not well established whether the decomposition of the
incoming dislocation is self-fed or   whether an external point defects
source is required. Both mechanisms might occur simultaneously depending
on the experimental conditions.

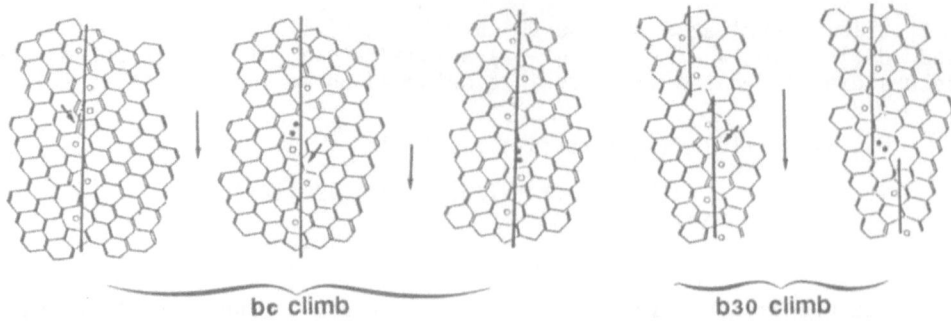

bc climb   b30 climb

Fig.3. Climb of the b_c and b_{30} GB dislocations. This mechanism consumes point defects which are drawn as black dots. The b_c climb involves 4 interstitial columns whereas the b_{30} involves only two for the dislocation to move over one period of the GB.

Under high compressive strains and at temperature higher than 850°C the GB is found to be filled with an increasing number of "b_c" GBD's. The b_c is characterized by a simple structural unit : a boat shaped six-atom ring (16). Due to its symmetry this defect is not associated with a GB step. These defects are distributed along the interface more or less periodically. This makes the misorientation angle increase in compression. Consequently, when climb is possible a grain boundary goes through a series of different coincidence configurations as the external strain is increased and more and more extrinsic dislocations from the grains enter the interface and become intrinsic. Fig.4 shows a Σ=11 GB that has been obtained from a Σ=9 one by straining the sample to 7% at 950°C in compression which increases the misorientation angle from θ=38.94° for Σ=9 to θ=50.5° for Σ=11. The structure of this Σ=11 GB corresponds to the insertion of one b_c per half period of the initial Σ=9 GB. As a matter of fact the structure found after deformation does not correspond to the structure of the Σ=11 GB found in an as-grown bicrystal(17). The interesting question remains if and how such different phases of crystallographically equivalent interfaces can transform into each other.

Fig.4. Evolution of the initial GB from the Σ=9 coincidence position to the Σ=11 orientation by a 7% strain in compression at 950°C. Due to the increasing number of incoming dislocations, the number of "b_c" residues increases leading to an increase in the misorientation angle from θ=38.94° for Σ=9 to θ=50° for Σ=11.

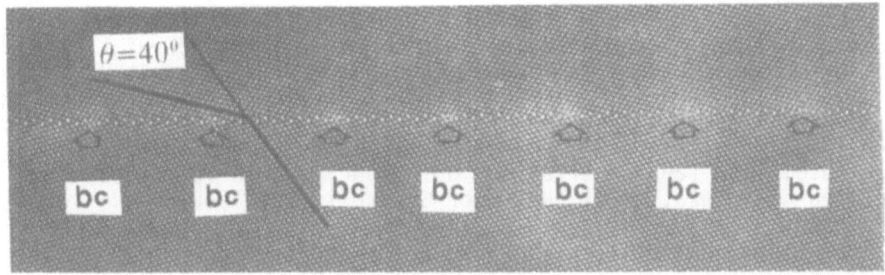

Fig.5. Subgrainboundary superimposed to the initial Σ=9 GB for a 1.3% strain in compression at 1200°C. The number of incoming dislocations is relatively low as compared to the one in fig.4. but the climb is large enough for the GB to reach a good recovery.

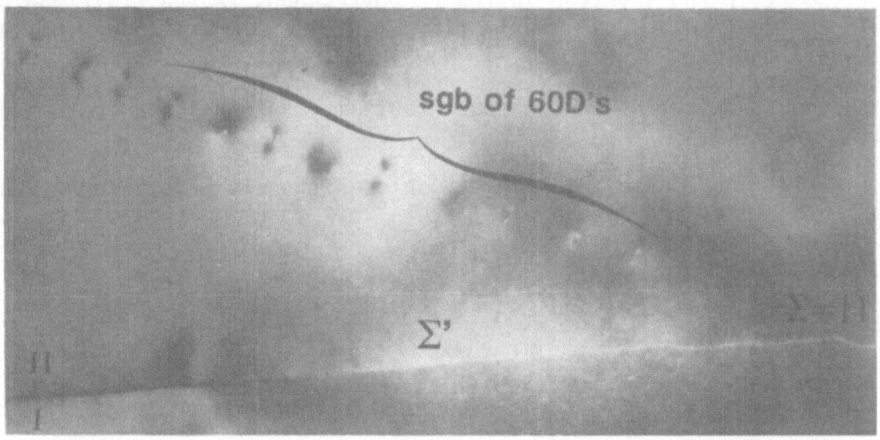

Fig.6. Recrystallization of the deformed bicrystal. A subgrain boundary-(SGB) was formed in one of the grain due to non symmetrical non homogeneous deformation conditions. (7% in compression at 950°C). On both sides of the SGB the GB became Σ=11 and Σ' whose coincidence index is not well defined because of the small angle of the sgb(θ=1.2°).

At 1200°C where the diffusion is large, considerable recovery of the GB could take place : a perfect subgrain boundary (SGB) of b_c dislocations is superimposed to the initial GB (fig. 5) even at low strain, i.e. even if only few dislocations integrate the GB. For large strains the symmetrical deformation is difficult to be maintained. As a consequence the grains are not equally stressed. At 950°C subgrain-boundaries can be formed in the grains being stabilized by the internal stresses (fig.6).

DISLOCATION-DISLOCATION INTERACTIONS

In the grain

A number of locks and dipoles has been detected in the vicinity of the GB. They are due to the reaction between 60° dislocations gliding on different (111) planes. The GB stops one or the other dislocation parti-cipating in the reaction, which is the reason why they are preferably observed near the boundary.

Lomer Cottrell lock.(Fig.7a) Theoretically, this lock $a/2[01\bar{1}]$ is expected to be dissociated in an asymmetrical configuration (18) with two intrinsic stacking faults whose widths are respectively $d_1 = 0.9$nm and

$d_2 = 3.4$ nm. The stair-rod would be $1/6[0\bar{1}1]a$ whereas the two other partials would be two 90° Shockley partials. However weak beam (19) and high resolution (20) observations both concluded in favour of a non dissociated configuration of the Lomer-Cottrell (LC) lock.

B2 lock. 60° dislocations can react and give a B2 lock parallel to [011] whose total Bürgers vector is [010]a. This lock is found asymmetrically dissociated (fig.7b), as predicted (18). The partials involved in this dissociation are two 30° Shockley partials and the stair-rod is $a/3[01\bar{1}]$. It seems that the LC dissociation is impeded by a core problem which does not occur for the B2 dissociation. As a matter of fact the stair-rod involved in the LC corresponds to the stair-rod which would be involved in the interstitial Frank loop formation during interstitials agglomeration (§ 2.1). However in the case of the LC lock the two stacking faults are intrinsic unlike in the case of the interstitial Frank loop nucleation where one fault is intrinsic and the other extrinsic.

Hirth lock. (Fig.7c). This lock is extremely rarely found. Unlike the theoretical predictions (18), this lock is asymmetrically dissociated on only one 111 plane such as:

$[\bar{1}00]a \rightarrow 1/6[\bar{4}11]a + 1/6[\bar{2}1\bar{1}]a.$

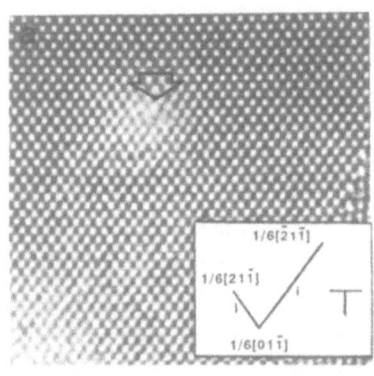

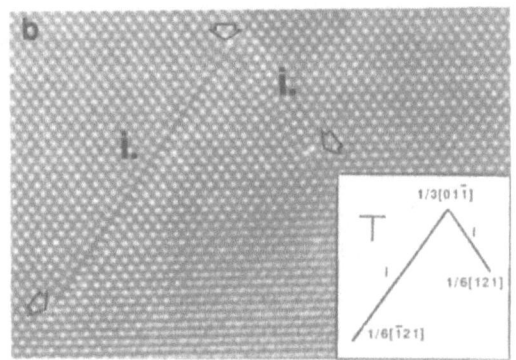

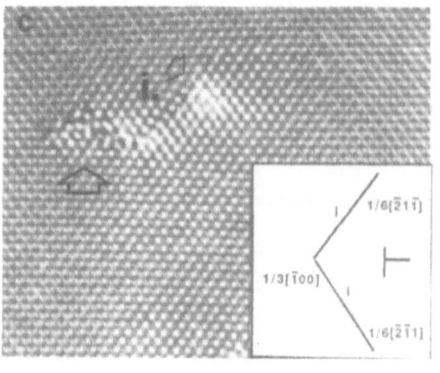

Fig.7. 60° dislocation interactions. a)The Lomer Cottrell lock is found in a non dissociated configuration unlike the theory whereas the B2 lock(b) dissociation corresponds to the predictions. The Hirth lock (c) is dissociated only on one 111 plane and a large precipitate is present in the core of the $a/6[\bar{4}11]$ partial. (b) micrograph by M. Elkajbaji.

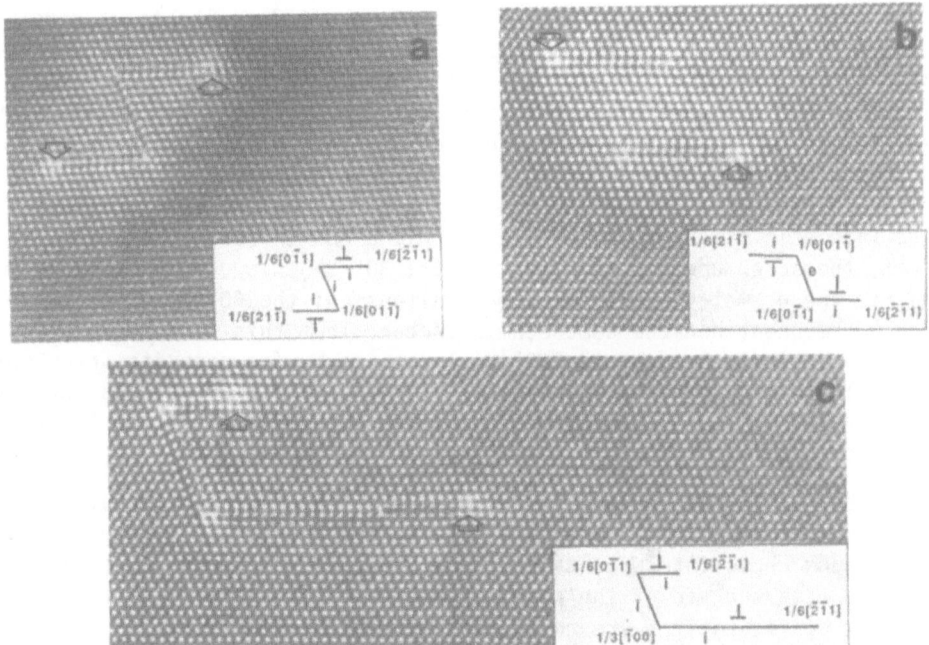

Fig.8. 60° dislocation reactions. a) Z faulted dipole b) S unfaulted dipole. The theoretically expected configurations are shown in the windows. c) Antidipole : two 60° dislocations of the same Burgers vector gliding on two parallel 111 planes reacted and gave a sessile configuration by the emission of a Shockley partial on the other 111 plane.

The stacking fault is intrinsic and both partials have an edge character. Moreover a large precipitate grew in the core of the first partial 1/6[4$\bar{1}$1]a. This precipitate looks very similar to hexagonal silicon described in (21). The same configuration has been found in subgrain boundaries (22) : the same [100] dislocation core being filled with a large precipitate. The non observation of the expected symmetrically dissociated configuration (18) is certainly due to the presence of the precipitate. As mentioned in (21), the supersaturation of interstitials at temperature below 800°C induces a phase transformation from cubic to hexagonal silicon, this phase being stabilized by the pressure in the dislocation core. Futhermore the edge component of the partial 1/6[411]a favours greatly the growth of these "411" defects (22).

60° dipoles. Two types of dipoles could be found : these are the S and the Z faulted dipoles as predicted in (24). As mentioned in (25,26) the S dipoles (fig. 8a) were never found in the faulted configuration unlike the Z dipoles (fig. 8b). As a matter of fact the stair-rod involved in the S configuration would be a/6[01$\bar{1}$] at the edge of one intrinsic and one extrinsic fault forming an obtuse angle like in the §2.1. In the Z faulted dipole the same stair-rod is at the edge of two intrinsic stacking faults forming an acute angle and is equivalent to the one involved in the LC lock. At present, the fact that the Z dipoles are observed with this stair-rod and that the LC lock is not observed in the asymmetrically dissociated configuration involving the same stair-rod is not understood.

Antifaulted dipoles. Close to the GB, two 60° dislocations gliding on two different but parallel (111) planes are sometimes trapped in a rather surprising configuration that we could call an antidipole (fig.8c). This configuration is often observed in the samples deformed in tension at 750°C. The 30° leading partials reacted together by the emission of a 30° Shockley partial. The global configuration became sessile. The projection of the Burgers vectors of the two stair-rods formed at the two edges of the intrinsic stacking faults are $1/6[01\bar{1}]a$ and $1/3[\bar{1}00]a$. As a matter of fact, the screw component of the initial 30° Shockley could be $\pm a/4[011]$ due to the geometry of the deformation tension: the 60°D with an opposite screw component had the same Schmid factor. Thus this kind of reaction might occur only between two 60° dislocations having opposite screw component. The widths of the two parallel stacking faults are always very dissimilar. It is due to the strong repulsive interaction between the two 90° trailing partials.

The configurations presented above are the most representative and the simplest of all. Close to the GB numerous complex configurations were also found whose interpretation is not so straightforward but could be explained with the aid of the previous configurations. Furthermore, these complex configurations are not detected at higher temperatures. In this condition the possibility for the dislocations to largely climb might aid to release these complex hardening configurations.

In the boundary

The GB dislocations move along the GB by glide and climb. They can then react and give new residual GBD's (9). Two basic reactions will be described here.

As far as the deformation remains compatible, the number of dislocations which integrates the GB from grain I and grain II is equal. The residues $b_{30}^{1-}(h=1.75\ h_0)$ and $b_{30}^{2+}(h=1.75\ h_0)$ coming respectively from the entrance of dislocations from I and II can react and give:

$$b_{30}^{1-}/h = 1.75\ h_0) + b_{30}^{2+}\ (h=-1.75\ h_0) \to b_c\ (h=0)$$

whereas the glissile residues b_g and $-b_g$ emitted at the time of the dislocation entrance annihilate together. The final result of the process is a group of 3 b_c (fig.9a). Consequently the problem of bond reconstruction which occurs in the core of the b_{30} GB dislocations disappears. Due to its structure the bond reconstruction of the b_c core might be easier. Each b_c moved apart by climb. A glissile residue can react with one of the three b_c by forming a b_{90} residue (fig.9b):

$$b_c + b_g \to b_{90}^1.$$

As mentioned at the end of § 3.1 numerous complex configurations were detected at temperatures below 850°C. The history of the configurations is relatively easy to reconstruct when the strain is low,i.e.,when the number of incoming dislocations is low. Otherwise the interpretation leads to an increasing number of possible paths. As already shown on fig.5 large recrystallization takes place in the GB at temperatures higher than 900°C. In these case the dislocation interact via point defects.

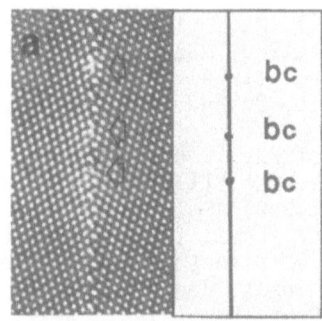 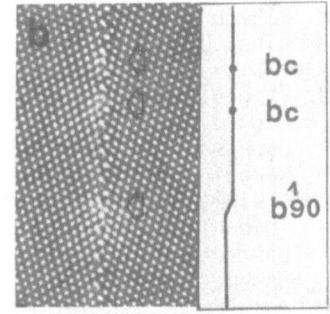

Fig.9. Two basic reactions between grainboundary dislocations. a) in compression and after the integration of primary dislocations coming from grain I and grain II the residues b_{30}^1 and b_{30}^2 reacted and gave a group of three b_c. b) one of the b_c reacted with a glissile residue b_g leading to a b_{90}^1 dislocation associated with a step.(courtesy Elkajbaji)

CONCLUSION

Both in the grain and in the boundary numerous complex configurations are found at moderate temperature (T<850°C). They involve generally point defects-dislocations-GB interactions. When the temperature increases, climb becomes important and can release some sessile configurations or on the contrary can permit an easier integration of the incoming dislocations into the GB and then impedes the plastic flow.

The main idea to emphasize here is the strong interaction between defects even at T ~ 750°C which can affect both the mechanical and the electrical properties of the material. At higher temperature, when climb takes place extensively, the configurations are generally simpler both in the grain and in the GB. However it must be reminded that numerous complex interactions occured before the stabilization of these simple configurations and a large number of impurities might also be dragged as the different kinds of defects move and interact. Futhermore the moving dislocations might retain a lot of intrinsic defects such as point-defects, kinks, jogs which might affect greatly the properties of the material.

REFERENCES

1. W. Bollmann, B. Michaut, G. Sainfort (1972) Phys. Stat. Sol. (a) 13, 637.
2. J. Hirth, R. Balluffi (1973) Acta Metallurg. 21, 929.
3. R. Pond, D.A. Smith (1977) Phil. Mag. 36, 353.
4. W. Bollmann, Crystal defects and crystalline interface. Springer Berlin (1970).
5. X. Baillin, J. Pelissier, J.J. Bacmann, A. Jacques, A. George, Phil. Mag (1987) A55, 143.
6. A. Jacques, A. George, X. Baillin, J.J. Bacmann, Phil. Mag. (1987) A55, 165.
7. Z. Shen, R.H. Wagoner, W.A. Clark, Scripta Met. (1986) 20, 921.
8. A. Jacques, X. Baillin, A. George, Proceeding 8th Int. Conf. on Strength of metals and alloys. Pergamon Press, Tampere-Finland (1988) vol.1, p. 245.

9. M. Elkajbaji, J. Thibault-Desseaux, Phil. Mag. (1988) A58, 325.
10. A. King, D.A. Smith, Acta Crystallogr. (1980) A36, 335.
11. A.W. King, Fu-Rongchen, Mater.Scien. Ing. (1984) 66, 227.
12. J. Thibault-Desseaux,H.O.K. Kirchner,J.L. Putaux, Phil. Mag. (accepted).
13. A. Ourmazd, D. Cherns, P.B. Hirsch, Inst.Phys.Conf. (1981) 60, 39.
14. Y. Komen, P. Petroff, R. Balluffi, Phil. Mag. (1972), 239.
15. A.H. King, D.A. Smith, Phil. Mag. A42 (1980) 495.
16. J. Thibault-Desseaux, M. Elkajbaji, Jour. Phys. (to be published).
17. A. Bourret, J.J. Bacmann, Grainboundaries structure and related phenomema.Proceedings of JIMIS4(1986). Trans. Jap. Inst. Met. Suppl.
18. A. Korner, H. Schmid, F. Prinz, Phys. Stat. So. (a) 51 (1979) 613.
19. A. Korner, M. Martinez-Hernandez, A. George, H. Kirchner, Phil. Mag. Letters, (1987) 55, 105.
20. M. Elkajbaji, J. Thibault-Desseaux, H. Kirchner, Phil. Mag. (1988) 57, 631.
21. A. Bourret, Inst. Phys. Conf. (1987) 87, 39.
22. A. Bourret,J.Thibault-Desseaux,F. Lancon,Jour. Phys. (1983)C4,44,15.
23. A. Seeger, G. Wobser (1966), Phys. Stat. Sol. a 18, 189.
24. J. Spence, M. Kolar (1979) Phil. Mag. A39, 59.
25. S. Chiang, C.B. Carter, D. Kohlstedt (1980) Phil. Mag. A42, 103.

THE INFLUENCE OF RESIDUAL CONTAMINATION ON THE STRUCTURE AND PROPERTIES
OF METAL/GaAs INTERFACES

Zuzanna Liliental-Weber

Center for Advanced Materials
Lawrence Berkeley Laboratory
1 Cyclotron Rd., Berkeley, CA 94720

INTRODUCTION

Reliable and reproducible metal contacts to semiconductors are
necessary if electronic devices are to function properly. Two types of
contacts are required to make these devices work properly: (1) ohmic
contacts, and (2) rectifying gate contacts, which in some applications
must survive annealing at 800°C or higher to activate implanted dopants.

Despite the wide spread use of rectifying contacts to GaAs, two
important issues remain to be resolved. One -- the basic mechanism
responsible for the observed Schottky barrier heights (which is still
being debated) and the second -- reproducibility and stability of
electrical performance during annealing and aging.

Most experimental data agree that the barrier heights for all metals
measured on GaAs fall within a few tenths of an eV in the midgap region,
indicating strong Fermi level pinning mechanism at metal/GaAs
interfaces. The measurements of barrier heights for many metals
deposited in situ on ultrahigh-vacuum- (UHV) cleaved GaAs, as determined
by Newman[1] using I-V and C-V characteristics, seem to be very
consistent. They show the same ideality factor n = 1.05 independent of
the reactivity of the particular metal. The lowest barrier height found
on n-GaAs was for Cr (Φ_b = 0.67 eV), and the highest was for Au
(Φ_b = 0.92 eV). To explain these results which are still
controversial several models have been proposed.[2] In some of them the
Fermi level pinning was ascribed to the inherent properties of ideal
metal/GaAs interfaces (metal-induced gap states[3,4]), in others to
native defects[5] which are formed upon metal deposition due to the
energy which was released during metal solidification or to
work-function differences between GaAs and microscopic interfacial
inclusions of arsenic, metal arsenides, or impurities (the Effective
Work Function model[6]). Part of our work is directed towards
contributing to this fundamental problem[7-11], however in this chapter
we will concentrate on our contribution to the second issue. We will
show, that surface contamination is indeed a major cause for thermal and
electrical instability connected with degradation of the
metal/semiconductor interfaces.

Our approach was to compare the structure, electrical properties and
stability of electrical properties of "ideal" contacts deposited in-situ

in UHV on cleaved GaAs (110) with contacts deposited on air-exposed GaAs (110) and in some cases with technologically prepared GaAs (100) contacts. We will describe in detail our results for four different systems: Au, Ag, Al and Cr on GaAs.

Experimental Procedure

To remove any unnecessary variables (e.g., impurities on the GaAs surface), the diodes used in this study were produced on clean GaAs surfaces formed by cleaving in an ultrahigh vacuum (UHV) with the metal deposited in situ.[12] Bulk n-GaAs bars (Si doped) were placed in an UHV chamber that was baked out to obtain a vacuum of $\sim 2 \times 10^{-10}$ torr. The bars were cleaved along their {110} planes. Metals were then deposited in situ using a resistance-type evaporator without breaking vacuum or additional heating. (During deposition the vacuum was kept $< 4 \times 10^{-10}$ torr.) In order to observe the influence of contamination on the properties of the electrical and structural contacts, a second batch of metal diodes was deposited on samples cleaved in air in the same vacuum chamber. In order to assure that the air-exposed surfaces were not subjected to any unnecessary heat before metalization, a chamber bakeout was not performed. For the diodes produced on the air-exposed surfaces, the pressure during the metal deposition was approximately 10^{-7} torr. The metal thickness for these two kind of diodes was ~ 100 nm. These two kinds of samples were annealed for 10 min at 405°C in a N_2 atmosphere.

In order to compare these diodes with diodes formed using typical commercial GaAs processing technology, a third batch of samples was prepared by deposition of Au layers on chemically cleaned samples. Au was chosen as the most frequently used metalization element. Electron-beam evaporation was used in the contact-fabrication process. The annealing was performed in a 95% Ar – 5% H_2 environment at the same time and temperature as for the first two batches of samples.

The structure and electrical properties of all contacts on GaAs were investigated by analytical and high-resolution transmission electron microscopy (TEM), combined with electrical characterization.

Electron microscopy was performed at LBL's National Center for Electron Microscopy in Berkeley using the JEOL JEM 200CX electron microscope equipped with a high-resolution pole piece (~ 0.25 nm point-to-point resolution) and the 1-MeV Atomic Resolution Microscope (ARM), which has a point-to-point resolution of ~ 0.16 nm.

Au Contacts[7-14]

The as-deposited Au layer was found to be polycrystalline in all three cases, with grain diameters in the 10-50 nm range. The largest grain size was found in UHV-deposited Au samples. Some of these grains were twinned along (111) planes. Such unannealed Au layers observed in cross sections show atomically flat interfaces with GaAs. Some of these grains, particularly in UHV-cleaved samples, were in the (211) or (011) orientation parallel to the (011) GaAs substrate orientation, but generally the grains were randomly oriented, resulting in diffraction patterns with textured rings.

Significant differences between these samples occur after annealing in N_2 at 405°C for 10 min (Fig. 1). For the UHV-cleaved samples, the interface remains flat and abrupt despite the annealing process [Fig. 1(a)]. The entire Au layer was almost monocrystalline, with the smallest grain size ~ 500 nm. Most of the grains were elongated along

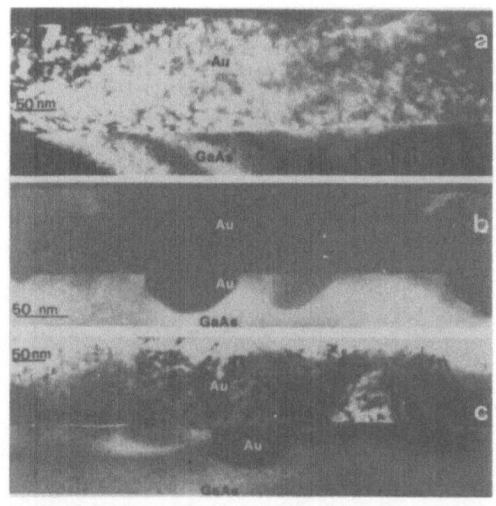

Fig.1. Cross sections of annealed Au/GaAs interfaces. (a) Au deposited *in situ* on (b) Au deposited UHV-cleaved GaAs, on GaAs cleaved in air, (c) Au deposited on chemically clean GaAs.

$[011]_{GaAs}$. Their orientation relationship towards the substrate was $(211)_{Au}$ slightly tilted from $(011)_{GaAs}$ e.g. $(522)_{Au}$ being parallel to $(011)_{GaAs}$. A perfect match between the substrate and the Au layer is expected for each fifth $\{200\}_{GaAs}$ plane with each sixth $\{111\}_{Au}$ plane. In many areas triangular features elongated along $[011]_{GaAs}$ were observed. These triangular features are just cross sections of prism-shaped features observed in plan-view samples (Fig. 2). These features probably voids were formed in the Au layer directly adjacent to the GaAs substrate [Fig. 2(a)]. High-magnification images using the ARM showed that these areas consisted of amorphous material with embedded gold particles.

The same annealing treatment for the Au samples deposited on GaAs cleaved in air resulted in the formation of metallic protrusions at the interface [Fig. 1(b)]. Many small grains, highly twinned and dislocated with irregular shapes, were observed in a plan view of annealed air exposed samples. The larger grains had two different shapes, triangular and rectangular.

In cross sections two different shapes of protrusions were found extending into the GaAs [(Fig. 1(b)]: (1) triangular protrusions whose sides are delineated by GaAs $\{111\}$ planes, and (2) multifaceted protrusions delineated by GaAs $\{111\}$, $\{110\}$, and $\{100\}$ planes. These two different protrusion shapes are probably related to the two different grain shapes visible in plan-view samples. Such protrusions were observed in the past for annealed Au-Ni-Ge contacts[15,16] and Au contacts,[17] and it was concluded that elevated temperatures are a sufficient condition for their formation. It is interesting to note that those protrusions were elongated along $[011]_{GaAs}$ [Fig. 2d], as were void-like features in the samples deposited in UHV and subsequently annealed [Figs. 2b and 2c]. Because the protrusions and void-like features are elongated in the same crystallographic direction, they can be easily misinterpreted by SEM type of studies. For many grains in air-exposed samples the $(011)_{Au} \| (011)_{GaAs}$ relationship was observed, but for some other grains $(\bar{4}11)_{Au} \| (011)_{GaAs}$ was found as well. Details about the relationship will be discussed in the next chapter.

Even more complicated interfaces were observed in annealed Au/GaAs samples formed on chemically prepared GaAs surfaces. The gold layer was separated from the GaAs substrate by a thin oxide band. Oxygen was

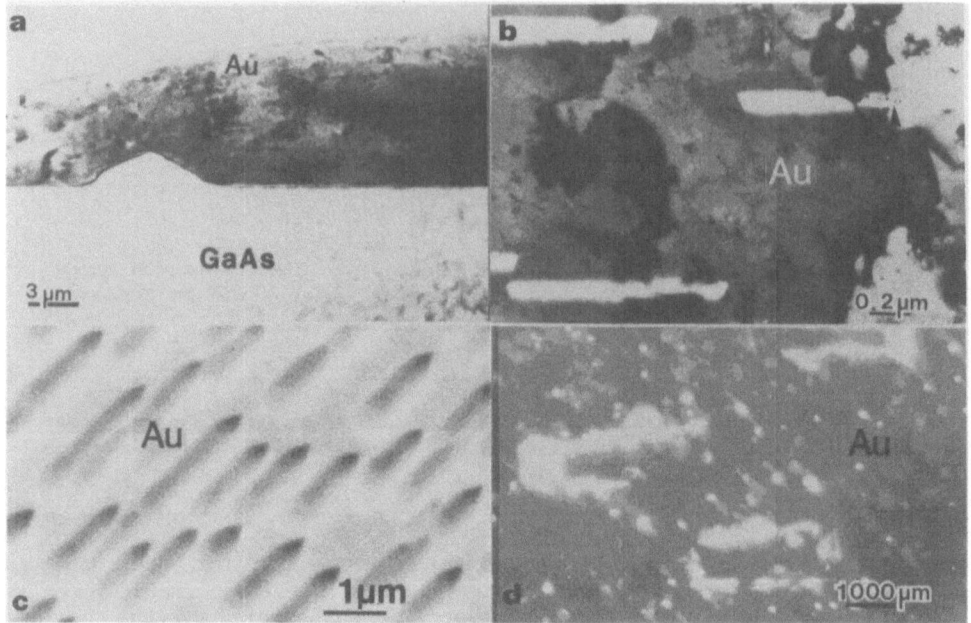

Fig. 2.(a)TEM micrograph of cross section of Au/GaAs interface from the sample prepared _in situ_ in UHV on cleaved GaAs after annealing in N_2 at 405°C for 10 min. The image was taken in the [011] zone axis. Note void-like triangular features formed in Au adjacent to the GaAs substrate. A high-magnification image of these triangular areas shows embedded Au particles inside the triangular areas. The same triangular features elongated along [011] are shown in plan view by TEM (b) and by SEM (c); (d) Plan-view image of Au sample prepared by deposition on an air-exposed cleaved GaAs surface annealed under the same conditions. Note the long grains elongated along [011]. These grains are the protrusions embedded in the GaAs substrate shown in Fig. 1(b).

detected on the interface by energy-dispersive x-ray spectroscopy (EDX) in chemically prepared samples and in the samples cleaved in air.[14] Oxygen was not detected in samples where Au was deposited _in situ_ on the UHV-cleaved surface. In many areas, the interface was found to be very flat and abrupt [Fig. 1(c)]. However, islands of gold with a wide range of shapes were found below the oxide layer as well. These islands were epitaxially regrown, with a much smaller defect density than in the layer above the oxide. The observation of separated islands below the oxide layer would suggest that Au has diffused through already exiting pinholes in the oxide. The orientation relationship for those grains was similar as for air-exposed samples, e.g. $(011)_{Au} \| (011)_{GaAs}$.

The Au layer above the oxide layer has many defects, and the grain size is much smaller than the annealed Au deposited in UHV.

These observations show that GaAs is very sensitive to oxidation and that the morphology of the interface is strongly influenced by the surface preparation prior to Au deposition. This demonstrates that the formation of protrusions is not the result of annealing at elevated temperatures alone but is clearly affected by the semiconductor surface-preparation technique prior to metal deposition.

As determined from I-V characteristics, there was not a large difference in barrier height for the Au diodes deposited in situ on UHV-cleaved GaAs samples (0.92 eV) and deposited on the samples cleaved in air (0.83 eV). After annealing, the barrier height decreases to 0.72 eV for both kinds of samples. A very important observation[18,19] is that those samples that were air exposed before Au deposition were found to age with time and/or exposure to electrical measurements where large bias voltages were used, whereas UHV-cleaved samples were stable. This observation is a very important consideration for the reliability of practical devices built on oxidized surfaces.

Ag Contacts[18-22]

Two kinds of Schottky diodes, similar to the first two kinds of Au diodes, were prepared for Ag: deposited on clean, UHV-cleaved (110) GaAs, and deposited on air-exposed cleaved (110) substrates.

Significant differences in the structures of the two kinds of contacts were observed. The metal/substrate interface was flat in both cases [Figs. 3(a) and 3(c)] for as-deposited samples; however, in the air-exposed samples an oxide layer ~ 40 Å thick was present on the GaAs surface. This oxide layer varied in thickness along the interface. The air-exposed diodes contained a higher density of twins and much smaller Ag grains in the metal layer than did the samples deposited in UHV conditions.

As in Au case, a difference in interface morphology was observed for these two kinds of samples after annealing. The interface remained flat for the UHV samples, and high-resolution electron microscopy showed that that $\{111\}_{Ag}$ planes were rotated slightly toward the $\{200\}_{GaAs}$ planes [Fig. 3(b)]. Large protrusions were formed at the interface of the air-exposed samples [Fig. 3(d)]. The faceted Ag protrusions grew into the GaAs. One of the facet planes of such a protrusion was always parallel to $\{111\}_{GaAs}$ and the other plane varied, but in many cases the other facet plane was parallel to $\{122\}_{GaAs}$.[21] When plan-view samples were prepared by ion-milling from the substrate side with partial Ag removal by short ion-milling from the metal side, it was clear that all these protrusions were embedded into the GaAs substrate, as was expected from cross-section samples, and that these protrusions were elongated along $[011]_{GaAs}$.[21] Also, as was the case for the annealed Au samples deposited on the air-exposed substrate, the orientation relationship for air-exposed samples was $[011]_{GaAs} \parallel [011]_{Ag}$ with $(011)_{GaAs} \parallel (011)_{Ag}$.

After annealing, voids formed at the metal/GaAs interface in many areas, and a large portion of the Ag layer peeled off. These void formations were observed in as-deposited air-exposed samples, but adhesion decreased after annealing. Occasionally adhesion problems occurred in UHV-deposited samples as well, but the problems were not so drastic as they were with the air-exposed samples after annealing. Many more problems occurred in those samples that remained in air longer before metal deposition and subsequent annealing.

The Schottky barrier height measured from I-V curves was 0.96 eV for as-deposited air-exposed samples, higher by 70 meV than that of UHV-cleaved diodes.[19] After annealing, a slight increase in barrier height (0.91 eV) was observed for UHV-deposited samples, while a large decrease (0.79 eV) and leakage was observed for air-exposed samples (Fig. 4). The large leakage current often reported in the literature[21-23] can be correlated with the adhesion problems in those samples.

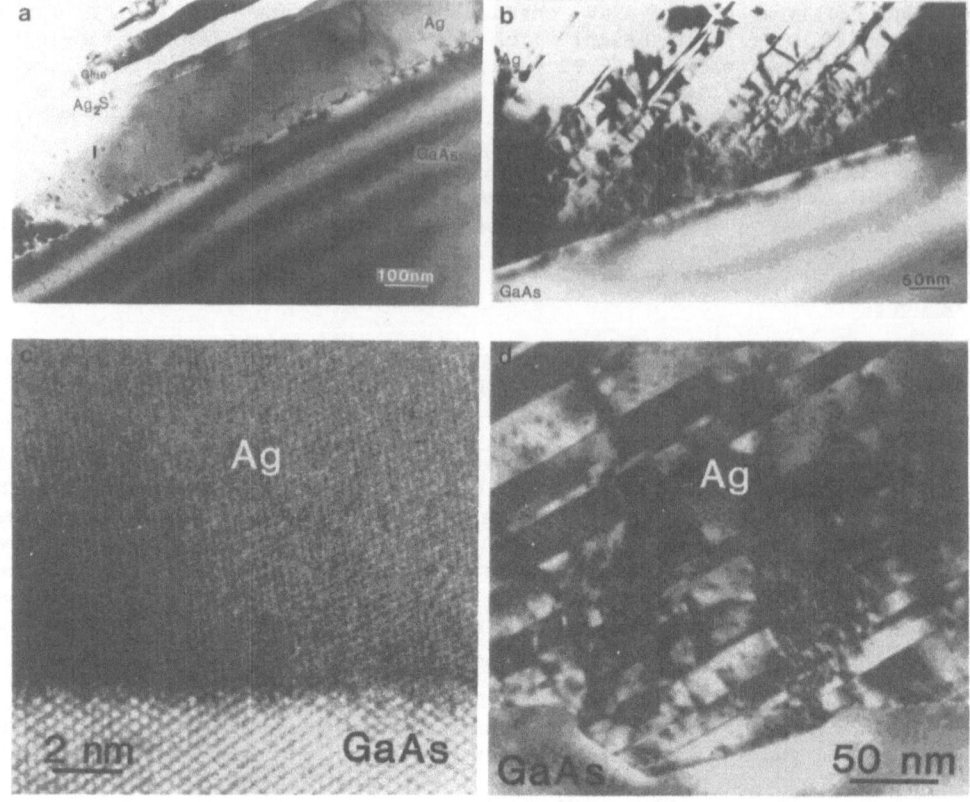

Fig. 3. Cross-section micrographs of Ag/GaAs interfaces. (a) Ag deposited _in situ_ on UHV-cleaved (110) GaAs. Note the very large Ag grain size. (b) High-resolution image of sample prepared under the same conditions as (a) after annealing in N_2 at 405°C for 10 min. (c) Ag deposited on air-exposed cleaved (110) GaAs, showing a high density of twins and an oxide layer formed on the interface. (d) Sample (c) annealed at 405°C in N_2. Note protrusion at the interface, a twinning of the Ag layer.

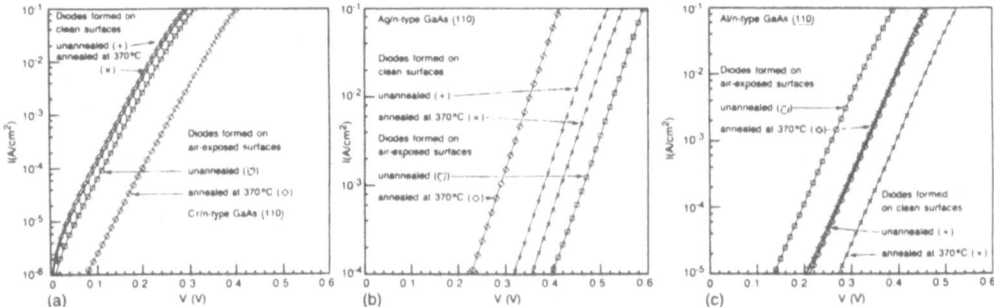

Fig. 4. Typical current-voltage (I-V) measurements for diodes a) Cr, b) Ag, c) Al formed on clean n-type GaAs (110) surfaces prepared by cleavage in UHV and on air-exposed surfaces prepared by cleavage and exposure to the atmosphere for ~ 1-2 hrs.

Electrical aging was performed for as-deposited air-exposed and UHV-cleaved samples.[18-20] For UHV-cleaved Ag diodes, electrical aging was performed with current densities from $2 \times 10^{-2} A/cm^2$ (0.60 V) up to 1.4 A/cm^2 for reverse bias (-19 V). For the UHV-cleaved Ag diodes, no significant change in barrier height and ideality factor was found after electrical aging under these conditions for more than 7 hrs. By contrast, for air-exposed Ag/GaAs diodes, 50 min at $4.3 \times 10^{-5} A/cm^2$ (-14 V) was sufficient to decrease the barrier height by 20 meV. More severe conditions of $2.3 \times 10^{-3} A/cm^2$ (-17 V) for the same 50-min period decrease the barrier height by 75 meV (Fig. 5).

Electrical aging reduced the barrier-height difference between the two kinds of diodes without significantly changing the ideality factor of either kind of diode (n = 1.06-1.085). Thus electrical aging caused changes in barrier height but did not significantly deteriorate the near-ideal Schottky characteristics of the diodes. It was also observed that the changes in barrier height were not stable: the Schottky barrier height returned almost to its initial value (before current stressing) within about five days. Light or forward current accelerated this recovery effect.

Current stressing of UHV-cleaved diodes did not result in any structural changes, but the air-exposed contacts showed a significant change.[18-20] Current stressing caused a decrease in the size of the Ag grains, the formation of voids separating these grains, and poor adhesion of the metal overlayer. Local electromigration of Ag resulted in Ag accumulation in parts of the contact and thinning or void formation in other parts[20]. Electromigration of Ag in the air-exposed diodes may be the result of large local current densities due to an inhomogeneous interfacial oxide layer, which acts to block current flow over part of the area of the contact.

The observations of void formation, enhanced electromigration combined with the formation of new compounds, can explain why Ag Schottky contacts, which are known to be generally leaky, have not been applied successfully in GaAs device technology. However, this study has shown that very stable and reliable Ag contacts can be obtained if the Ag is deposited on atomically clean surfaces, such as the UHV-cleaved surfaces used for these observations.

Al Contacts[9,18,19,24]

For Al deposited on UHV-cleaved GaAs, the grain size of 100-300 nm was observed. The interface with GaAs remains flat and the Al (111) planes form a small angle with the GaAs (111) planes (Fig. 6). This angle remains constant for grains with different orientations. Upon annealing at 375°C in N_2 for 10 min, the interface remains flat and

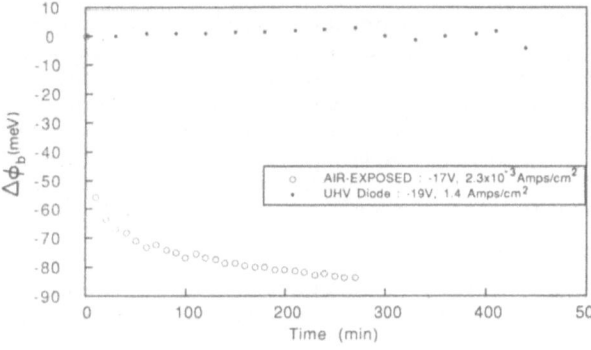

Fig. 5. Typical results of electrical aging for Ag/n-type GaAs (110) diodes formed on air-exposed and UHV-cleaved GaAs (110) surfaces. The change in barrier height is plotted as a function of the time the diodes were exposed to electrical aging.

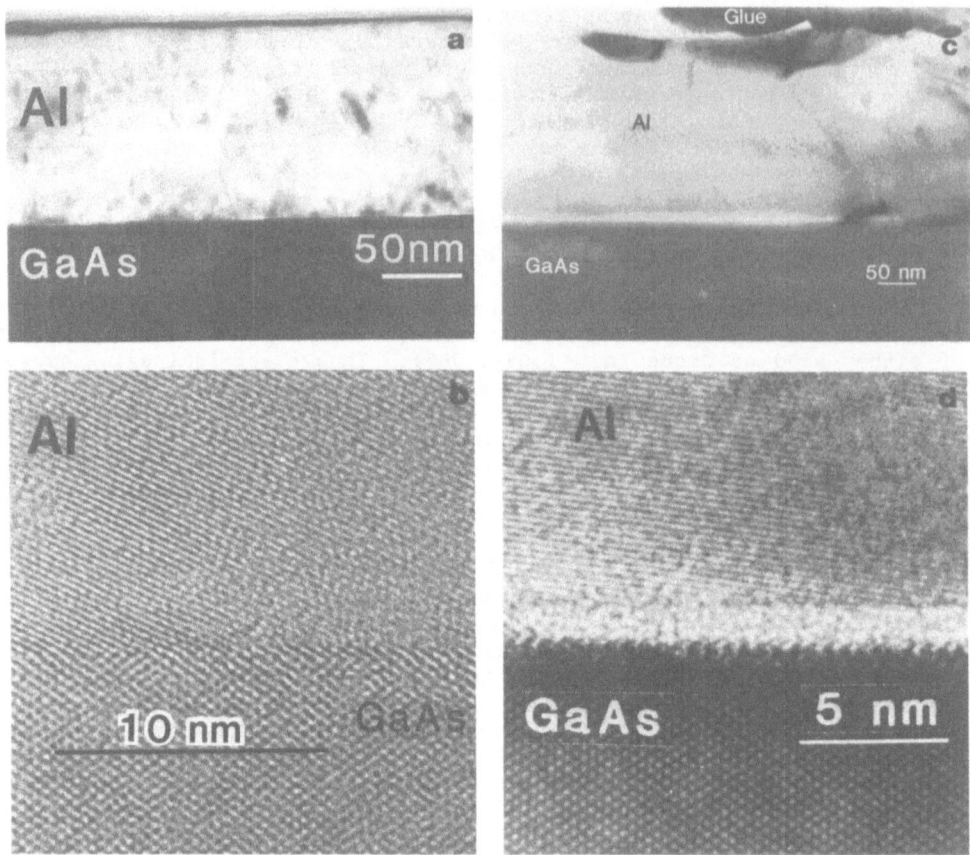

Fig. 6. TEM micrograph of cross sections of Al/GaAs interfaces (a) from the sample prepared on UHV-cleaved substrate; (b) high-resolution image of the same sample annealed at 405°C in N_2; (c) from the sample prepared on air-exposed GaAs; (d) high-resolution image of the same sample annealed under the same conditions. Note amorphous layer at the interface.

the grain size does not increase. In some areas a very thin layer (50-100 Å) of AlGaAs was formed. The formation of AlGaAs did not occur uniformly. There were large areas where this phase was not detected.[9,24] Individual grains of Al above AlGaAs or in intimate contact with GaAs did not change the orientation relationship upon annealing.

For the samples cleaved in air, the interface remained flat before and after annealing, but a significant decrease in Al grain size was observed in these samples. In some areas of the annealed air-exposed samples, a thin layer of AlGaAs was detected as well. For Al metalization, in contrast to the previously described metals (Au and Ag), no protrusions at the interface were observed for cleaved air-exposed samples after annealing. This probably can be explained by the formation of an AlGaAs phase in intimate contact with GaAs and no As outdiffusion from the systems.

In all observed cases, Al (or AlGaAs) was always found in intimate contact with the GaAs substrate. Void formation was not observed, either in as-deposited samples or in annealed ones. Aging of these contacts did not influence the interface structure.

Al deposited in situ on UHV-cleaved GaAs forms Schottky contacts with a barrier height of 0.83 eV (Fig. 4c). After ex situ annealing for 10 min in N_2 atmosphere at 360°C or above, the barrier height increases to 0.90 eV.[19,25] This is in contrast to the behavior of the Au diodes, where the barrier height decreases upon annealing. For as-deposited samples cleaved in air, the barrier height was lower (0.76 eV) than for UHV-cleaved samples, but a similar increase of 70 meV was observed after annealing.[19,25] The increase of barrier height for AlGaAs upon annealing has frequently been attributed to the formation of the interfacial AlGaAs with a larger bandgap.[26,27] However, recent studies showed, that the barrier height of Al on n-GaAs (110) added up together with the barrier height of Al on p-GaAs to the GaAs bandgap.[28] The observed changes in barrier height are thus due to a downward shift of the Fermi level pinning position rather than the formation of AlGaAs. We have attributed this shift of the Fermi level pinning position to a change of stoichiometry due to the replacement of Ga by As.[9,10]

Air-exposed Al/GaAs diodes were aged at -9.7 V for more than 7 hrs with a reverse current flow of 1.3 A/cm^2. There was a very small, almost insignificant, increase of ~ 9 meV in the barrier height after electrical aging. No change in barrier height was noticed for UHV-deposited samples with the same aging parameters. This study shows that Al contacts are stable upon annealing. Strong adhesion between Al and the substrate exists for both UHV-deposited and air-exposed samples. No protrusions were found at the interfaces upon annealing.

Cr Contacts[18,19,29]

The TEM study of Cr layers deposited on clean UHV-cleaved GaAs surfaces consistently showed a columnar structure in the Cr layer. These columns were inclined 80° to the interface, and this inclination was probably related to the deposition direction. The size of the columns was in the range of 4-12 nm. Voids up to 5 nm wide were formed between some of the columns. The void formation initiated in the Cr layer, about 10-15 nm from the interface with the GaAs [Fig. 7(a)].

High-resolution images of the interface taken in the ARM (1 MeV) in [100] and [110] projections show that the interface with GaAs was flat on an atomic scale. The individual columns were misoriented with respect to each other by a few degrees [Fig. 7(c)].

The orientation relationship between the GaAs substrate and the Cr layer was $\{100\}_{Cr}\|\{100\}_{GaAs}$ with $[011]_{Cr}\|[022]_{GaAs}$. Cr matches GaAs almost perfectly to GaAs because the Cr lattice parameter (a = 0.288 nm) is almost exactly half that of GaAs (a = 0.565 nm).

For the samples with Cr deposited on the air-exposed GaAs surface, the interfaces were flat, similar to the UHV-deposited samples. A columnar structure of the Cr overlayer was observed as well, but the columns were almost randomly oriented, with a high void density between them [Fig. 7(b)]. The size of the columns of these samples was less than 2 nm. Extra spots of chromium oxide were detected in these samples.

Annealing for 10 min in N_2 at atmospheric pressure at 370°C did not cause the formation of a new phase in either the UHV samples or the air-exposed samples.

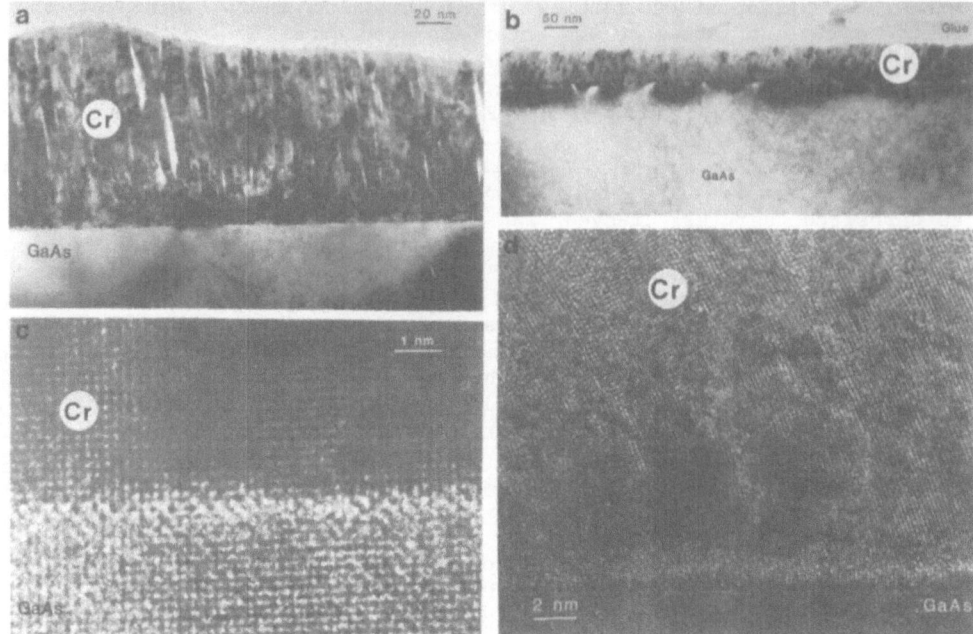

Fig. 7. TEM micrographs from Cr/GaAs interfaces: (a) low-magnification micrograph showing columnar structure of Cr with voids between columns for UHV as-deposited samples. Note that columns are almost parallel to each other and inclined ~ 80° toward the interface with GaAs; (b) low-magnification micrograph showing columnar structure of Cr with columns inclined in a different direction to the substrate for air-exposed samples; (c) high-resolution micrograph of annealed samples deposited in UHV. Note two perpendicular (110) planes in a Cr column to the left and only one set of lattice images for the planes parallel to the substrate in the second column (darker image); (d) high-resolution image of annealed air-exposed samples. Note thick oxide layer at the interface and increased buckling of lattice planes toward the top of the layer.

High-resolution images from annealed UHV samples show that the structure and interface abruptness remained stable after annealing. Individual columns remained slightly misoriented to one another [Fig. 7(c)].

A high-resolution image of the air-exposed samples showed an oxide layer about 1 nm thick at the interface [Fig. 7(d)]. Voids between columns remained after annealing. Individual columns and lattice planes were found to change their inclination direction continuously with increasing layer thickness. The top of the layer consisted of small polycrystalline Cr grains.

I-V and C-V characteristics taken on both types of as-deposited diodes showed a barrier height of 0.66 eV for UHV-cleaved samples and 0.68 eV for air-exposed samples (Fig. 4). This small difference was within measurement error. It shows that an oxide layer at the interface does not influence the barrier height for as-deposited samples. Similar values (0.69 eV) were reported for UHV-cleaved samples by McLean and

Williams,[30] and slightly higher values (0.73 eV) were reported for Cr/GaAs (100) by Waldrop.[31]

The barrier height (0.68 eV) and ideality factor (n = 1.06) of the UHV-deposited contacts, as determined by I-V electrical measurements, did not a change upon annealing at these temperatures (see Fig. 4).

For air-exposed samples the barrier height increased (Fig. 4) from 0.68 eV to 0.76 eV after annealing. The low barrier height of Cr (0.66 eV) for as-deposited samples (compared to other metals, e.g., 0.92 eV for Au) may be associated with the accumulation of As near the interface.[7,10]

Aging of Cr diodes at -19 V for more than 6 hrs with a reverse current flow of 3 A/cm^2 did not change the barrier height by more than 6 meV (within experimental error) for either UHV-cleaved samples or air-exposed ones.

Orientation Relationship in Metal/GaAs Interfaces[9]

The orientation relationship between the metals investigated (Au, Ag, Al, and Cr) for as-deposited samples was almost random for all samples. When the metals were deposited in situ on UHV-cleaved samples, the metal grains were always larger than those of the metals deposited on the air-exposed GaAs substrate.

The difference in orientation relationship between differently prepared samples occurred after annealing. These differences are shown for Au in Table 1 and described in detail in Ref. 9.

Table 1. Orientation relationship between Au and GaAs.

Crystallographic axis	x	y	z
GaAs	011	100	$0\bar{1}1$
Type I	011	100	$0\bar{1}1$
Type IIa	$\bar{4}11$	122	$0\bar{1}1$
Type IIb	$\bar{4}11$	$0\bar{1}1$	$\overline{122}$
Type III	$\bar{5}22$	455	$0\bar{1}1$

Type I orientation relationships were observed for all air-exposed samples (Au, Ag, Cr) except Al. This type of orientation relationship was explained for Au by Yoshiie and Bauer[32] as the epitaxial relationship to the newly formed Au-Ga phase, e.g., $(0\bar{1}1)_{GaAs}\|(1\bar{1}0)_{AuGa}\|(0\bar{1}1)_{Au}$ with $[01\bar{1}]_{GaAs}\|[001]_{AuGa}\|[01\bar{1}]_{Au}$. However, formation of an Au-Ga phase is not necessary to fulfill the minimum mismatch on the interface. The mechanism is probably more general. Our data show that the Au orientation relationship for cleaved (110) GaAs surfaces depends on both the environment in which the GaAs surface was prepared before annealing and the annealing conditions, and not necessarily on the Au-Ga phases formed. The type I orientation relationship exists in annealed Au even when an Au-Ga phase is not formed, but it is characteristic for annealed Au deposited on air-exposed GaAs and exists for other metals like Ag, where this phase is not formed. A possible explanation for this behavior is that the γ-Ga$_2$O$_3$ grows epitaxially[33,34] as a type I orientation: $(011)_{Ga_2O_3} \parallel (011)_{GaAs}$ with $[100]_{Ga_2O_3} \parallel [100]_{GaAs}$. This oxide provides an excellent lattice match to Au: d_{400} (Ga_2O_3) = 0.205 nm, as compared with d_{200} (Au) = 0.203 nm (with similar d values for Ag and Cr) and d_{044} (Ga_2O_3) = 0.145 nm, with d_{022} (Au) = 0.149 nm. This observation would suggest that as soon as GaAs is exposed to air, epitaxial γ-Ga_2O_3 is formed, and the deposited metal epitaxially relates to the oxide already existing at the interface.

The oxide on GaAs is not a continuous layer. In the areas where the oxide is not present, twinning takes place, giving a better match at the interface, leading to a type IIa orientation relationship. A type IIb orientation was observed in most cases for Au deposited on UHV-cleaved GaAs with _in situ_ postannealing.

The orientation relationship in UHV-deposited Au samples annealed _ex situ_ in N_2 (405°C, 10 min, as done for air-exposed samples) was completely different and was described as type III. This type of orientation relationship was observed not only in Au/GaAs samples but also in Al/GaAs and Ag/GaAs samples. A small rotation angle (~ 10°) between $(111)_{Au,Ag,Al}$ planes and $(111)_{GaAs}$ planes was characteristic for all three metals (they have similar lattice parameters) deposited _in situ_ on UHV-cleaved GaAs and annealed in N_2. This behavior was explained by As accumulation at the interface.[9,10] The metal probably tries to accommodate to the accumulated As or to the As plane in the GaAs substrate. This discussion shows, that the macroscopic orientation relationship of metal grains on GaAs is very sensitive to interfacial contamination. Comparing of the crystallographic orientation relationships of metal grains on GaAs (110) revealed distinct differences between UHV-deposited and air-exposed samples. In fact, the observation of the orientation relationship after annealing for metals with similar lattice parameters can be used as an additional tool in recognizing how clean the GaAs surface was before metal deposition. All metals deposited on air-exposed substrates follow the orientation relationship of the oxide present on the semiconductor surface.

CONCLUSIONS

This study shows that interface morphology, orientation relationship, and formation of new phases strongly depend on the surface preparation of GaAs before metal deposition and/or on the annealing environment. The metals investigated (Au, Ag, and Al, with lattice parameters close to each other) deposited _in situ_ on a UHV-cleaved GaAs surface show very similar relationships with GaAs upon annealing. This relationship changes when GaAs is exposed to air before metal deposition. All metals investigated, when deposited on a UHV-cleaved GaAs substrate, are stable upon annealing. The interface between metal and GaAs remains abrupt upon annealing. In the case of Cr almost perfect matching to GaAs was observed for UHV deposited samples, but random orientation for air-exposed samples.

This study shows that impurities at the semiconductor surface can affect the stability of the barrier height of Schottky contacts. These changes in barrier height depend on the metal used, and on the intensity and direction of the potential and current during the electrical aging. The dramatic example of Ag contacts and their change upon current stressing only for air-exposed samples confirms the importance of surface preparation before metalization.

A great part of frequently observed problems with reproducibility and stability of Schottky barrier heights on GaAs can be ascribed to insuffient cleaning of water surfaces before metal deposition. Comparison of the results of current processing with UHV–prepared samples allows in a unique way to define the goals available for non–contamination technology.

ACKNOWLEDGMENTS

This work was supported by the Materials Science Division of the Department of Energy under Contract DE–AC03–76SF00098 and partially by SDIO managed by ONR under Contract N00014–86–K–0668.

REFERENCES

1. N. Newman, M. Van Schilfgaarde, T. Kendelewicz, M. D. Williams, and W. Spicer, Phys. Rev. B 33, 1146 (1986).
2. J. Vac. Sci. Technol. B, Vol. 5, No. 4, Jul/Aug 1987.
3. V. Heine, Phys. Rev. A 138, 1689 (1965).
4. J. Tersoff, Phys. Rev. Lett. 52 465 (1984).
5. W. E. Spicer, Z. Liliental–Weber, E. Weber, N. Newman, T. Kendelewicz, R. Cao, C. McCants, P. Mahowald, K. Miyano, and I. Lindau, J. Vac. Sci. Technol. B6, 1245 (1988).
6. J. L. Freeouf and J. M. Woodall, Appl. Phys. Lett. 39, 727 (1981).
7. Z. Liliental–Weber, N. Newman, W. E. Spicer, R. Gronsky, J. Washburn, and E. R. Weber, in Thin Films--Interfaces and Phenomena, edited by R. J. Nemanich, P. S. Ho, and S. S. Lau (Materials Research Society, Pittsburgh, 1986), Vol. 54, p. 415.
8. Z. Liliental–Weber, R. Gronsky, J. Washburn, N. Newman, W. E. Spicer, and E. R. Weber, J. Vac. Sci. Technol. B4, 912 (1986).
9. Z. Liliental–Weber, J. Vac. Sci. Technol. B5, 1007 (1987).
10. Z. Liliental–Weber, E. R. Weber, N. Newman, W. E. Spicer, R. Gronsky and J. Washburn, "Defects in Semiconductor" in Material Science Forum Vol. 10–12, 1986 p. 1223 edit. H. J. von Bardelebon, Trans. Tech. Publ. Ltd., Switzerland.
11. W. E. Spicer, Z. Liliental–Weber, E. R. Weber, N. Newman, T. Kendelewicz, R. Cao, C. McCants, P. Mahowalk, K. Miyano, and I. Lindau, J. Vac. Sci. Technol. B6, 1245, (1988).
12. N. Newman, W. S. Petro, T. Kendelewicz, S. H. Pan, S. J. Eglash, and W. E. Spicer, J. Appl. Phys. 57, 1247 (1985).
13. D. Coulman, N. Newman, G. Reid, Z. Liliental–Weber, E. R. Weber, and W. E. Spicer, J. Vac. Sci. Technol. A5, 1521 (1987).
14. Z. Liliental–Weber, J. Washburn, N. Newman, W. E. Spicer, and E. R. Weber, Appl. Phys. Lett. 49, 1514 (1986).
15. Z. Liliental, R. W. Carpenter, and J. Escher, Ultramicroscopy 14, 135 (1984).
16. T. S. Kuan, P. E. Batson, T. N. Jackson, H. Ruppreicht, and E. L. Wilkie, J. Appl. Phys. 54, 6952 (1983).
17. T. Yoshiie, C. L. Bauer, and A. G. Milnes, Thin Solid Films 111, 149 (1984).
18. A. Miret, N.Newman, E. R. Weber, Z. Liliental–Weber, J. Washburn, and W. E. Spicer, J. Appl. Phys. 63, 2006 (1988).
19. N. Newman, Z. Liliental–Weber, E. R. Weber, J. Washburn, and W. E. Spicer, Appl. Phys. Lett. 53, 145 (1988).
20. Z. Liliental–Weber, A. Miret–Goutier, N. Newman, C. Jou, W. E. Spicer, J. Washburn, and E. R. Weber, Mat. Res. Soc. Symp. Proc. 102, 241 (1988).
21. C. J. Jou, J. Washburn, Z. Liliental–Weber, and R. Gronsky, J. Electrochem. Soc. (1988) (in press).
22. W. E. Spicer, N. Newman, T. Kendelewicz, W. G. Potro, M. D. Williams, C. E. McCants, and I. Lindau, J. Vac. Sci. Technol. B3, 1178 (1985).
23. R. Ludeke and G. Landgreen, J. Vac. Sci. Technol. 19, 667 (1981).

24. Z. Liliental-Weber, C. Nelson, R. Gronsky, J. Washburn, and R. Ludeke, Mat. Res. Soc. Symp. Proc. 77, 229 (1987).

25. N. Newman, W. E. Spicer, and R. R. Weber, J. Vac. Sci. Technol. B5, 1020 (1987).

26. S. P. Svensson, G. Landgreen and T. G. Andersson, J. Appl. Phys. 54, 4475 (1983).

27. N. Newman, K. K. Chin, W. G. Petro, T. Kendelewski, M. D. Williams, C. E. McCants and W. E. Spicer, J. Vac. Sci. Technol. A3, 996 (1985).

28. N. Newman, W. E. Spicer and E. R. Weber, J. Vac. Sci. Technol. B5, 1020 (1987).

29. Z. Liliental, N. Newman, J. Washburn, and E. R. Weber, Appl. Phys. Lett. (1988) (in press).

30. A. B. McLean and R. H. Williams, J. Phys. C. Sol. State. Phys. 21, 783 (1988).

31. J. R. Waldrop, J. Vac. Sci. Technol. B2, 445 (1986).

32. T. Yoshiie and C. L. Bauer, J. Vac. Sci. Technol. A1, 554 (1983).

33. T. T. Sands, J. Washburn, and R. Gronsky, Mater. Lett. 3, 247 (1985).

34. O. R. Monteiro and J. W. Evans, J. Vac. Sci. Technol. (1989) (in press).

HIGH-RESOLUTION ELECTRON MICROSCOPY OF TWIN-FREE (111)
CdTe LAYERS GROWN ON VICINAL (001) GaAs SURFACES

G. Feuillet, Y. Gobil [*], J. Cibert[*], S. Tatarenko[*], and
K. Saminadayar

CEA/CENG, DRF/SPh/PSC
[*]Lab. de Spectrométrie Physique, CNRS, 38041 Grenoble
Cédex

ABSTRACT

We report here on a High-Resolution Transmission Electron
Microscopy investigation of (111) epitaxial CdTe layers on (001) GaAs
substrates where twinning is impeded by a suitable misorientation of
the substrate surface. It is also found that the layer (111) growth
planes are slightly tilted with respect to the (001) GaAs planes.
This tilt is explained in terms of preferential nucleation at surface
steps and of mismatch accomodation on the vicinal interface.

INTRODUCTION

CdTe epitaxial growth has been achieved on a variety of substrates, either almost lattice matched such as $Cd_{0.96}$ $Zn_{0.04}$ Te (Magnea et
Al., 1987) or non lattice matched such as GaAs (Otsuka et Al., 1985).
(001) CdTe growth on (001) GaAs is feasible but implies that the
interface has to undergo a 14,6 % mismatch in all directions, and one
is left with a high-density of grown-in defects (Feuillet, 1988). On
the contrary, it appears easier to grow CdTe in the <111> orienta-

tion. In this case, the [$1\bar{1}2$] CdTe direction is aligned with the

[$1\bar{1}0$] GaAs direction and the misfit in this common direction is reduced to 0.7 % ; in the direction at 90° from the previous one, i.e.
when [110] CdTe is parallel to [110] GaAs, the misfit is again
14.6 %. But, due to the low stacking-fault energy in CdTe, twinning
occurs parallel to the interface, the density of twins decreasing
when going from the interface towards the top surface of the layer.
We shall report on preliminary results obtained when (111) CdTe
growth is initiated on vicinal GaAs (001) surfaces and demonstrate
the possibility of eliminating twinning.

Observations

The vicinal GaAs surfaces, tilted by 6° around [$1\bar{1}0$] were prepared as detailed elsewhere (Cibert et Al, 1988). This tilt leads to

exposing Ga dangling bonds at the terrace edges as deduced from previous polarity determination experiments by ion channeling (Chami et Al, 1988). T.E.M. samples were prepared in cross sections by Ar^+ -ion milling and subsequently observed in a JEOL 200CX microscope operating at 200 KeV.

Fig. 1 is a high-resolution lattice image of a typical portion of the CdTe/GaAs interface viewed along the common [1$\bar{1}$0] direction. There are no more twins parallel to the (111) growth planes. In addition there appears to be extra half-planes (as indicated by arrows) parallel to the (111) growth planes, ending up at the interface. In addition, the (111) CdTe planes are no longer parallel to the (001) GaAs plane, the rotation angle being of the order of 5°. Other "inclined" (11$\bar{1}$) extra-half-planes are detected that correspond to the usual misfit dislocations.

Fig. 1. <110> lattice image of a vicinal (111) CdTe/(001) GaAs interface.

Discussion

The way epitaxial layers "rotate" when grown on vicinal surfaces has been observed a number of times, especially in the case of low misfit systems (cf, for instance (Ohki et Al, 1988)). In these cases, one has to accomodate the misfit within the interfacial inclined plane which, in short, can be written :

$$\frac{a_L}{\mathrm{Sin}(\theta+\delta)} = \frac{a_s}{\mathrm{Sin}\,\theta},$$

where θ is the misorientation angle of the substrate, δ the layer "tilt" angle, a_L and as the respective layer and substrate lattice parameter in the growth direction. This means that lattice accomodation is possible by a rotation of the layer lattice planes without resorting to misfit dislocations even for layers with overcritical thickness. Coming back to our case, $a_L \simeq a_{CdTe}\,/\sqrt{2} = 3.74$ Å, $a_s = a_{GaAs}/2 = 2.82$ Å, $\theta = 6°$ yielding $\delta=2°$, which is much less than the observed 5°. In order to understand this discrepancy one has to keep in mind that the misfit involved is very large (14,6 %) and moreover, that the epilayer does not retain the substrate orientation. Thus one has to look into the details of (111) CdTe nucleation at (001) GaAs surface steps.

In these regards, one notices that an interfacial coïncidence exists if the (113) GaAs planes are made parallel to the (110) CdTe planes as depicted in Fig. 2a. In this case, the (111) CdTe planes are inclined by 10° to the (001) GaAs planes. It is noteworthy that the (113) GaAs planes are precisely the planes where two missing As atoms would sit on either side of a monomolecular step on the vicinal GaAs surface (Fig. 2b). These vacant sites can be occupied by Te atoms which, if they belong to (110) CdTe planes, will initiate (111) CdTe growth (Cf. the growth model of Cohen-Solal et Al, 1986) but with a 10° tilt.

Thus, since on the terraces CdTe tend to grow with its [111] direction parallel to the [001] GaAs one, and on the steps with a 10° tilt, (111) extra-half-planes are introduced as definitely found in Fig. 1.

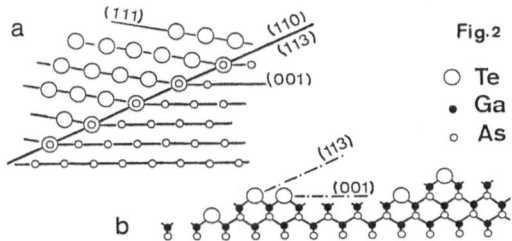

Fig. 2. (a) GaAs(113)/CdTe(111) interface. The CdTe(111) planes form an angle of 10° with the GaAs(001). (b) At the GaAs(113) planes, two missing As atoms would sit on either side of a monomolecular step on the vicinal GaAs surface.

Conclusion

Misorienting the (001) GaAs substrate appears to be a very conclusive way for eliminating twinning in (111) CdTe epilayers. The layer tilt with respect to the substrate can be understood if one refers to the nucleation mechanisms of CdTe on the terraces and at steps. We will further study the epilayer rotation as a function of cutting angle and substrate preparation.

A detailed analysis of the high-resolution T.E.M. pictures will also allow us to visualize the height of the steps on the vicinal GaAs surface and to relate these to the possible twins, when they occur.

Acknowledgements

The authors would like to thank J.L. Pautrat and Y. Merle d'Aubigné as heads of the mixed CEA/CNRS II-VI team.

References

1. N. Magnea, F. Dal'bo, C. Fontaine, A. Million, J.P. Gaillard, Le Si Dang, Y. Merle d'Aubigné, S. Tatarenko, J. of Crystal Growth, $\underline{81}$, 501 (1987).
2. M. Otsuka, L.A. Kolodziejski, R.L. Gunshor, S. Datta, R.N. Bicknell and J.F. Schetzina, A.P.L. $\underline{46}$, 860 (1985).
3. J. Cibert, Y. Gobil, K. Saminadayar, S. Tatarenko, C. Chami, G. Feuillet, Le Si Dang, E. Ligeon, Accepted for publication A.P.L. (1989)
4. A.C. Chami, E. Ligeon, R. Danielou, J. Fontenille, A.P.L. $\underline{52}$ (18), 1502 (1988).
5. G. Feuillet, Proceedings of the Nato Workshop "Evaluation of Advanced Semiconductors by Electron Microscopy" Bristol, 1988, Plenum Press - To be published.
6. A. Ohki, N. Shibata, S. Zembutsu, J.A.P., $\underline{64}$ (2), 694, 1988.
7. G. Cohen-Solal, F. Bailly, M. Barbé, A.P.L., $\underline{49}$, 1519, 1986.

STM AND RELATED TECHNIQUES

D. W. Pohl

IBM Research Division
Zurich Research Laboratory
8803 Rüschlikon, Switzerland

1. INTRODUCTION

STM has raised a high degree of attention in the world of microscopy and surface science over the past couple of years.[1,2] What are the reasons? In my opinion, there are four major points:

1. *Resolution:* STM is capable of imaging individual atoms on a surface. There are few other techniques that provide similar resolution, showing single atoms only under very particular conditions. In field ion microscopy, for instance, only atoms located at the apex of the field-emission tip can be imaged individually.

2. *New information:* Tunnel electrons arrive at the sample surface with millivolt to few-volt energies. They gently stroke the surface instead of striking it like in a conventional electron microscope. As a consequence STM provides preferentially information on the most softly bound electrons — the electrons near the Fermi level, which happen to be the most important for many chemical and electrical properties and processes. Furthermore, having such a small energy, object damage due to electron irradiation is practically impossible. This can be important for the study of low-stability materials such as biological tissue.

3. *New approach:* The STM as an instrument has more in common with a profilometer than with a conventional microscope. In any conventional microscope, a beam of particles/waves is directed onto or emitted from the surface, creating an image by diffraction, deflection, or projection. In STM, a stylus of finite, actually huge mass (compared to the structures to be analyzed) is physically moved across the surface — a few years ago, probably no one would have expected that such a scheme could be used successfully to obtain atomic resolution.

4. *Nano-engineering:* The surprising results of STM indicate a possibility to manipulate macroscopic bodies with atomic precision. As we are penetrating deeper into the world of nanometer dimensions — be it in microelectronics, be it in biology — such a capability is of utmost interest not only for surface characterization but also for modification, processing and alignment on the mesoscopic, say, 0.3 to 30 nm scale.

2. PRINCIPLE OF STM

STM bears considerable similarity to a profilometer but is much more sensitive. The essential difference is that the stylus comes to a halt *before* "touching" the sample surface (...but what does that mean at a few Angstrom tip/sample separation?! A brief discussion is found in Sect. 6.2). This implies that no or only very little surface deformation takes place in the interaction zone. To achieve this goal, an interfacial interaction has to be exploited which is highly distance dependent — and easy to detect.

The tunnel current between two conductors fulfills these conditions in an excellent way. It decays more or less exponentially for all the media studied so far, with a 1/e decay length of ~1 Å on clean surfaces (in UHV) and somewhat larger at ambient conditions. The variation in tunnel current can be readily monitored and used in a feedback loop for distance adjustment to a set value, for instance 1 nA at 0.1 V between tip and sample, corresponding to a gap of typically 3 to 10 Å. For this purpose, the tunnel tip is mounted on an actuator which expands/contracts according to the control-loop output.

At the set distance, the STM stylus can be raster-scanned across the sample surface; whenever there is a step or just a protruding electron orbit, the current tends to in-/decrease, causing a corrective action of the control loop. The record of the control-loop output represents the surface of the object or, more precisely, the topography of its Fermi surface in real space.

3. INSTRUMENTAL ASPECTS

The fact that the STM scheme is so simple represents a great challenge for experimentalists with instrumental inclinations. In fact, a considerable number of designs have been created over the last couple of years. Although looking differently, the basic scheme is the same, viz. to supply as much scanning/positioning flexibility as possible while maintaining a high degree of stiffness of the whole setup (as a rule of thumb, the lowest elastic resonance should not occur below 3 kHz).[3] The main structural elements are (Fig. 1): (i) the "scanner head", in general made of piezoelectric ceramic material, providing 3-dimensional motion of the tip with Å-precision however over a small range only; (ii) the sample holder, mounted opposite to the scanner head, usually on a stage which allows rough approach to the tip. Very often, the stage can be translated parallel to the sample surface at least in one dimension in order to select a suitable sample area for STM characterization. (iii) A frame connecting head and holder. In general, it rests on a base which provides damping, usually in the 100 to 400 Hz range.

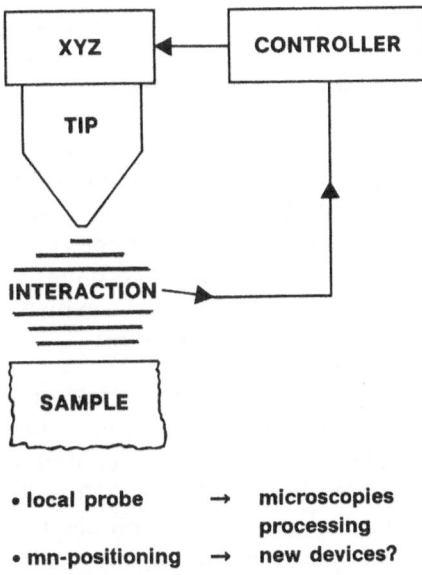

- local probe → microscopies
  processing
- mn-positioning → new devices?

Figure 1. STM principle

Various permutations between the different design elements are feasible, and today the instrumental situation resembles to some extent that of the early days of aviation: There were single-, double- and triple-deckers in the air, having propellers in front or at the stern, all of them capable of flying but far different from what would be called an airplane nowadays. In analogy, the present variety may be considered a symbol for the stimulating pioneering era in the STM field; in a few years, it may well be replaced by the streamlined, optimized designs of a mature though probably less exciting noncontact-mesoscopic-stylus-profilometry-and-microscopy.

4. IMAGING WITH STM

STM is best known for its capability to provide atomic-resolution images. A famous example is the Si(111) surface with its typical 7×7 reconstruction. This surface has been investigated by many groups, originally by the STM Inventors Binnig and Rohrer[1] in their effort to establish the limits of resolution of their new instrument. Figure 2 depicts the regular arrays of the topmost atoms as well as defects in this superstructure[4] which are of great importance for nucleation, growth and dynamics of deposits on the surface.

Figure 3 shows the same type of surface on a larger scale which includes a number of atomic steps. Here the interplay between the atomic 7×7 and the "mesoscopic" step structure is of particular interest. It is seen that the step does not cut arbitrarily through the 7×7 pattern but aligns to one of the symmetry directions through the midpoints of the structure. Results of similar quality have been obtained for (100) and (110)-oriented surfaces of Si, GaAs,[5] and other semiconductors. The excellent resolution is a consequence of the well localized and oriented bonds at the semiconductor surface.

With regard to atomic resolution the situation is less favorable for metals. Their local electron distribution at the Fermi level is smooth in general, showing little corrugation on a plane surface. Adsorbates, however, can establish prominent features in the STM image of such a surface. For instance, clusters of Cu-phthalocyanine molecules on a silver substrate have been imaged successfully[6] with individual molecules showing up as balls which form an ordered micro-crystal of several hundred angstrom in diameter.

With such dimensions, we have left the strictly atomic scale and now want to contemplate a few more mesoscopic structures. Figure 4 sketches a series of images from a polycrystalline silver surface[7] (coated with a monolayer of lead on the right hand side) over an area of 200 by 200 Å. The exciting aspect of this picture: this is the surface of an electrode in an electrolytic cell observed under operating conditions! The insets in the upper part sketch the experimental

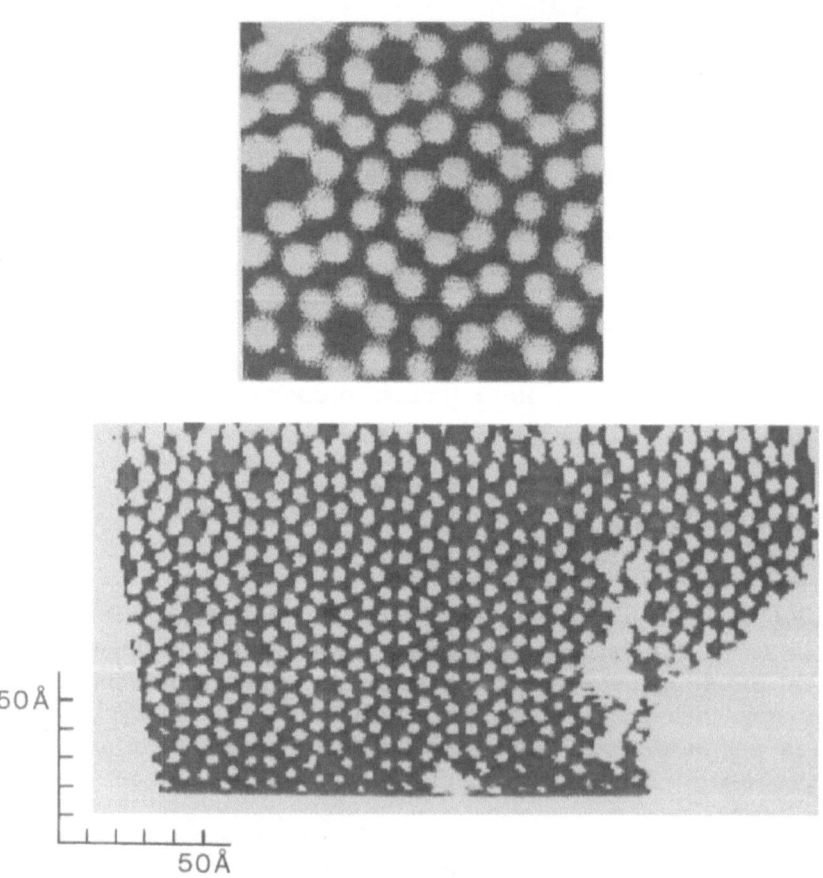

50Å

50Å

Figure 2. Silicon (111) surface showing 7×7 reconstruction with defects [top: from Ref. 4b; bottom: from Ref. 4a, Copyright 1986 by International Business Machines Corporation; reprinted with permission].

Figure 3. Silicon (111) surface showing 7x7 reconstruction with steps [from Ref. 4b].

arrangement and the I/V cycle indicating the conditions for lead deposition and dissolution. The ionic current between tip and sample was kept sufficiently low to still regulate the distance-dependent electron tunneling. This finding opens a completely new form of electrochemistry allowing, for the first time, to characterize deposition processes in situ, down to the atomic level, and to investigate the all-important double layer at the electrode in a completely new way.

Figure 5 finally brings us into the exciting world of biophysics.[8] Although biological molecules usually lack conductivity, it turns out that STM on sufficiently thin layers is possible and provides useful information. The figure shows patches of a lipid membrane (HPI) on a flat gold film. When zooming in on an isolated patch, one recognizes a regular molecular arrangement which forms a sort of a two-dimensional crystal in this case. Although still quite at the beginning and from its theoretical basis not well understood, the use of STM in certain areas of biology appears to open new horizons.

The last examples show that a good deal of interesting research can be done without exploiting the full resolution capability of STM; by doing so, one gains in experimental simplicity and can work at ambient (in vivo!) conditions, but has to deal with contaminated, ill-defined surfaces. Atomic resolution imaging as discussed at the beginning of this section, on the other hand, in general requires UHV conditions which make setups expensive and elaborate, but provide well-controlled surface conditions.

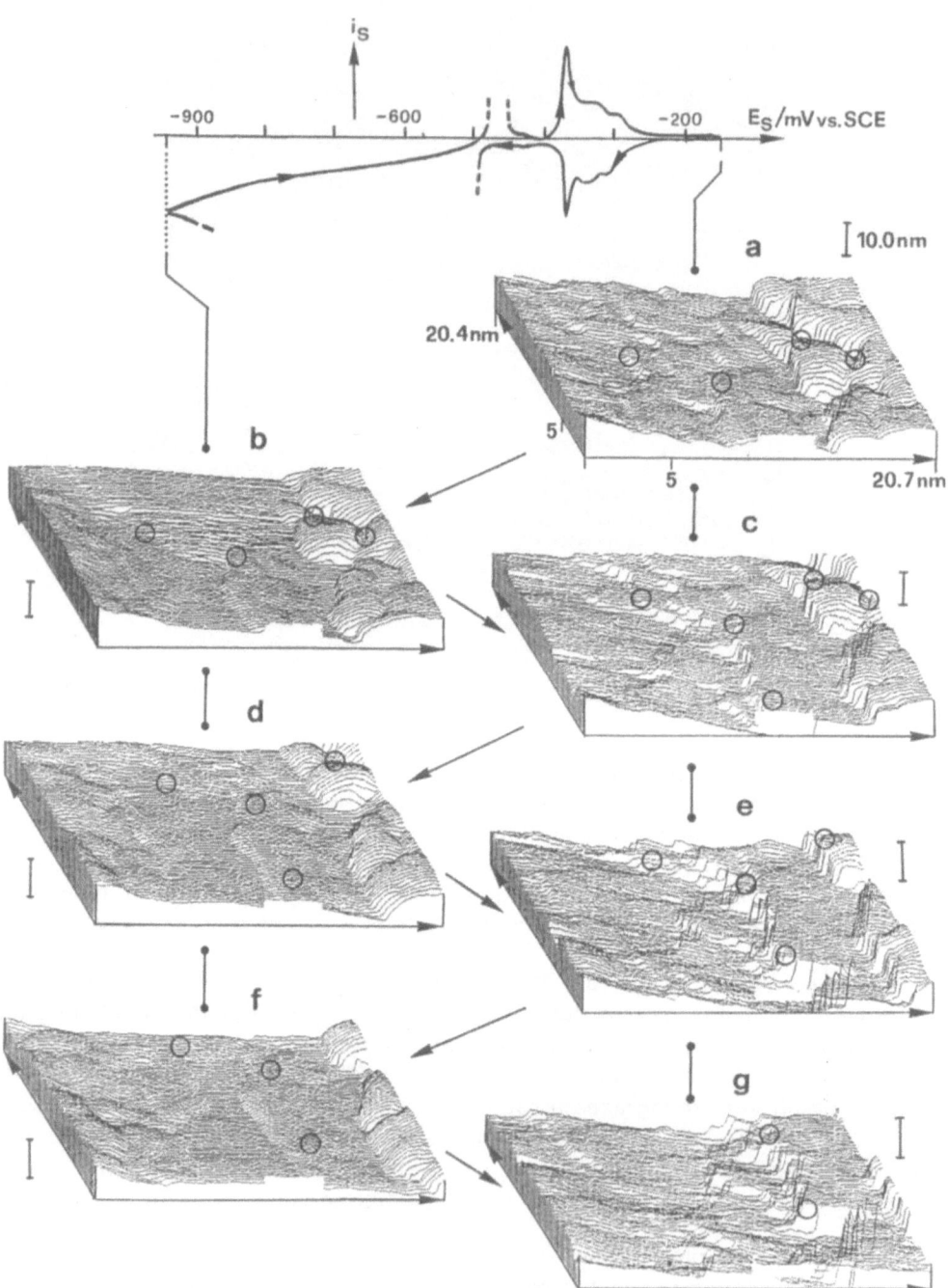

Figure 4. STM imaging of an electrode surface in electrolytical medium [from Ref. 7].

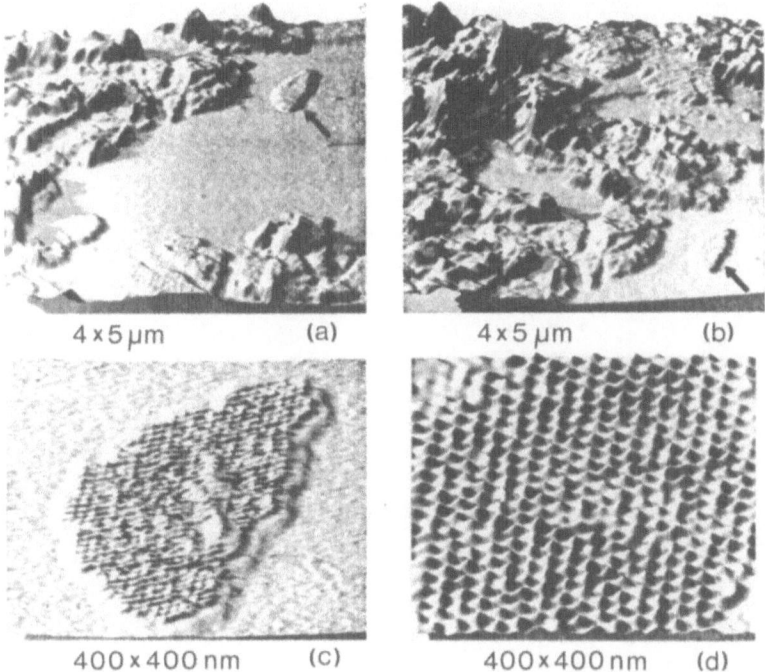

5. DIFFERENT MODES OF OPERATION

Although one could spend one's lifetime with mapping the topographies of
all sorts of interesting material surfaces, it is intriguing to look out for other
information possibly obtainable with an STM. For this purpose, let us consider
the parameters upon which the tunnel current I_t depends (Fig. 6).

The *lateral position* (x, y) influences I_t not only via topography $z(x, y)$ but
also through possible chemical changes of the sample surface. The latter make
themselves felt in the value of the local *workfunction* $\phi(x, y)$ which in turn influences the decay length of the tunnel probability. This effect can be probed by
modulating the gap width and recording the response in I_t (caution: interpretation is problematic for contaminated or soft media). The (position-dependent)
chemistry makes itself also felt in the current-voltage characteristic $I_t(V)$. This
dependence leads to the important fields of *tunnel spectroscopy*[9] and *potentiometry*,[10] which will be dealt with in the following paper by H. Salemink.

Another aspect of interest is the way in which x and y are varied in an
STM experiment. Figure 7 presents an alternative to the usual raster-scan
mode:[11] the tip is x- and y-modulated with small amplitude (resulting in a cir-
cular motion). The resulting modulation in the tunnel current is used to move to
the tip to an extreme point of the surface, following the line of steepest inclination and finding the bottom of the small crater in the figure where it stays
locked for an arbitrary length of time. In a similar way, one might track local-

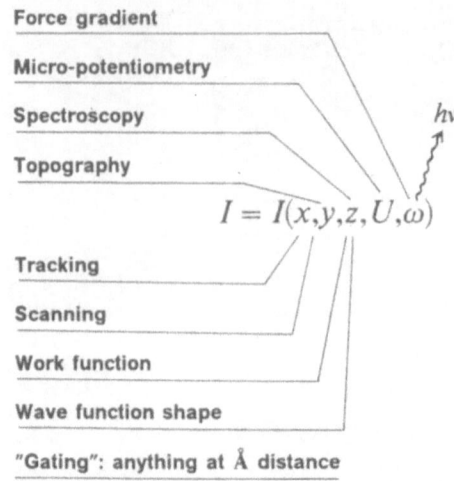

Figure 6. Different modes of STM operation

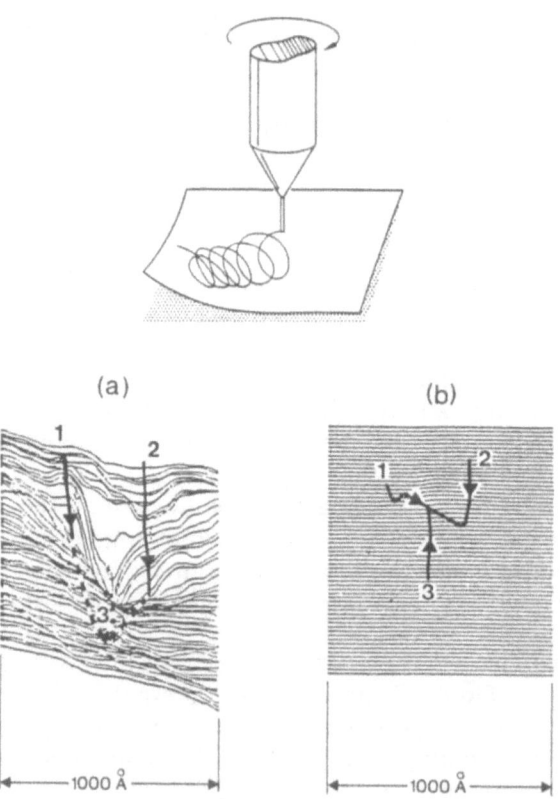

Figure 7. Tracking tunneling microscope [from Ref. 11, © 1988 American Institute of Physics].

ized (point) defects on a surface. The tip alternatively can be made to move on a line of constant height by the same technique, changing the role of two lateral feedback loops. This "tracking" mode can be useful for the observation of surface dynamics (e.g., filling up the well) or for tracing out certain features of a surface such as a grain boundary or a semiconductor junction.

When tunneling across the gap, electrons gain the energy $e U_t$. For $U_t = 2$ to 3 V, the tunnel electrons may dissipate their energy by *emission of visible light*, in particular if the surface plasmon energy falls into this range (Ag, Au, Cu, etc.). This effect, well known for MIM sandwich structures, was recently also found in an STM arrangement.[12,13] A surprisingly high yield, $\sim 10^{-4}$, provides light levels sufficient for detection with an optical multichannel analyzer.

Figure 8 shows the emission spectra from a silver surface with the tunnel voltage as parameter.[13] The peak at 2.3 eV can be associated with the plasmon resonance of the STM tip-sample geometry. Interestingly, the light emission strongly varies with position, for instance when moving from one grain to another on a polycrystalline silver film and possibly also during passage over adsorbate molecules. The results have been associated with inelastic tunneling processes [Fig. 8(a)].[14]

Instead of studying the response to a lateral translation, one may obtain interesting information also by moving the tip normal to the surface, i.e., by systematically varying the gap width Δz. For increasing Δz, one first of all finds a decay of I_t which is exponential over several decades, indicating a large range

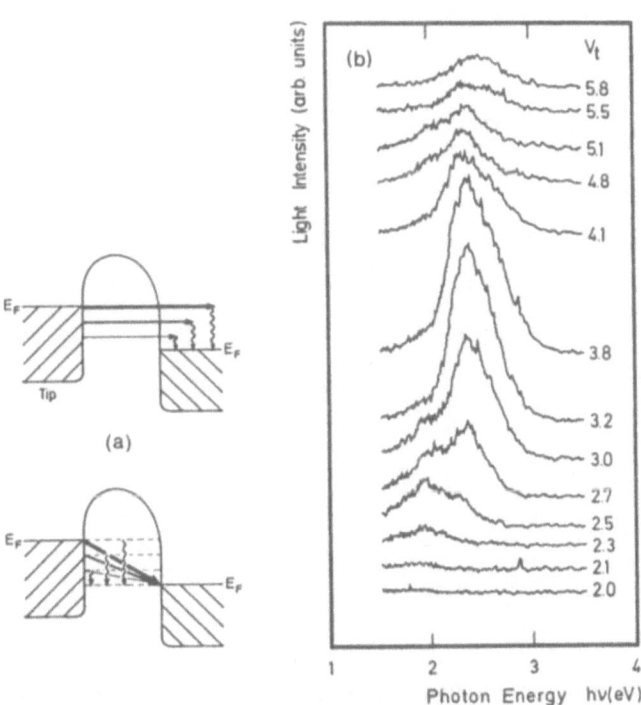

Figure 8. (a) Elastic (top) / inelastic (bottom) tunneling processes: theory favors the latter;[14] (b) experimental light emission spectra. [From Ref. 13, © Les Editions de Physique 1989].

of validity for the usual assumptions in tunnel theory.[9] If U_t is increased appropriately, the transition from pure *tunneling to field emission* can be conveniently investigated, including the resonances due to image and Gundlach states.[15]

At the gap widths of a few angstrom with which we are dealing in STM, *considerable forces* can be expected to build up between tip and sample. We were recently able to demonstrate the metallic binding force quantitatively in the system Ir(tip)/Ir(sample).[16] For this purpose, the sample was cut as a small strip from an Ir foil and mounted to form an elastic cantilever beam [Fig. 9(a)]. In the proximity of the tip, the gradient of the interaction force influences the elastic resonance ω_{CB} of the beam as if an additional spring were installed between tip and sample. ω_{CB} can be readily detected by gently shaking the sample with the help of a piezoelement and recording the modulation $I_t(\omega)$.

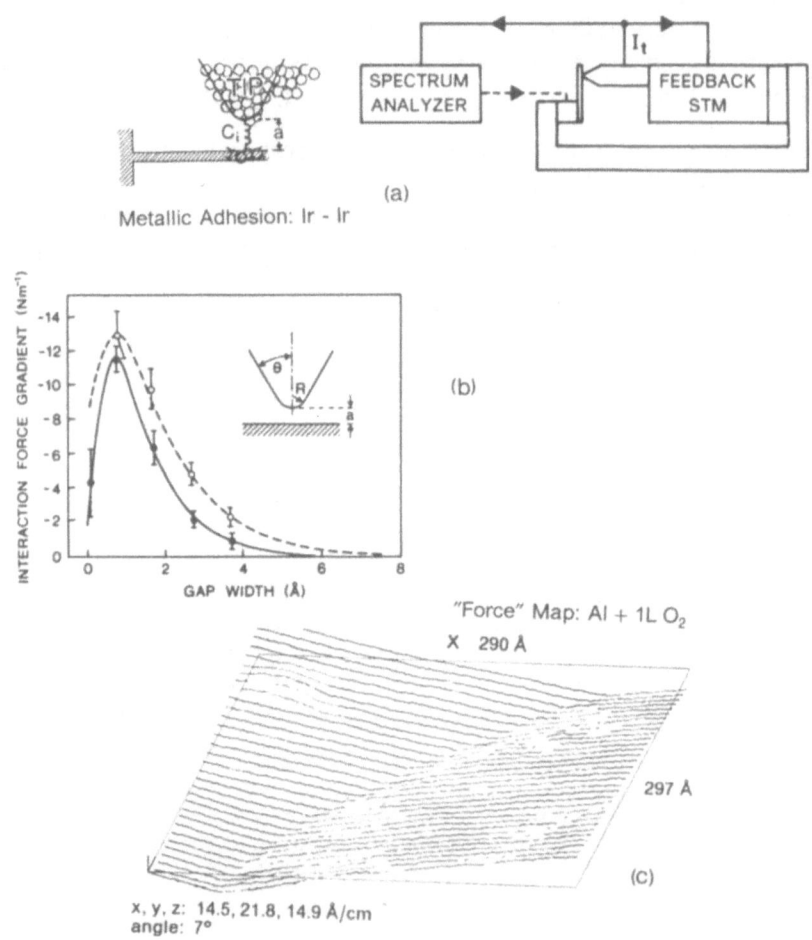

Figure 9. Forces during tunneling: (a) setup, (b) force vs. distance (experimental points and fit to two slightly different models) [from Ref. 16, © 1988 The Royal Microscopical Society], and (c) force map across an Al grain boundary (left: attraction, surface clean; right: repulsive, probably covered with monolayer of oxygen).[16]

Figure 9(b) was derived from such measurements. The data points can be well fitted to a jellium model (solid and dashed curves). The significance of this finding lies in the fact that chemical binding between two atoms, formation of an electrical contact, adhesion between two materials and many other important properties are determined by interfacial interactions on the sub-10 Å scale. Interfacial forces at such small distances, on the other hand, neither were accessible to experiment before, nor was the validity of the jellium model in this range absolutely clear. The interaction force turns out to depend on surface composition, becoming, for instance, completely repulsive in the presence of oxygen adsorbates. Similar experiments were performed recently also at ambient conditions,[17] revealing that the establishment of a tunnel contact in the presence of air, i.e. not perfectly clean conditions is much more complex than in UHV.

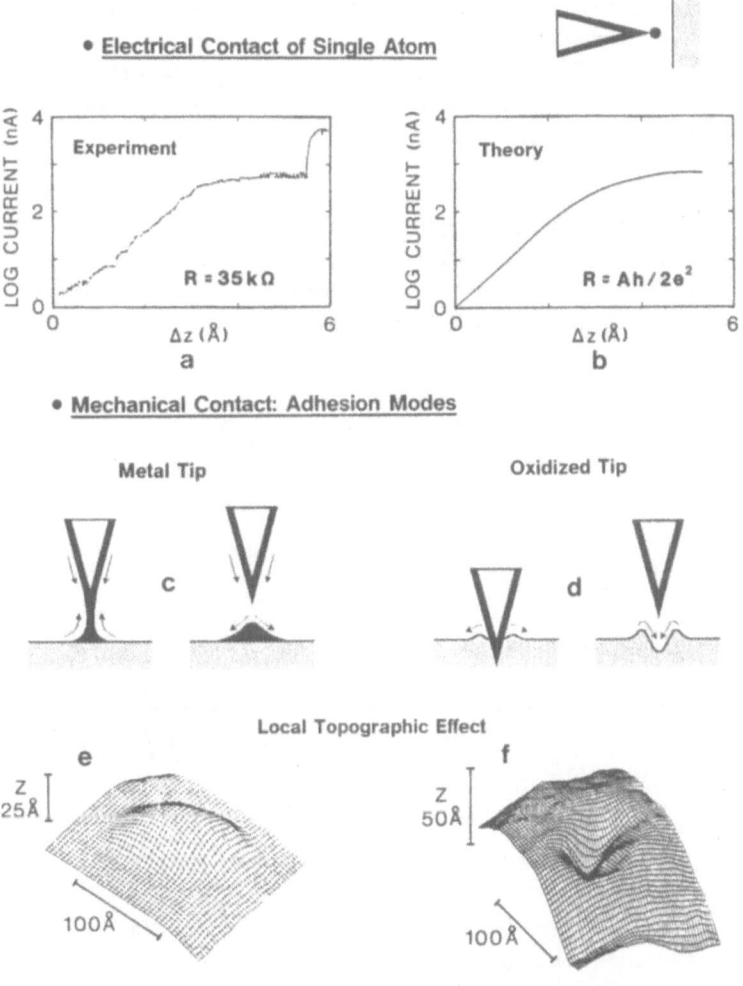

Figure 10. From tunneling to point contact: (a) experimental[18] and (b) theoretical[19] current vs. gap width dependence [from Ref. 19, © 1987 The American Physical Society]; (c,d) clean metal tip touching clean/oxidized sample [from 18a, © Elsevier Science Publishers B.V.], leaving (e) a bulge or (f) a crater after retraction [from 18b, © 1987 The American Physical Society].

When the gap is closed gradually and in a well-controlled way (starting from an STM set point $\Delta z \simeq 0.5$ nm), the transition from *tunneling to point contact* can be studied in detail, a problem of great practical interest with regard to the role of electrical switches in everyday life and many kinds of machinery. Figure 10 [18,19] depicts experimental (a) and theoretical (b) dependences $I_t(\Delta z)$ for an atomically sharp tungsten tip on a clean silver surface. The most interesting features are a saturation of I_t followed by a vertical step. The latter may represent the sudden closure of the gap by the formation of a bulge on the soft silver surface [Fig. 10(c)] under the influence of the metallic binding force. This idea is supported by topographic STM images taken before and after contact, which indeed show such an effect on the previously flat sample area [Fig. 10(d)].

The plateau before contact was interpreted by Lang[19] to be caused by the breakdown of the tunnel barrier to a narrow channel on the tip/sample axis. The channel behaves like a short open duct for electrons exhibiting a resistance of several times $\hbar/2e^2 = 12.7$ kΩ. The experimental data appear to support this assumption, which relates I_t to fundamental constants in an intriguing way. Figures 10(e) and 10(f) depict the result of the same experiment with an adsorbate-covered sample. A crater is found after touching the surface, indicative of a repulsive rather than attractive interaction.

Figure 11, last not least, shows a first step towards *spin-polarized tunneling* on magnetic media.[20] Secondary electrons are created with an STM operated in the field-emission mode and gap widths of a few millimeters. They appear considerably polarized when analyzed with a Mott detector. Although a far cry from true STM operation, the result indicates potential magnetic characterization by STM.

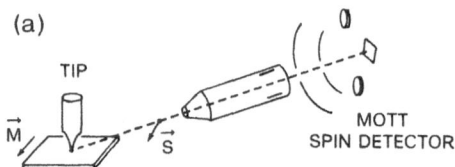

(a)

- First result: -160 V / 0.8 nA / 3 mm

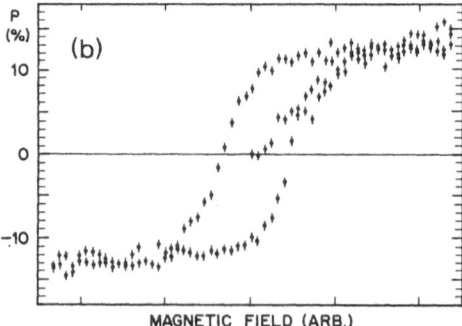

Figure 11. Magnetic effects and STM:[20] (a) setup and (b) experimental secondary electron spin polarization upon variation of sample magnetization [from Ref. 20, © 1989 American Institute of Physics].

6. ALTERNATIVE INTERACTIONS

It was stated in the introduction that in STM, unlike in conventional profi-lometers, the probe stylus is kept at a fixed small separation from the sample by detecting an interaction and regulating it to a fixed set value with the help of a feedback circuit (Fig. 1).   After the great success with the tunneling, it is intriguing to catalogue other possible interactions (X) and try to exploit them in an analogous way ("SXM"). Figure 12 lists a number of such interactions.

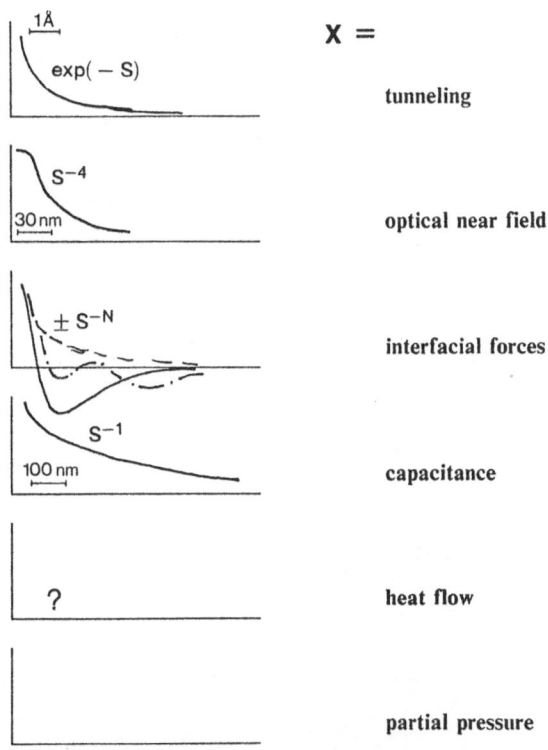

Figure 12. Alternative interfacial interactions.

6.1. X = Near-Field Optics: SNOM

Optical near fields represent the evanescent waves that build up around strongly curved parts of an illuminated medium.   Their extension is roughly equal to the radius of curvature, and they become strongly modified upon approaching another medium. The modification is specific for the optical prop-erties of the second medium and may be detected in the far field radiated for the curved part. The latter may be considered an "optical antenna," "scattering center" or simply as the "tip" of a scanning optical near-field microscope (SNOM).[21]

In Figure 13(a) the first practical SNOM is sketched; Fig. 13(b) shows the variation of far field (observed in transmission) upon sample approach, and Fig. 13(c) depicts an NO image of a semitransparent metal film with 100-nm holes, using green light of 488 nm wavelength.[22] This SNOM micrograph, one of the first (ever) made (and therefore lacking any image processing), displays features < 20 nm in size, for instance details of the structure at the rim of the holes. To our knowledge, it is the highest resolution optical image ever made.

With its resolving power, SNOM comes close to conventional SEM, but is not restricted to conducting media nor does it require operation in vacuum. Our results were recently confirmed by a group at Cornell University using a similar scheme.[23] It has also been possible to operate SNOM in reflection[24] and in a topographic mode, achieving similar resolution, in particular when exploiting plasmon effects.[25]

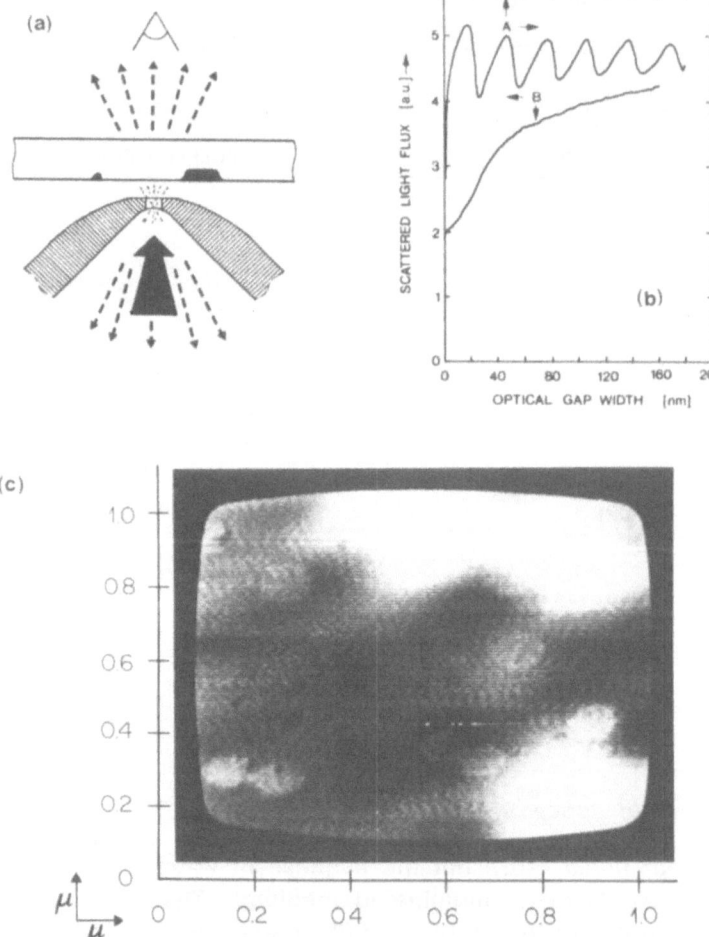

Figure 13. Scanning optical near-field microscopy:[22] (a) setup, (b) variation of transmitted light intensity with gap width, (c) SNOM image of metal film with 100 nm holes, λ = 488 nm [from Ref. 22a, © 1985, The Society of Photo-Optical Instrumentation Engineers].

6.2. X = Forces: SFM

The existence of interfacial interaction forces has already been discussed in context with STM. To become an independent method, it is important to sense an interaction force also without a tunnel current associated with the tip-sample arrangement. Two schemes invented for this purpose are sketched in Fig. 14.

In 14(a) [26], the force tip is mounted on a deflectable cantilever beam whose metallic rear side is part of a rudimentary STM. Any deflection caused by the approach of the sample (from the left side) results in a variation of the tunnel current on the right, which is used to keep the force at a set small value (usually repulsive) of the order of 10^{-8} N (this is to be compared with the load exerted by the conventional profilometers of 10^{-5} to $> 10^{-3}$ N).

The arrangement in Fig. 14(b)[27] exploits the sub-angstrom sensitivity of optical heterodyne interferometry for detection of cantilever deflection. In the particular setup shown here, the cantilever consists of an iron wire, which can be magnetized to detect magnetic interactions with high sensitivity.

SFM has already proven very useful for surface topographic characterization on the mesoscopic scale, as shown in Fig. 15.[27] The sample is a silicon surface with grooves 2 μ apart (a,b). Part (c) depicts a small section of the slope of a groove which reveals topographic details, for instance facets and steps along crystallographic orientations.

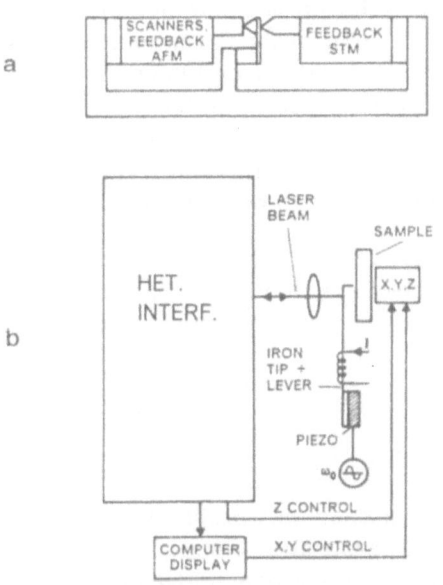

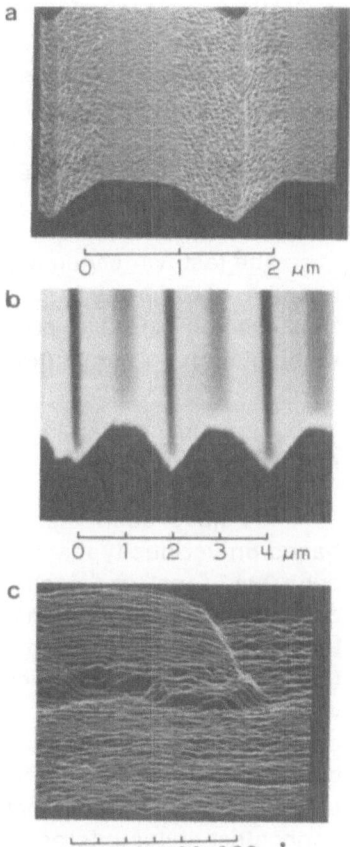

Figure 14. Scanning force microscopy: detection of cantilever beam deflection by (a) tunneling [from Ref. 26, © 1986 The American Physical Society], or by (b) heterodyne interferometry.[27]

Figure 15. (a) Profile of V-shaped grooves in silicon wafer, (b) corresponding SEM image, and (c) high resolution blow-up from (a). [From Ref. 27, © 1987 American Institute of Physics].

197

The ultimate resolution of SFM (also called "AFM", atomic force microscope) is not yet well established. So far, atomic resolution has been found for graphite[28] and other layer compounds. These soft materials, however, behave strangely in STM and SFM; it is not yet clear whether these results are representative also for other materials. If so, SFM will certainly become of great fundamental importance since it can be applied to insulating surfaces where electron-beam techniques fail.

6.3. X = Other Interactions

Williams et al.[29] developed an intriguing thermal profiler where the tunnel current is replaced by a heat current through an air gap. The flow is sensed by means of a tiny thermoelement integrated into the probe tip. Resolutions of less than 100 nm were obtained; moreover, changes in the surface temperature of a few mK can be imaged. This opens interesting perspectives for in-vivo observation of biological samples and processes occurring therein.

The capacitance microscope developed by Matey and Blanc[30] also falls into the category of noncontact stylus microscopes. The scheme recently was taken up again,[31,32] with first results indicating a resolving power of < 50 nm, even if the sample is a dielectric medium.

My last two examples of microscopes analogous to STM are based on material transport between "tip" and sample. The tip is a micropipette with an internal diameter of less than one micron. In one of the arrangements,[33] it is moved along a membrane with microscopic pores. Its far end is connected to a mass spectrometer which records the amount of gas entering the pipette during the scanning process. In the other setup, the micropipette is immersed into an electrolytic solution.[33] The resistance between its interior and a reference electrode increases when the end approaches the surface to be investigated.

The last two schemes, though somewhat exotic in their present state, might become of interest in combination with other techniques, in particular with "mesoscopic" surface processing and the further development of nanometer manipulation and technology in general.

7. SUMMARY

We have seen that noncontact stylus microscopes/profilometers may be based on various types of interactions. With appropriate sensors, the probe tips can "feel," "see" or "smell" that they are next to the surface of investigation. The corresponding SXMs hence provide different information about a surface, and they collect it at different resolutions ranging from atomic to sub-micron scale (Fig. 16). The various techniques thus supplement each other in a convenient way, in particular when integrated into the same instrument. Simultaneous detection of forces and of near-field optics with tunneling has already been demonstrated; other combinations appear similarly feasible. Hence, in an outlook, the multifunctional SXM of the future may become a general-purpose noncontact stylus microscope which could be applied to a host of tasks not conceivable with STM alone.

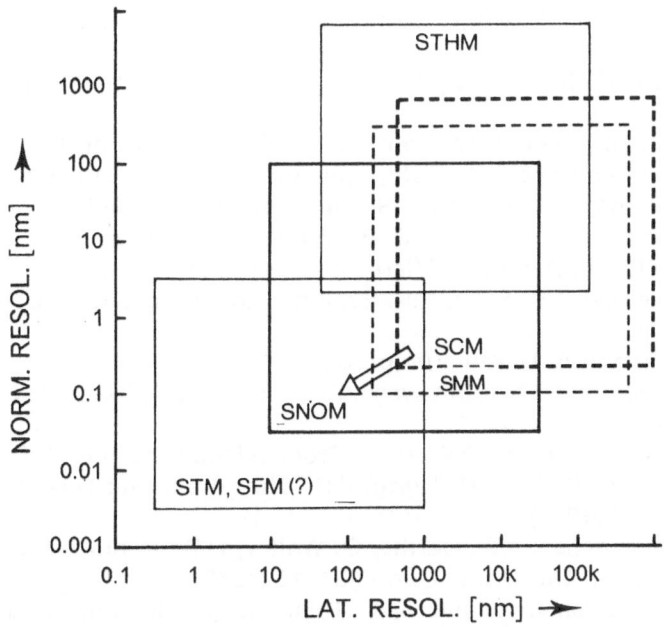

Figure 16. Resolving power of different SXMs lateral and normal to the sample surface (STHM: thermal profiler, SCM: capacitance microscope, SMM: micropipette microscope.

While the above aspects may open doors for X intriguing developments in the future, most of the ongoing work concentrates on STM and the physical insights to be gained with tunneling. The activity in the other techniques is still small though growing. Instrumental developments are being pursued mostly by a few small startup companies whose products are not yet far from the experimental level. The beginning tide of publications in the open literature indicates that the new form of microscopy is well on its way, being established at more and more research institutes around the globe.

REFERENCES

1. G. Binnig and H. Rohrer, Physica 127B, 37 (1984).
2. P.K. Hansma, J. Appl. Phys. 61, R1 (1987).
3. D.W. Pohl, IBM J. Res. Develop. 30, 417 (1986).
4. a) R.J. Hamers and J.E. Demuth, IBM J. Res. Develop. 30, 396 (1986); b) U. Köhler, R.J. Hamers and J.E. Demuth (in prep.).
5. R.M. Feenstra, J.A. Stroscio, J. Tersoff and A. Fein, Phys. Rev. Lett. 58, 1192 (1987).
6. J. Gimzewski, E. Stoll and R. Schlittler, Surf. Sci. 181, 267 (1987).
7. R. Christoph, H. Siegenthaler, H. Rohrer and H. Wiese, Electrochimica Acta (to be publ.).

8. B. Michel and G. Travaglini, Proc. 1988 STM Conf., Oxford: J. Microsc. (in press).

9. A. Baratoff, G. Binnig, H. Fuchs, F. Salvan and E. Stoll, Surf. Sci. 168, 734 (1986).

10. P. Muralt, H. Meier, D.W. Pohl and H. Salemink, Appl. Phys. Lett. 50, 1352 (1987); P. Muralt and D.W. Pohl, Appl. Phys. Lett. 48, 514 (1986).

11. D.W. Pohl and R. Möller, Rev. Sci. Instrum. 59, 840 (1988).

12. J.H. Coombs, J.K. Gimzewski, B. Reihl, J.K. Sass and R.R. Schlittler, Proc. 1988 STM Conf., Oxford: J. Microsc. (in press).

13. J.K. Gimzewski, J.K. Sass, R.R. Schlittler and J. Schott, Europhys. Lett. 8, 435 (1989).

14. B. Persson and A. Baratoff (in prep.).

15. J.H. Coombs and J.K. Gimzewski, Proc. 1988 STM Conf., Oxford: J. Microsc. (in press).

16. U. Dürig, O. Züger and D.W. Pohl, Proc. 1988 STM Conf., Oxford: J. Microsc. 152, pt. I, 259 (1988). U. Dürig, J.K. Gimzewski and D.W. Pohl, Phys. Rev. Lett. 57, 2403 (1986).

17. M. Nonnenmacher, J.W. Bartha, O. Wolter, D.W. Pohl and R. Kassing, Beitr. Elektronenmikr. Direktabb. Oberfl. 21, 13 (1988).

18. a) J.K. Gimzewski, R. Möller, D.W. Pohl and R.R. Schlittler, Surf. Sci. 189/190, 15 (1987), b) J.K. Gimzewski and R. Möller, Phys. Rev. B 36, 1284 (1987).

19. N.D. Lang, Phys. Rev. B 36, 8173 (1987).

20. R. Allenspach and A. Bischof, Appl. Phys. Lett. 54, 587 (1989).

21. for a recent review see D.W. Pohl, U.Ch. Fischer and U.T. Dürig, Proc. SPIE 897, 84 (1988).

22. a) D.W. Pohl, W. Denk and U. Dürig, Proc. SPIE 565, 56 (Micron and Submicron Integrated Circuit Metrology 1985); b) U. Dürig, D.W. Pohl and F. Rohner, J. Appl. Phys. 59, 3318 (1986).

23. E. Betzig, M. Isaacson, H. Barshatzky, A. Lewis and K. Lin, Proc. SPIE 897, 91 (1988).

24. U.Ch. Fischer, U.T. Dürig and D.W. Pohl, Appl. Phys. Lett. 52, 249 (1988).

25. U.Ch. Fischer and D.W. Pohl, Phys. Rev. Lett. 62, 458 (1989).

26. G. Binnig, C.F. Quate and Ch. Gerber, Phys. Rev. Lett. 56, 930 (1986).

27. Y. Martin, C.C. Williams and H.K. Wickramasinghe, J. Appl. Phys. 61, 4723 (1987).

28. See, for instance, G. Binnig, Ch. Gerber, E. Stoll, T.R. Albrecht and C.F. Quate, Europhys. Lett. 3, 1281 (1987).

29. C.C. Williams and H.K. Wickramasinghe, Appl. Phys. Lett. 49, 1587 (1986).

30. J.R. Matey and J. Blanc, J. Appl. Phys. 57, 1437 (1985).

31. H.P. Kleinknecht, H. Meier and J. Sandercock, Beitr. Elektronenmikr. Direktabb. Oberfl. 21, 19 (1988).

32. H.K. Wickramasinghe (priv. communic.).

33. J.A. Jarell, J.G. King and J.W. Mills, Science 211, 277 (1981).

34. P. Hansma, Proc. 1988 STM Conf., Oxford: J. Microsc. (in press).

TUNNELING SPECTROSCOPY AND III-V SEMICONDUCTOR INTERFACES

O. Albrektsen* and H.W.M. Salemink

IBM Research Division, Zurich Research Laboratory

8803 Rüschlikon, Switzerland

The (110) cleavage plane of GaAs is of high interest for the characterization of semiconductor multilayers grown along the <001> direction. After cleaving in ultra-high vacuum (UHV), the (110) GaAs face exhibits a simple and regular, unreconstructed 1×1 surface unit cell with both Ga and As atoms showing up in the terminating surface,[1] fig. 1. For the clean surface, electronic surface states are found outside the semiconductor bandgap, in the valence and conduction band.[2] Hence, the electronic spectroscopy at the surface has a similarity to that of the bulk material. Such a situation is not found on the silicon surfaces, where the surface states dominate in the semiconductor bandgap.[3] As the (110) plane is an orthogonal cross section to the <001> growth direction, the tunneling spectroscopy on this face might be used to study the electronic properties along the growth direction, in a similar sense that TEM reflects the structural

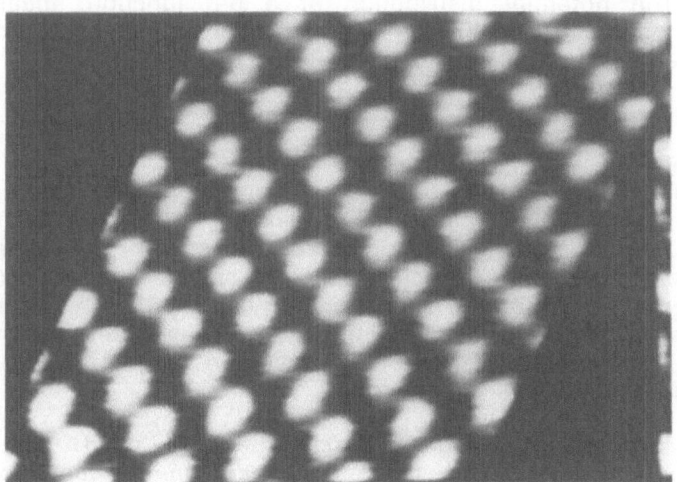

Figure 1. Clean UHV cleaved GaAs (110) surface; sample energy is positive, tunneling into empty states on Ga atoms (white). Unit cell dimensions are 4.0×5.6 Å. Slight chaining along the <001> direction.

coordination in the interfaces.[4] The instrument used for the experiments described here is a usual 'pocket size' scanning tunneling microscope (STM), which is mounted inside a UHV electron microscope;[5] in a connected UHV chamber, the GaAs samples are cleaved and the tunneling tips are prepared by sputtering and heating. Both samples and tips are transferred in UHV to the analysis chamber where the SEM is used to move the tunneling tip to the layers of interest. A description of the operation of the STM can be found in the literature.[6-8]

Figure 2 shows the potential step at a p-n junction in an AlGaAs heterostructure along the <001> direction, as measured in direct space with a tunneling potentiometric technique;[9,10] the topographic structure in this area is atomically flat, but the lateral resolution on the (001) face was too low to resolve the atomic structure of fig. 1. From such curves, the spatial extent of the electric field at a material junction can be estimated.[10,11]

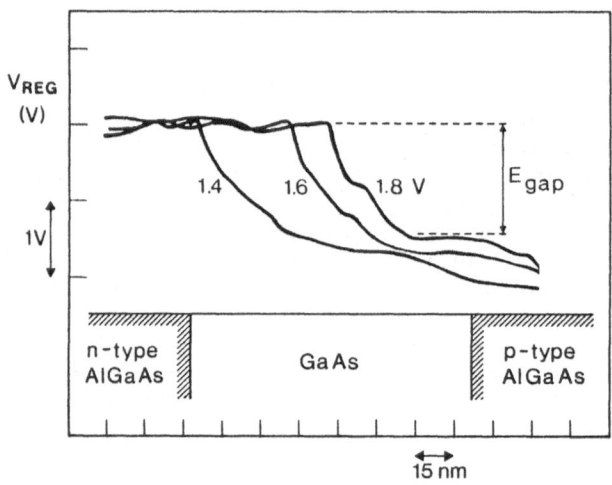

Figure 2. The potential distribution at a p-n junction, measured by a tunneling potentiometric technique over the cleavage plane (from Ref. 10, © 1987 American Institute of Physics).

A direct measurement of the electron depletion and confinement regions in GaAs-AlGaAs interfaces is shown in fig. 3; with the STM, current-voltage characteristics are obtained across from the interface, at fixed gap distance. In fig. 3b, the current at a selected voltage (= electron injection energy) is plotted as a function of position; the local conductivity at the interface is traced, reflecting the typical depletion regions in such interfaces.[12] For the given structure, a depletion length of 15 nm is found, which is in agreement with calculations.

The local current voltage characteristics are used to measure the energy position of conduction and valence band in the semiconductor[1] and the band shifts in the vicinity of adsorbed atoms.[13] Similarly it is used to find the valence band offset in a material interface.[14] Usually, this quantity is derived from photo-emission data, which cover macroscopic areas and therefore include sample inhomogeneities. Two such spectroscopy curves are shown for the AlGaAs-GaAs interface in fig. 4. The derived valence band offset (0.35 eV) agrees very well with the most reliable photo-emission data (0.38 eV)[15] for this interface. In this particular case, the result implies a relative change $\Delta Ec/\Delta Eg = 0.25$ for the conduction band for xAl $= 0.50$. Similar spectroscopic data has been obtained by Feenstra et al. on the changes in the GaAs band structure, due to localized, individual adsorbed oxygen atoms.[13] A charge-screened area of several unit cells is found on n-type GaAs, indicating a negative charge on the surface, whereas the state is found to be neutral on p-type.

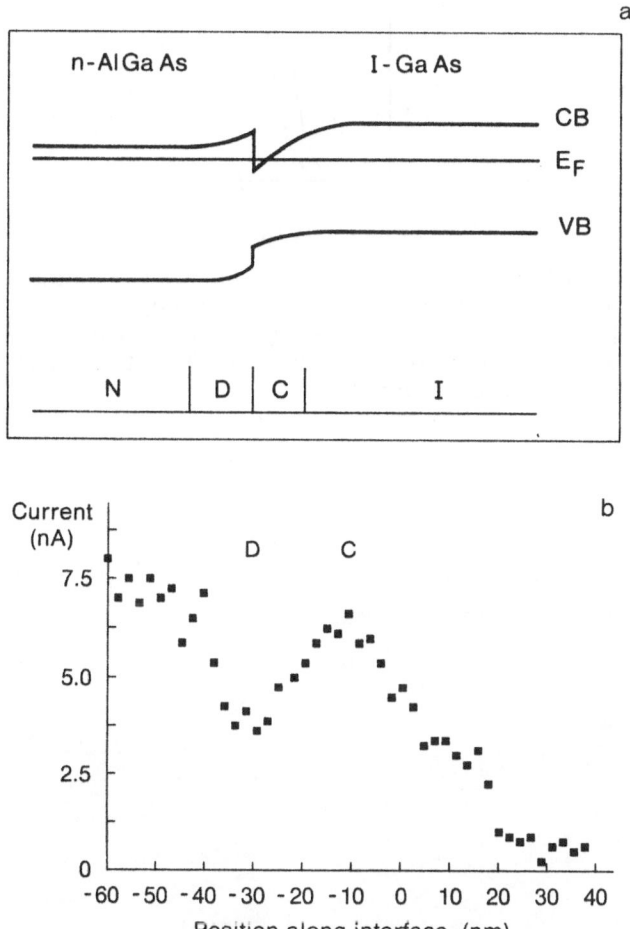

Figure 3. (a) Schematic energy band diagram of the $Al_xGa_{1-x}As$-GaAs heterojunction. n-type AlGaAs ($x_{Al} = 0.50$, Si 1×10^{18} cm^{-3}), depletion layer (D), electron confinement layer (C) and semi-insulating GaAs (I).
(b) Local conductivity in central part of (a) near interface D-C. Tracing of tunnel current at sample voltage of $+3$ V as a function of relative position along the interface. These data are derived from a set of 45 I-V curves. taken along the interface at 2 nm separation. The local maximum and minimum reflect the enhanced or reduced tunneling conductivity into the confined (C) or depleted (D) interfacial zone (from Ref. 14, © 1989 American Institute of Physics).

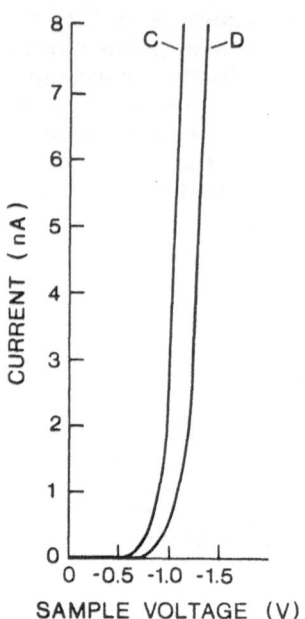

Figure 4. Typical local I-V curves for the electron-depleted (D) and confined (C) region of Fig. 3. Spatial separation of both curves is 25 nm. Tunneling is out of the valence band. The valence band offset of 0.35 eV is derived from the difference in the extrapolated gaps in the I-V curves and from the voltage shift in the curves (from Ref. 14).

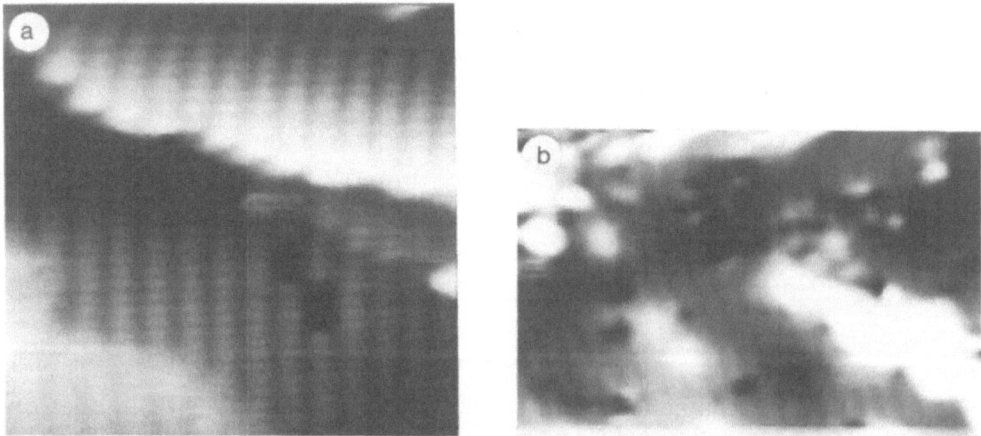

Figure 5. **(a)** GaAs (110) surface, tunneling out of As filled states. Note in the As rows the missing individual atoms, with no appreciable charge screening in the defect vicinity. Image is 5.8 nm wide and surface corrugation in center part is 0.015 nm.
(b) GaAs(110) surface after 70 hours exposure to system pressure of 5E-11 mbar. Contamination is visible in the form of white clusters of adsorbed atoms. Apparent individual atoms are also seen. Between the clusters, atomically clean surface is found at this coverage. Image is 9.7 nm wide.

Two examples of local defects on the GaAs(110) face are given in fig. 5. In fig. 5a the tunnel voltage is set to emphasize the filled states on the As atoms; the two small defects near the center are 3 and 4 missing As atoms, respectively, in the surface registry: no observable re-distribution of charge apparently results from the missing atoms and only single As atomic rows are affected. This can be understood if the charge density on the As atoms is sufficiently localized, so that a missing atom does not largely affect its neighbors. After prolonged exposure to the environmental pressure of 5E-11 mbar (in this case for 70 hours), the initially clean GaAs (110) surface contaminates, and exhibits a typical image as in fig. 5b. The Auger spectrum made in UHV after this STM image showed minor contributions from oxygen and carbon. The surface contamination is seen as white clusters and individual atoms; still clean atomic As structure can be observed between these adsorbed atoms. Also some missing As atoms like in fig. 5a are visible.

Final Remarks

Tunneling spectroscopy as described in this paper and the relevant references can be regarded as a very local, low-energy electron analysis technique with atomic resolution capability. It may therefore be expected that it will be applied to electronic structures with interesting effects in the vicinity of the Fermi level. The GaAs (110) cross-sectional face to the preferred growth direction is an important test case in this respect, as this material is used to construct many multilayer structures.

References

*   Institute of Applied Physics, Technical University of Copenhagen, DK-2800, Lyngby, Denmark
1.   R. Feenstra, J.A. Stroscio, J. Tersoff and A.P. Fein, *Phys. Rev. Lett.* 58:1192 (1987).
2.   B. Reihl, T. Riesterer, M. Tschudy and P. Perfetti, *Phys. Rev.* B38(13):456 (1988-II)
3.   J.A. Stroscio, R.M. Feenstra and A.P. Fein, *Phys. Rev. Lett.* 57:2579 (1986).
4.   P. Muralt, H. Meier, D.W. Pohl and H. Salemink, *Superlattices and Microstructures* 2:519 (1986).
5.   C. Gerber, G. Binnig, H. Fuchs, O. Marti and H. Rohrer, *Rev. Sci. Instrum.* 57:221 (1986).
6.   Proc. STM Workshop, Oberlech, Austria (1985): *IBM J. Res. Devel.* 30(4,5) (1986).
7.   N. Garcia (ed.) Proc. STM '86, Santiago de Compostela, Spain, *Surf. Sci.* 181, Special Issue (1986).
8.   R. Feenstra (ed.) Proc. STM '87, *J. Vac. Sci. Tochnol.* A6(2) (1988).
9.   P. Muralt and D.W. Pohl, *Appl. Phys. Lett.* 48:514 (1986).
10. P. Muralt, H. Meier, D.W. Pohl and H.W.M. Salemink, *Appl. Phys. Lett.* 50:1352 (1987).
11. J.R. Kirtley, S. Washburn and M.J. Brady, *Phys. Rev. Lett.* 60:1546 (1988).
12. N. Ashcroft and N. Mermin, "Solid State Physics," Holt, Rinehard and Winston, New York (1976) p.596; G. Duggan *in:* "Heterojunction Band Discontinuities, Physics and Device Applications," F. Capasso and G. Margaritondo, ed., Elsevier Science Publishers, Amsterdam (1987) p.207.
13. J.A. Stroscio, R.M. Feenstra and A.P. Fein, *Phys. Rev.* B36:7718 (1987).
14. H.W.M. Salemink, H.P. Meier, R. Ellialtioglu, J.W. Gerritsen and P.R. Muralt, *Appl. Phys. Lett.* to appear March 1989.
15. M. Heiblum, M.I. Nathan and M. Eizenberg, *Appl. Phys. Lett.* 47:503 (1985).

SEM STUDIES OF INDIVIDUAL DEFECTS IN SEMICONDUCTORS

D.B. Holt

Department of Materials
Imperial College of Science
Department of Technology and Medicine
London SW7 2BP

INTRODUCTION

Studies of the physical properties of individual defects in semiconductors are essential to resolve important current questions in this field. These questions outlined here are the role of impurity dislocation interactions, the glide or shuffle set character of dislocation core structures and the influence of the polar character of defects in semiconducting compounds. The principles underlying the use of scanning electron microscopy methods such as EBIC (electron beam induced current) and CL (cathodoluminescence) and the analogous scanning light (laser) microscopy methods for such studies, OBIC (optical BIC), IRBIC (infrared BIC) and SRPL (spatially resolved photoluminescence) are outlined. The significance of the many types of resolution available is emphasised as is the fact that spatial and signal resolution are reciprocally related. The application of the phenomenological (Donolato) theory of dark EBIC (and CL) contrast makes it possible to determine the strengths (recombination velocities) of defects. The physical theory of dislocation recombination (Wilshaw) derives this strength in terms of more basic parameters. It is emphasised that some defects give rise to bright CL contrast or bright-dark EBIC contrast which can provide information on other properties.

The electrical effects of dislocations in semiconductors were first studied by measuring the resistivity and Hall effect in plastically deformed Ge, Si and InSb (Alexander and Haasen 1968, Labusch and Schroter 1978). The results proved difficult to interpret unambiguously for two main reasons. (i) Plastic deformation introduces enormous numbers of dislocations of many types: edge, screw and 60°; constricted and dissociated; dipoles and loops together with much point defect "debris" and it is their combined effects that were measured. (ii) Deformation at the relatively high temperatures that were then believed to be essential, allows impurity diffusion with the likelihood of decoration of the dislocation lines and drastic alteration of their core structures, electronic states and physical properties.

Techniques have been introduced to avoid these complications. These are (i) lower temperature deformation techniques, (ii) physical measurements such as ESR that are more directly related to defect electronic states and most

recently, (iii) microscopy methods to analyse individual defects. The first
two are dealt with elsewhere in this volume by H. Alexander and this paper
introduces the microscopy of individual defects. Weak-beam transmission
electron microscopy (TEM) and atomic resolution TEM can resolve the short-
range defect structure (e.g. unit or closely dissociated dislocations, clean
or precipitate decorated defect character) and the atomic core structures of
defects respectively. These techniques are dealt with in the contributions
of J. Heydenreich, A. Ourmazd and J. Thibault Dessaux. Scanning beam methods
have emerged that enable the electronic energy level or band properties of
individual defects to be measured.

This paper is intended to provide an introduction for those of C. Donolato
and P. Wilshaw, concerned with dislocation EBIC (electron beam induced
current) contrast theory. This is the basis for interpreting the results of
the most developed of the scanning beam methods for the measurement of the
electronic properties of individual defects. An attempt will also be made to
set in perspective the OBIC (optical beam induced current) technique (to be
dealt with by P. Gondi and IRBIC (infrared modified optical beam induced
current) method dealt with by A. Cavallini. Some results of recent SEM
studies of defects will also be briefly reviewed.

Before introducing these scanning beam techniques, it is desirable to
outline the type of question about defect core structures and properties that
such methods are needed to resolve. These include the role of the intrinsic
defect core and of impurity decoration of microstructural defects, whether
dislocation cores have glide or shuffle set atomic structures and the defect
polarity in semiconducting compounds.

Intrinsic Defect Core Effects and Impurity Decoration

Dislocation electronic properties are often dominated by impurity
decoration, especially after device processing. It is generally found that
below about 10^3 to 10^4 dislocations cm^{-2} in wafers of Si or GaAs, no gains in
yield or performance are obtained by further reducing dislocation densities.
Moreover large numbers of dislocations may be introduced in processing without
necessarily rendering device performance unacceptable. This is apparently
due to defect passivation by impurity atoms, analogous to the hydrogen
passivation used to render the grain boundaries in polycrystalline Si solar
cells innocuous. Introduction of hydrogen to saturate dangling bonds is also
successfully used to render non-crystalline ("amorphous") Si capable of being
doped n- and p-type. It is found that dislocations have strong electrical
effects when they are either relatively clean, as in fundamental dislocation
studies or are decorated by "heavy metal" (e.g. Cu or Fe) precipitates so they
act as short circuits or microplasma breakdown sites. It is, therefore, the
metal precipitates that generally cause yield problems and technological
processes of gettering were long ago developed to avoid their occurrence in
device processing. In the intermediate situation, when a relatively few
impurity atoms have diffused to the dislocation line, bond saturation and
electrical passivation apparently occurs. The literature on these points was
reviewed elsewhere (Holt, 1979). It remains to determine the mechanisms and
kinetics of the processes of impurity modification of defect properties and to
establish intrinsic defect properties. This requires property measurement
techniques capable of resolving individual defects of well-defined core
structures i.e. microcharacterization techniques.

Glide or Shuffle Set Cores?

Recent plasticity and TEM evidence suggests that dislocation core
structures are of glide set types but older evidence on defect polarity makes
better sense on the shuffle set interpretation. The case for the glide set
core hypothesis has been persuasively presented by Hirsch (1980, 1985) and

need not be repeated here. It depends essentially on the weak beam TEM observation that dislocations in e.g. Si and GaAs move in dissociated form, the belief, based on core structure models, that shuffle-set partial dislocations would be less mobile than those with glide set core structures and some suggestive atomic resolution TEM pictures. Blanc (1975) pointed out that the one form could transform to the other by absorbing or emitting native point defects and thermodynamically, a mixture of the two is to be expected.

There is some old evidence suggesting that dislocations have shuffle set cores which depends on observations of differences in chemical and electronic behaviour between dislocations introduced by polar bending of III-V compounds. The electronic (chemical) character of III and V atom dangling bonds is established by evidence concerning {111} surfaces. Then polar bending experiments, which were interpreted as introducing an excess of either α or β dislocations, are found to produce predictable effects on the original shuffle set interpretation but the glide set interpretation reverses the sign of the dislocations, making their properties difficult to understand.

Opposite (111)A and ($\bar{1}\bar{1}\bar{1}$)B faces of {111} slices of semiconducting compounds with the sphalerite structure have gross, macroscopic physical and chemical differences (for a recent review see e.g. Holt, 1988). In oxidizing reagents, for example, {111} slices are more rapidly attacked on the ($\bar{1}\bar{1}\bar{1}$)B faces which are therefore polished, while the more slowly dissolved (111)A faces develop pits. There is direct, Rutherford scattering evidence that the B faces consist of triply bonded higher valence atoms (of the VI element in II-VI compounds) and of triply bonded low valence (II) atoms in the case of the A faces i.e. the polar character of the two types of face is proven. The III atom (111)A faces of III-V compounds, which can be thought of as having approximately empty dangling bonds, thus are known to be slowly attacked by oxidizing reagents whereas the B (V atom) faces, with relatively full dangling bonds are etched rapidly.

Positive or negative edge dislocations were believed to be selectively introduced by plastic bending of opposite sign as shown in Figure 1.

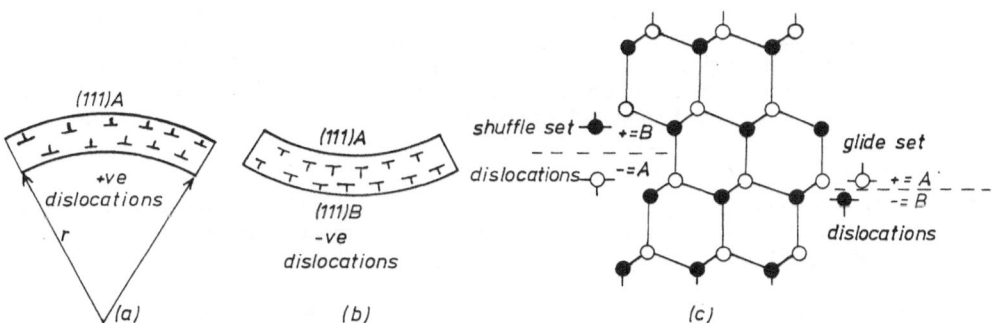

Fig. 1. Bending crystals in opposite senses introduces predominantly (a) positive and (b) negative edge dislocations. In polar crystals such as semiconducting compounds with sphalerite structure these dislocations are of opposite polarity but (c) the positive dislocations, for example, are β-type (B atom edged) if of the shuffle set but of α type (A atom edged) if of the glide set core structure.

The two signs of dislocation differ in polarity i.e. one sign of polar bending introduces α dislocations which have A atoms along the edge of the extra half plane and so have relatively empty dangling bonds in the core while bending in the opposite direction introduces β dislocations that are B atom edged and so have relatively full dangling bonds. The positive and negative edge dislocations have their chemical characters reversed on moving from models of the shuffle set to those of the glide set (Figure 1(c)). Thus by polar bending, as shown in Figure 2, a majority of either A atom or B atom dangling

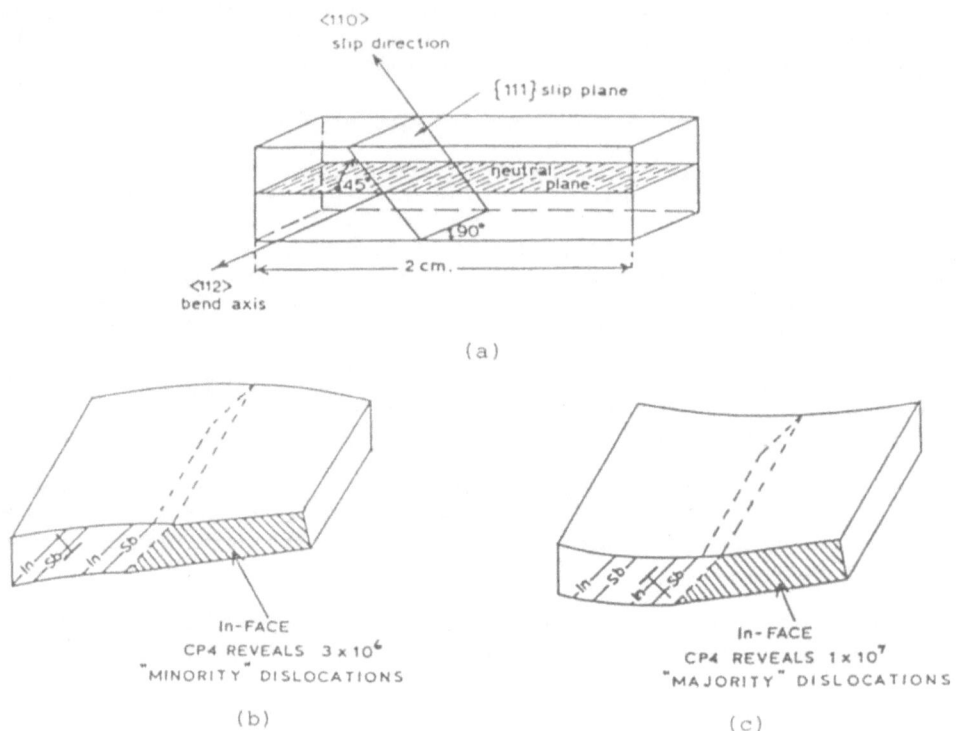

Fig. 2. The crystallography of single slip bending of InSb crystals. (The view here is rotated through 180° relative to those (b) and (c). A specimen bent upwards in the middle to introduce Sb-edge dislocations assuming the shuffle set form and (c) the opposite bend to introduce a majority of In-edged dislocations on the same assumption. An In (111) face was cut for etch pit measurements in each case and the In-dislocation densities for 6 cm. radius bends of opposite signs are given in (b) and (c). (After Bell et al, 1966).

bond dislocations were thought to be introduced into InSb crystals with an orientation near {111}, chosen to produce deformation on only one slip system, by Bell et al (1966) and Bell and Willoughby (1970).

   The net density of dislocations required to accommodate plastic bending to a radius r is given by the difference of the majority and minority densities:

$$\rho_{Sb} - \rho_{In} = \frac{1}{r\ b\ \cos 45°} \tag{1}$$

for this geometry and one particular sign of bending. Bending the opposite way introduces an equal net density of dislocations of the opposite chemical character. The absolute polarity of the crystals was established, before bending, by etching. (111) In faces were cut so dislocation etch pitting could be used to check the numbers and polar type of the dislocations introduced using selective etchants that had been carefully "calibrated" by Bell and Willoughby (1966). They obtained results in agreement with equation (1) if, as expected In-edged dislocations are revealed by the etchant used, on the assumption that the dislocations were of the type to be expected on the shuffle set assignment.

Confirmatory evidence was that the dislocations identified as In-edged on the basis of the shuffle set assumption introduced acceptor centres into both p- and n-type InSb. Bending to produce a majority of Sb-edged dislocations however introduced acceptor centres into n-type InSb but in p-type InSb their introduction led to no increase in the acceptor concentration and in one case led to a small decrease i.e. to an introduction of donor centres. This would be expected on the basis of largely empty dangling bond states along In-edged dislocations which can only accept but relatively full dangling bonds along Sb-edged dislocations which can accept or donate depending on the position of the Fermi level. If a glide set identification were made, the polarity of all the dislocations would be reversed and the chemical character of the dangling bonds would not correspond to the observed etching or electrical properties. This work is reported here both because of its interest and because it was carefully done. Later work has not confirmed its conclusions, however.

Only a little work on polar bent specimens has been reported subsequently. It has raised doubts about the introduction of dislocations of opposite sign by polar bending and has not found differences in properties between the dislocations introduced (for a review see Farvacque et al 1983). As in all deformation experiments, large numbers of dislocations of many types are introduced by polar bending and their overall effects have been observed. It is obviously desirable to repeat this type of study using techniques for microcharacterizing individual dislocations. There is some evidence suggesting polar property differences between orthogonal grids of misfit dislocations (see Figure 3 below). The most strongly polar of defects are antiphase boundaries (APBs) which occur commonly, for example, in GaAs grown on (100) Si.

Defect Polarity

Antiphase or polarity reversal domains are volumes in a semiconducting compound between which the occupation of the two sublattices is reversed. That is, if the sphalerite structure has A atoms at the points of an f.c.c. lattice and B atoms at 1/4, 1/4, 1/4 then on the other side of an antiphase boundary the B atoms occupy the lattice sites and the A atoms occupy the 1/4, 1/4, 1/4 type sites. All the cross boundary bonds in APBs are necessarily wrong bonds. Both A APBs in which the bonds are all A-A ones and B APBs containing all B-B bonds are geometrically possible (see e.g. Holt, 1969). It is found that, although APBs are relatively high energy defects, they can form profusely in epitaxial films of polar sphalerite structure compounds grown on non-polar diamond structure elements in (110) and (100) orientations (Morizane 1977). Recent studies of APBs in GaAs/Ge(100) (Gowers 1984 and Neave et al 1983) and in GaAs/Si (100) (Kroemer 1983) suggested a growth mechanism for nucleating antiphase domains (APDs) and a method for avoiding them. Most recently SEM EBIC was used and showed dark contrast and therefore enhanced recombination at APBs in GaAs/Si. TEM showed that APBs moved from one crystallographic plane to another as they propagated up through the epitaxial film and gave evidence of Si segregation to the boundaries (Chu et al 1988). Polarity reversal can also occur at grain boundaries. For example, both upright and inverted (polarity reversing) twins are crystallographically possible (Holt 1984). The study of such strongly polar defects presents the possibility of observing the effects of large numbers of wrong bonds and their modification by impurity decoration.

This again underlines the importance of microcharacterization techniques i.e. those with the spatial resolution to observe individual defects and the capability of determining their electronic and optical properties. Only such information can be related to the results of TEM analyses giving the core structure of the defects down to the atomic level. The electronic microcharacterization techniques rely on the use of small probes of electrons or light that are scanned over the surface, exciting small sub-surface volumes to produce signals that are detected and processed to obtain information on the response of the material and its alteration in the presence of a defect.

Principles of Scanning Beam Microcharacterization Techniques

Scanning electron microscopes (SEMs) and scanning laser (or light) microscopes (SLMs) operate on the same principles but differ in the probe used. We will begin by considering the longer established SEM which consists of two sub-systems. The first is an electron optical column to produce a finely focussed electron beam and scan it over the specimen surface. The other is a display system which detects one of six forms of energy induced by beam bombardment. These are employed as signals to form the basis for the six modes of the SEM. The display system processes this data and displays or reads it out in some form.

Continuous bombardment can generate five types of useful signal. (1) X-rays are the basis for electron probe microanalysis (EPMA), (2) emitted electrons are the basis of the emissive mode exhibiting topographic and voltage contrast, (3) beam induced potential differences can be detected as electron beam induced current (EBIC) or voltage (EBIV) signals, (4) light in and near the visible is known as cathodoluminescence (CL) and, if the specimen is thin, (4) electrons are transmitted and can be used in the STEM (scanning transmission electron microscopy) mode. If the beam is chopped at high frequency, the intermittent bombardment generates (6) ultrasonic waves which are the signals for SEAM (scanning electroacoustic microscopy). For a review of this new mode see Balk (1988).

It is well-known in the X-ray case that microcomputers can rapidly provide quantitative data. In EPMA this gives the compositions, calculated from X-ray line intensities, of volumes of the order of a μm cubed. In each mode two types of output can be obtained. The first are point analyses in which the beam is stationary and some form of signal spectrum is recorded e.g. CL spectra or voltage waveforms in a circuit in the case of stroboscopic voltage contrast. The second is micrographic displays produced by selecting a particular signal level or wavelength, etc. and scanning a raster on the specimen. It is important to remember that the performance of an SEM mode is specified by both spatial resolution and signal (spectral) resolution values.

For the determination of the properties of defects in semiconductors, the most important modes are EBIC and CL. Voltage contrast and SEAM can also provide defect data but the first is only applicable to device operation and the second is new and as yet little used.

Scanning laser microscopes differ from the SEMs essentially only in the probe scanning sub-system. In SLMs either the specimen is scanned by a motorized stage under a stationary beam of light or the beam is scanned over the specimen surface by galvanometers or rotating mirrors. A discussion of the alternatives and their advantages and disadvantages is given by Wilke (1985). Some form of induced signal is detected, amplified and displayed or otherwise processed and recorded. The analogues of EBIC and CL are known as OBIC and SRPL. OBIC (optical beam induced current) or LBIC (laser BIC) is covered elsewhere in this volume as is the powerful variant known as IRBIC (infra-red

BIC). SRPL (spatially resolved photoluminecence) has been little used but Yamaguchi et al (1981), for example, observed misfit dislocation networks in InGaAsP/InP (100) this way. They found that of the two orthogonal grids of ⟨110⟩ dislocations, one gave much stronger contrast (Figure 3). This is one of the few observations of a polarity-related difference in dislocation electronic properties.

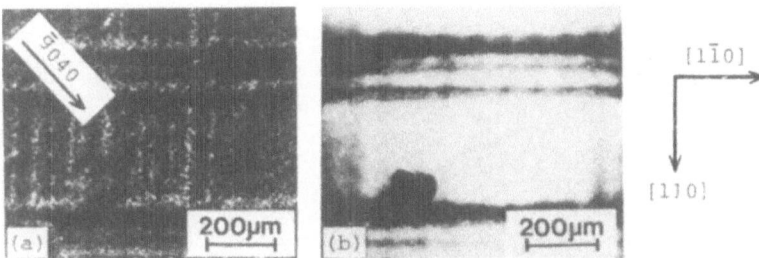

Fig. 3. Asymmetry in < 100 > misfit dislocations observed by SRPL. (a) X-ray topograph showing the dislocations in both grids. (b) SRPL image of the same region using a Nd:YAG laser to excite the InGaAsP. (After Yamaguchi et al 1981).

Obvious disadvantages of the SEM techniques are possible beam charging and damage and the necessity for a vacuum in the specimen chamber. SLMs do not produce surface charging or other electrical damage. It seems likely that OBIC will prove to be a truly non-destructive inspection technique for VLSI. A time-dependent theory of OBIC imaging of defects has recently been published (Pester and Wilson 1988).

Beam Energy Dissipation and the Calculation of Signal Strengths

To extract information from observed contrast, that is the variations of signal strength, it is necessary to have an expression for the excitation i.e. for the spatial distribution of the dissipated beam energy. The dissipated energy distribution can be represented by semi-empirical functions such as the polynomial depth dose function due to Everhart and Hoff (1972). The uniform dissipation sphere is often used for simplicity, to represent the beam energy dissipation volume. An important alternative is the method of Monte Carlo electron trajectory simulation. These programs are now available to run on desk top computers and can compute the results for a few thousand electrons in a few tens of minutes (Joy 1988, Napchan 1988). The energy deposited in the material is computed for each step of the trajectory and the results are stored as a matrix from which the programs will provide a variety of depth dose and lateral dose distributions, the number and total energy of the backscattered electrons, etc. The graphic display gives both the simulated trajectories and the depth dose function as shown in Figure 4. The Napchan program is adapted to run for multilayer materials such as epitaxial multilayers with top metal contacts.

Monte Carlo simulations can be used to provide the excitation energy distributions for use in computer calculations of contrast profiles for curve fitting to extract both materials, device and defect parameters. This has been done for EBIC only, so far (Joy 1986, Hungerford and Holt 1987, 1988, Napchan and Holt 1987), but the extension of the method to calculate CL line scan profiles should be straightforward. The convenience and speed of such

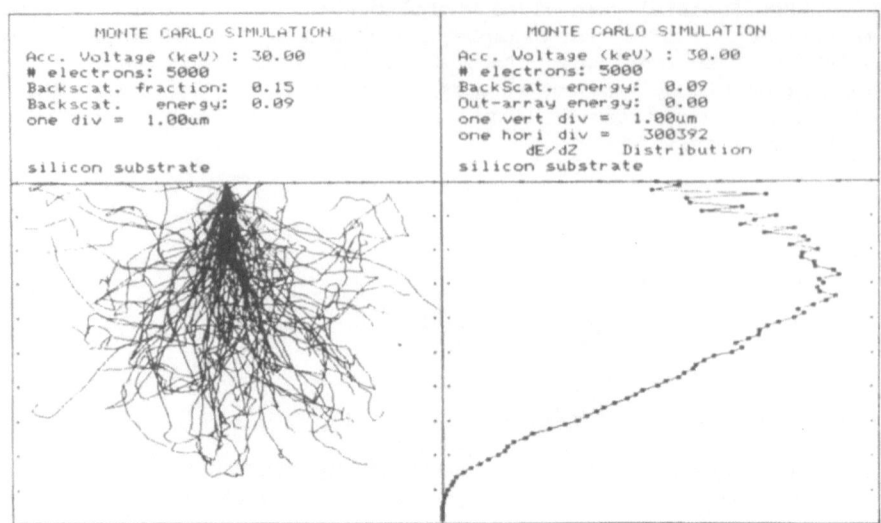

Fig. 4.   A plot of the trajectories of 5000 30 kV electrons incident on
silicon (at the left) and the energy deposited per unit depth
i.e. the depth dose distribution (at the right) resulting from a
microcomputer Monte Carlo simulation.   Computing the trajectories
of larger numbers of electrons obscures the pattern on the left
but gives a smoother depth dose function.   (After Napchan,
private communication).

simulations is such that curve fitting becomes a practical method for
measuring beam induced changes in surface recombination velocity in Si for the
first time, for example (Hungerford and Holt 1988).   It is to be expected that
as these programs are freely available from the authors they will be widely
adopted by experimentalists.   It is to be hoped that the Lambert's Law
determined depth dose for SLM excitation and the subsequent calculation of
OBIC profiles will similarly be microcomputerized.

Spatial and Signal Resolution

     To interpret scanning-beam-measured defect properties in terms of the
core structure, it is necessary to follow scanning beam examination by TEM-
type analysis.   This can be done in two ways.   (1) The bulk specimen may be
examined in an SEM or SLM, then thinned and examined in transmission or (2) the
specimen may be thinned first and scanning and transmission observations may
both be carried out in the same instrument of STEM or TEMSCAN type.   Each route
has advantages and drawbacks.   The method of thinning first ensures that the
same defect can be examined by both types of technique but it reduces the volume
and level of excitation, since most of the beam power will be transmitted.   In
general signal and spatial resolution are reciprocally related.

     To improve lateral spatial resolution the excited volume must be reduced.
This volume is determined in a bulk specimen by the electron penetration
(Gruen) range.   This is the approximate diameter of the volume filled by
electron tracks in plots like Figure 4 (left).   It depends on the beam voltage
and so can be reduced by turning this down in the case of bulk specimens.   The
excited volume is also reduced in thin specimens as the beam has a limited path

length along which it is scattered. In both cases the lateral spatial resolution improves at the expense of a fall in the total excitation energy and consequently in the intensity of both EBIC and CL signals.

The SEM provides lateral and depth spatial resolution. Depth resolution is obtained by varying the beam voltage and consequently the penetration range to excite or not excite a defect or an epitaxial layer at a particular depth. There are also specific forms of signal (spectral) resolution for each mode. Bimberg et al (1988) were able to image columnar areas of thickness uniform to one atomic step height in quantum wells by selecting thickness-dependent exciton lines. In addition, a stationary beam can be chopped so the decay of the EBIC or CL signal can be observed. From such "time resolved" observations minority carrier lifetimes or recombination times can be determined. The minimum time interval measurable is the time resolution. Thus the performance of an SEM is not specified by one resolution value.

The Phenomenological Theory of EBIC Defect Contrast

Many defects appear in dark contrast in EBIC micrographs and a successful treatment of this contrast was developed by Donolato and is covered later in this volume. Here it is only necessary to introduce the main ideas involved to prepare for a discussion of other aspects of the study of defects.

The Donolato theory models any defect as a volume of appropriate shape, F, in which the minority carrier lifetime is reduced from τ to τ' (Figure 5).

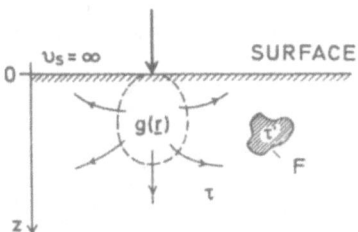

Fig. 5. The Donolato model for EBIC dark contrast. The incident beam produces a hole-electron pair density g at each point r througout a certain volume. Some of these carriers diffuse to the Schottky barrier at the top surface and are collected to produce the EBIC current. The defect is represented as a volume F in which the minority carrier lifetime is reduced from the bulk value τ to τ'. (After Donolato 1985).

Thus enhanced recombination (only) is assumed to characterize defects so reducing the EBIC signal locally. The theory computes the form of this reduction i.e. the EBIC line scan profile recorded across the defect and gives the contrast as the product of the EBIC strength or recombination velocity of the defect times a correction factor. The latter is a function of geometrical and beam parameters. By evaluating the correction factor for a particular case, the defect strength can be determined.

The Results of EBIC Dislocation Contrast Studies

Dislocation EBIC recombination velocities were first determined by Kittler and Seifert (1981). Pasemann (1981a, 1981b) developed a higher order version of the phenomenological theory and Pasemann et al (1982) applied it to experimental observations of dislocation EBIC contrast. This showed that the strengths of all the 14 dislocations measured lay within a few percent of one of the values; 0.29 for 60° dislocations (assumed to be constricted on the basis of the work of Ourmazd et al 1981), 0.68 for 60° dislocations assumed to be dissociated and 0.02 for one screw dislocation. These results suggest that the dislocation strength is at least not totally dominated by impurity effects.

Recently Donolato (1985) extended his treatment to enable the surface recombination velocities of grain boundaries to be determined. Measurements on polycrystalline Si solar cells showed that passivation reduced the recombination strength so that in some cases the EBIC dark contrast became unobservable. Observations in our laboratory suggest that the recombination velocities of coherent boundaries are lower than those of incoherent ones (B. Reza, private communication).

Bright-dark EBIC Grain Boundary Contrast

The phenomenological theory was proposed to model dark contrast by assuming that defects provide faster recombination. However, this is not the only form of defect EBIC contrast. Bright, dark and bright-dark contrast has been observed at grain boundaries in ZnS (Russell et al 1980), ZnSe (Russell et al 1981) and GaP (Ziegler et al 1985). The observation of bright contrast indicates the presence of a local charge collection barrier i.e. of a region with a built-in electrical field that separates hole-electron pairs.

Grain boundaries have long been modelled as charged surfaces (e.g. negative due to trapped electrons) with shielding space charge layers (Figure 6)

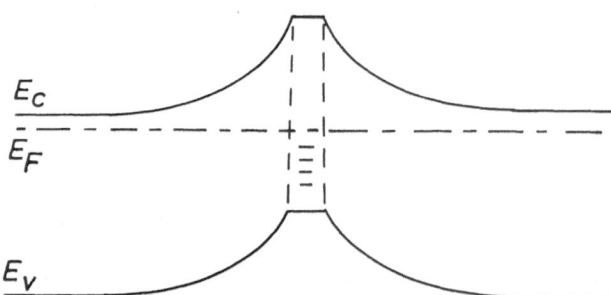

Fig. 6. Energy band diagram of a grain boundary with a negative surface charge.

resulting in band bending and opposite signs of field on either side. The oppositely directed charge separating fields on either side mean the EBIC current will reverse direction as the beam crosses the boundary. Consequently the EBIC line scan profile will consist of peaks of negative and positive EBIC current which produces bright-dark contrast in EBIC micrographs. Just such contrast was observed by Ziegler et al (Figure 7) who modelled the boundaries as pairs of Schottky barriers of different heights back-to-back. The reason that dark-bright constrast is not seen in EBIC

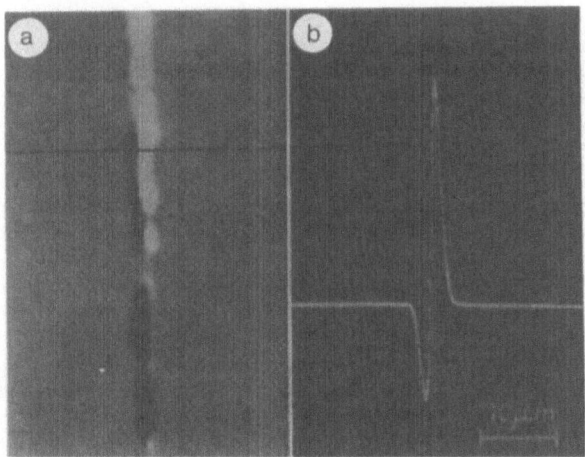

Fig. 7. A grain boundary in GaP observed in (a) an EBIC micrograph
showing the contrast to vary from bright at the top, through
dark-bright to dark at the bottom and (b) an EBIC line scan
profile across a position showing dark-bright contrast. (After
Ziegler et al 1982)

micrographs of polycrystalline Si is not known. OBIC line scans showing peak
and dip profiles were reported for grain boundaries in Ge by Figielski (1960).

A Physical Theory of Dislocation EBIC Contrast

Read (1954, 1955a, 1955b) modelled dislocations in semiconductors as a
line of acceptor states so the dislocation developed a negative line charge
with a shielding cylindrical positive space charge i.e. the band diagram of
Figure 8.

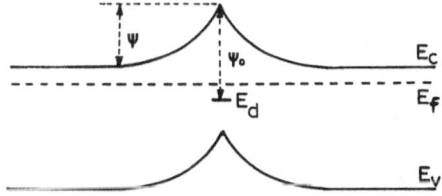

Fig. 8. Energy band diagrams of a dislocation with a single deep
acceptor level (after Wilshaw and Booker 1987).

Wilshaw and Booker (1987, 1988) applied this model to treat dislocation
recombination and EBIC dark contrast and this will be discussed later by
Wilshaw. Again the band bending suggests that bright-dark contrast should be
observable at dislocations by EBIC. Matare and Laakso (1969) reported

bright-dark dislocation contrast in what were apparently β-conductivity micrographs (observations without a charge collecting barrier but with an external bias applied between the contacts) but no recent reports of such behaviour are known to the author.

CL Studies of Dislocations

In panchromatic and near bandgap monochromatic CL micrographs defects generally exhibit dark contrast. That is, dislocations locally reduce the emission of radiation of photons of energies near the bandgap value. This suggests they locally reduce the lifetime as in the phenomenological theory. Direct, time-resolved-CL measurements of the lifetime at and away from dislocations in GaP by Rasul and Davidson (1977) showed that it was locally reduced just to the extent required to account for the observed dark contrast. The phenomenological theory was successfully extended to treat this contrast by Lohnert and Kubalek (1984), Jakubowicz (1986) and Pasemann and Hergert (1986, 1987). The essential difference is that the CL is self absorbed in getting out while there is no equivalent effect for the EBIC signal. With increasing depth, therefore, the CL contrast from dislocations is reduced while the EBIC contrast is not. The ratio of the EBIC and CL contrast can be shown to be a simple function of the dislocation depth and the experimental data is in agreement with the theory (Jakubowicz et al, 1987). It is useful to have a mathematical model for CL dark contrast but it is not likely to lead to any new information about the electronic properties of defects, unlike bright CL contrast.

Bright CL Defect Contrast

While dislocations generally act as non-radiative recombination centres for near bandgap radiation, there is no reason why they cannot produce radiative recombinaton across part of the gap. If that happens the defect will appear in bright contrast in monochromatic CL micrographs made by recording the intensity emitted at that wavelength. The importance of such bright contrast is that the CL spectrum can be studied to obtain information on the energy levels and radiative recombination mechanisms involved. Use of TEM Cl or STEM CL i.e. CL observations on thinned specimens with good spatial resolution (but low CL intensities) makes it possible to determine the variation of CL emission spectra with position along and away from e.g. a dislocation line.

The first reports of bright dislocation CL contrast concerned natural diamond. It was found that different dislocations emitted two different wavelength bands, one in the blue and the other in the yellow-green range (Hanley et al 1977). This is ascribed to the segregation of two different types of luminescence centres to the dislocations in the two groups. It is also found that the blue dislocation CL is linearly polarized (Kiflawi and Lang 1974, Yamamoto et al 1984).

It was shown by Myhajlenko et al (1984) using a spectroscopic CL system on a transmission electron microscope (TEM CL), that the emission from the dislocations could be distinguished from that due to decorating particles in ZnSe:Al as shown in Figure 9. Spectrum a is characteristic of the bulk and shows only the near bandgap emission. Spectrum b is characteristic of the dislocations and exhibits a new emission marked Y with an LO (longitudinal optical) phonon replica. Spectrum c arose from a barely discernable particle in the dislocation tangle. It contains an additional DAP (donor acceptor pair) emission band thought to involve the Al dopant in the crystal (J.W. Steeds, private communicaton). TEM CL also provided evidence of a concentration gradient in the Cottrell atmosphere around dislocation cores in GaAs which results in a "red shift" in the emission as the line is approached

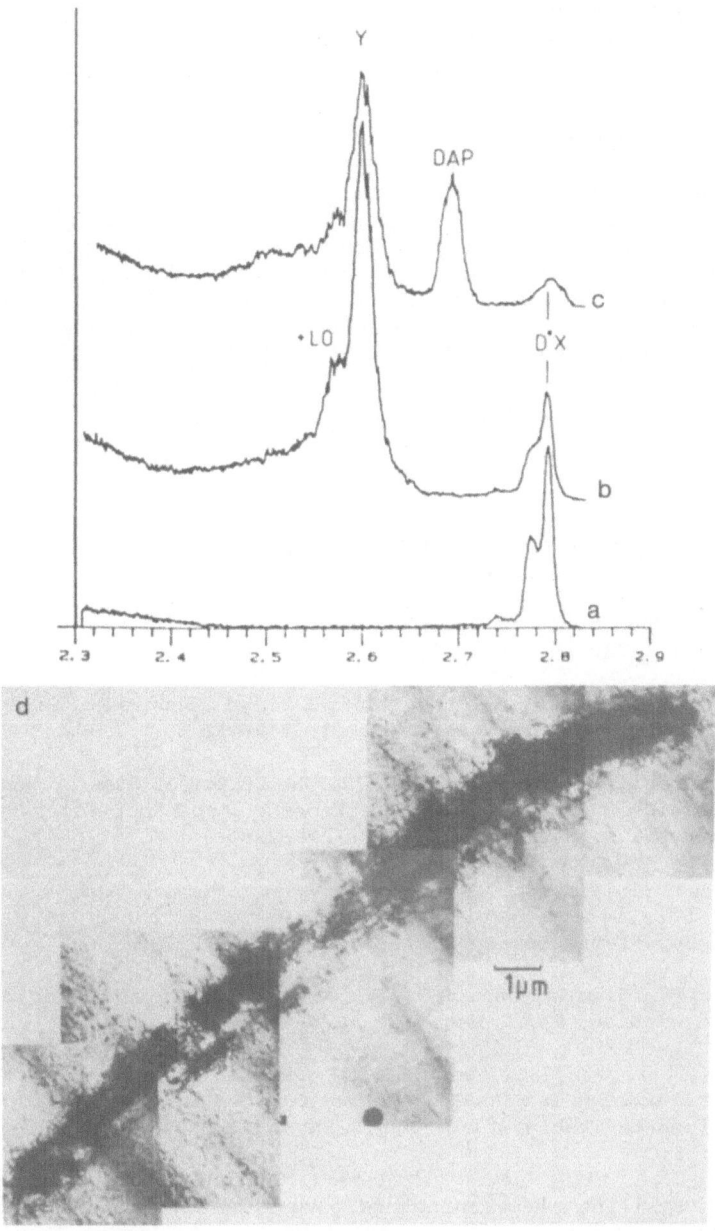

Fig. 9. TEM CL spectra recorded (a) in bulk ZnSe:Al, (b) at the dislocation
tangle shown in the TEM micrograph in (d) and (c) the spectrum
recorded at a small particle in the tangle (d). After Myhajlenko
et al (1984)

(J.W. Steeds, private communication). The general conclusion of Steeds (1988) review of dislocation bright TEM CL contrast studies is that the effects observed are due to impurity segregation to the dislocations. The effects were strongly dependent on the growth or heat treatment of the specimens but not on the dislocation character.

CONCLUSIONS

The work quoted above shows that spatially resolved physical measurement techniques such as EBIC and CL and their analogues OBIC and SRPL (and SEAM) can yield unique information on defect core properties. Most of the published EBIC and CL defect results indicate that the observed properties were impurity determined but some exhibit variations with dislocation type so this is not conclusive. Little recent work has been reported on polarity dependent properties e.g. of α versus β dislocations. In addition to its inherent interest, this could help determine whether dislocations have predominantly glide or shuffle set cores. It is clear that this field is likely to remain lively and productive.

REFERENCES

Alexander, H. and Haasen, P., 1968, Dislocations and Plastic Flow in the Diamond Structure, Sol. State Phys. 22:27-158

Balk, L.J., 1988, Scanning Electron Acoustic Microscopy, Adv. Eletron. El Phys. 71: 1-73

Bell, R.L. and Willoughby, A.F.W., 1966, Etch-pit Studies of Dislocations in Indium Antimonide, J. Mater, Sci. 1: 219-228

Bell, R.L. and Willoughby, A.F.W., 1970, The Effect of Plastic Bending on the Electrical Properties of Indium Antimonide Part 2 Four-Point Bending of n-type Materials, J. Mater. Sci. 1: 219-228

Bimberg, D., Christen, J., Fukunaga, T., Nakashima, H., Mars, D.E. and Miller J.N., 1988, Direct Imaging of the Columnar Structure of GaAs Quantum Wells, Superlattices and Microstructures 4: 257-263

Blanc, J., 1975, Thermodynamics of "Glide" and "Shuffle" Dislocations in the Diamond Lattice, Phil. Mag. 32, 1023-1032

Chu, S.N.G., Nakahara, S., Pearton, S.J., Boone, T. and Vernon, S.M., 1988, Antiphase Domains in GaAs Grown by Metalorganic Chemical Vapor Deposition on Silicon-on-Insulator, J. Appl. Phys. 64:2981-2989

Donolato, C., 1985, Beam Induced Current Characterization, in "Polycrystalline Semiconductors. Physical Properties and Applications"., G. Harbeke ed., Springer-Verlag, Berlin pp. 138-154

Farvacque, J.L., Ferre, D.and Vignaud, D., 1983, Experimental Studies of Energy States Associated with Dislocations in III-V Compounds, J. de Phys. Colloque C4 pp C4-115 to C4-123

Figielski, T., 1960, Electronic Processes at Intercrystalline Barriers in Germanium, Acta Phys. Polonica XIX: 607-630

Gowers, J.P., 1984, TEM Image Contrast from Antiphase Domains in GaAs:Ge (001) Grown by MBE, Appl. Phys. A34:231-236

Hanley, P.L. Kiflawi, I and Lang, A.R., 1977, On Topographically Identifiable Sources of Cathodoluminescence in Natural Diamonds, Phil Trans. R. Soc. Lond. 284: 329-368

Hirsch, P.B., 1980, Electronic and Mechanical Properties of Dislocations in Semiconductors, Proc. Mat. Res. Sci. Ann. Meeting, 1980, Boston, North-Holland, New York, 257-271

Hirsch, P.B., 1985, Dislocations in Semiconductors, Mat. Sci. Technol. 1:666-677

Holt, D.B., 1969, Antiphase Boudnaries in Semiconducting Compounds, J. Phys. Chem. Solids 30: 1297-1308

Holt, D.B. 1979, Device Effects of Dislocations, J. de Phys. Colloque C6 pp. C6-189 to C6-199

Holt, D.B., 1984, Polarity Reversal and Symmetry in Semiconducting Compounds with the Sphalerite and Wurtzite Structures, J. Mater. Sci. 19: 439-446

Holt, D.B., 1988, Surface Polarity and Symmetry in Semiconducting Compounds Part 1. Macroscopic Effects of Polarity, J. Mater. Sci. 23: 1131-1136

Hungerford, G.A. and Holt, D.B., 1987, Electron Dose Induced Variations in EBIC Line Scan Profiles Across Silicon p-n Junctions, in "Microscopy of Semiconducting Materials 1987. Conf. Series No. 87", A.G. Cullis and P. Augustus, eds., Inst. Phys., Bristol ppl 721-726

Hungerford, G.A. and Holt, D.B., 1988, Theoretical curve Fitting to EBIC Junction Profiles, in "EUREM 88. Conf. Series No. 93" P.J. Goodhew and H.G. Dickinson, eds., Inst. Phys., Bristol pp69-70

Jakubowicz, A., 1986, Theory of Cathodoluminescence Contrast from Localized Defects in Semiconductors, J. Appl. Phys., 59: 2205-2209

Jakubowicz, A., Bode, M. and Habermeier, H-U, 1987, Simultaneous EBIC/CL Investigations of Dislocations in GaAs, in "Microscopy of Semiconducting Materials 1987" Conf. Series No 87, A.G. Cullis and P.D. Augustus, eds. Inst. Phys., Bristol pp763-768

Joy, D.C., 1986, The Interpretation of EBIC Images Using Monte Carlo Simulations, J. Microscopy 143: 233-248

Joy, D.C., 1988, An Introduction to Monte Carlo Simulations, in "EUREM 88. Conf. Series No. 93". P.J. Goodhew and H.G. Dickinson, eds., Inst. Phys., Bristol pp23-32

Kiflawi, I. and Lang, A.R., 1974, Linearly Polarized Luminescence from Linear Defects in Natural and Synthetic Diamond, Phi. Mag. 30: 219-223

Kittler, M., and Seifert, W., 1981, On the Sensitivity of the EBIC Technique as Applied to Defect Investigations in Silicon, Phys. Stat. Sol (A), 66: 573-583

Kroemer, H., 1983, Heterostructure Devices: A Device Physicist Looks at Interfaces, Surf. Sci. 132:543-576

Labusch, R. and Schröter, W., 1978, Electrical Properties of Dislocations in Semiconductors, in "Dislocations in Solids Vol. 5," F.R.N. Nabarro, ed., North-Holland, Amsterdam.

Löhnert, K. and Kubalek, E., 1984, The Cathodoluminescence Contrast Formation of Localized Non-Radiative Defects in Semiconductors, Phys. Stat. Sol. a83: 307-314

Matare, H.F. and Laakso, C.W., 1969, Space-Charge Domains at Dislocation Sites, J. Appl. Phys. 40: 476-482

Morizane, K., 1977, Antiphase Domain Structures in GaP and GaAs Epitaxial Layers Grown on Si and Ge, J. Cryst. Growth 39:249-254

Myhajlenko, S., Batstone, J.L. Hutchinson, H.J. and Steeds, J.W., 1984, Luminescence Studies of Individual Dislocations in II-VI (ZnSe) and III-V (InP) Semiconductors, J. Phys. C17: 6477-

Napchan, E. and Holt, D.B., 1987, Application of Monte Carlo Simulations in the SEM Study of Heterojunctions, in "Microscopy of Semiconducting Materials 1987. Conf. Series No. 87", A.G. Cullis and P. Augustus, eds., Inst. Phys., Bristol pp. 733-738

Napchan, E., 1988, Electron and Photon-Matter Interaction: Energy Dissipation and Injection Level, J. de Phys. to be published, (Proc. BIADS 88)

Neave, J.H., Larsen, P.K., Joyce, B.A., Gowers, J.P. and Van der Veen, J.F., 1983, Some Observations on Ge:GaAs(001) and GaAs:Ge(001) Interfaces and Films, J. Vac. Sci. Technol. B1:668-675

Ourmazd, A., Weber, E., Gottschalk, G.R. and Alexander, H., 1981, The Electrical Behaviour of Individual Screw and 60° Dislocations in n-type Silicon, in "Microscopy of Semiconducting Materials 1981. Conf. Series No. 60", A.G. Cullis and D.C. Joy, eds, Inst. Phys. Bristol pp. 63-68

Pasemann, L., 1981a, A Contribution to the Theory of the EBIC Contrast of Lattice Defects in Semiconductors, Ultramicroscopy 6: 237-250

Pasemann, L., 1981b, On the EBIC Contrast of Dislocations, Cryst. Res. Tech. 16: 147-148

Pasemann, L. and Hergert, W., 1986, A Theoretical Study of the Determination of the Depth of a Dislocation by Combined Use of EBIC and CL Technique, Ultramicroscopy 19: 15-22

Pasemann, L. and Hergert, W., 1987, On EBIC and Cathodoluminescence Contrast from an Individual Dislocation, Izv. Akad. Nauk Ser. Fiz. 51: 1528-1534

Pasemann, L., Blumentritt, H. and Gleichmann, R., 1982, Interpretation of the EBIC Contrast of Dislocations in Silicon, Phys. Stat. Sol. A70:197-209

Pester, P.D. and Wilson, T., 1988, Time-dependent Theory of Optical-Beam-Induced Current Imaging of Defects in Semiconductors, J. Appl. Phys. 64: 1131-1135

Rasul, A. and Davidson, S.M., 1977, A Detailed Study of Radiative and Non-Radiative Recombination around Dislocations in GaP, in "Gallium Arsenide and Related Compounds (Edinburgh) 1976", Inst. Phys. Bristol pp.306-316

Read, W.T., 1954, Theory of Dislocations in Germanium, Phil. Mag. 45: 775-796

Read, W.T., 1955a, Statistics of the Occupation of Dislocation Acceptor Centres, Phil. Mag. 45: 1119-1128

Read, W.T., 1955b, Scattering of Electrons by Charged Dislocations in Semiconductors, Phil. Mag. 46: 111-131

Russell, G.J., Robertson, M.J. Vincent, B., and Woods, J., 1980, An Electron Beam Induced Current Study of Grain Boundaries in Zinc Selenide, J. Mater. Sci. 15: 939-944

Russell, G.J. Waite, P., and Woods, J., 1981, in "Microscopy of Semiconducting Materials 1981. Conf. Series No. 60", A.G. Cullis and D.C. Joy, eds., Inst. Phys., Bristol pp. 371-376

Steeds, J.W., 1988, Performance and Applications of a STEM-Cathodoluminescence System, J. de Phys. to be published (Proc. BIADS 88).

Wilke, V., 1985, Optical Scanning Microscopy - The Laser Scan Microscope, Scanning 7: 88-96

Wilshaw, P.R. and Booker, G.R., 1985, New Results and an Interpretation for SEM EBIC Contast arising from Individual Dislocations in Silicon, in "Microscopy of Semiconducting Materials 1985. Conf. Series No. 76", A.G. Cullis and D.B. Holt, eds., Inst. Phys. Bristol pp. 329-336

Wilshaw, P.R. and Booker, G.R. 1987, The Theory of Recombination at Dislocations in Silicon and an Interpretation of EBIC Results in Terms of Fundamental Dislocation Parameters, Izv. Acad. Nauk Ser. Fiz. 51: 1582-1586

Yamaguchi, A., Komiya, S., Ueda, O. Nakajima, K., Umebu, I. and Akita, K., 1981, Asymmetric Character of Misfit Dislocations in LPE DH InGaAsP/InP, in "Gallium Arsenide and Related Compounds 1981. Conf. Series No. 63", T. Sugano, ed., Inst. Phys., Bristol pp. 161-166

Yamamoto, N., Spence, J.C.H. and Fathy, D., 1984, Cathodoluminescence and Polarization Studies from Individual Dislocations in Diamond, Phil. Mag. 49: 609-629

Ziegler, E., Siegel, W., Blumtritt, H. and Breitenstein, O., 1982, Electrical and EBIC Investigations of GaP Grain Boundaries, Phys. Stat. Sol. a72: 593-605

QUANTITATIVE CHARACTERIZATION OF SEMICONDUCTOR DEFECTS BY

ELECTRON BEAM INDUCED CURRENT

C.Donolato

CNR - Istituto LAMEL
Via Castagnoli 1 - 40126 Bologna, Italy

INTRODUCTION

In the electron beam induced current (EBIC) mode, the finely focused electron beam of the scanning electron microscope is used to in-
ject locally excess carriers in a semiconductor. If the semiconductor
surface is provided with a rectifying contact, carrier collection will
occur and give rise to a signal that can be used to produce an image of
the specimen. EBIC imaging has been widely used to assess rapidly the
electrical activity of semiconductor defects; however, additional information can be obtained by analyzing quantitatively the defect contrast. The principles and applications of EBIC microscopy have been reviewed by Hanoka and Bell (1981), Leamy (1982), and Holt and Lesniak
(1985); review articles dealing more specifically with the related theory have been published (Jakubowicz, 1987; Donolato, 1988).

This paper illustrates current phenomenological models that aim at
describing quantitatively the EBIC contrast of semiconductor defects.
Both first-order and higher-order/exact treatments are discussed and
the notion of recombination strength of defects with different geometrical extension is introduced. Usual methods to recover the value of
this strength from contrast measurements are examined. An interpretation of the defect strength in terms of microscopical processes lies
beyond the scope of this paper; such connection is developed for the
case of dislocations in the paper by Wilshaw et al. in this volume. Recent studies dealing with generalizations or different formulations of
the EBIC theory are also discussed.

THE LINEAR MODEL OF THE EBIC CONTRAST

Charge Collection

Fig.1 shows a common specimen configurations used for EBIC
observations; the charge-collecting barrier is provided by a Schottky
diode. The electron beam penetrates the thin surface metallization and
generates in the semiconductor electron-hole pairs. In low injection
conditions, the resulting current flow can be discussed by considering
only the transport of excess minority carriers (see, e.g. Grove,
1967).

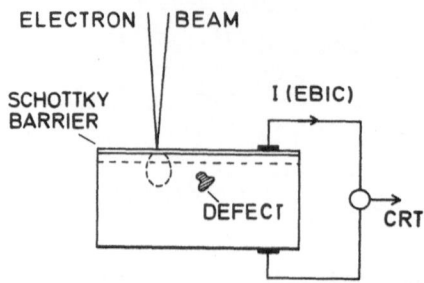

Fig. 1. Schematic illustration of the experimental arrangement in Schottky barrier charge-collection microscopy.

Minority carrier motion occurs by diffusion in the neutral semiconductor, whereas they undergo additionally drift in the field region.

The main characteristic of the electron beam injection is its high localization: injected carriers essentially spread out radially from the impact point of the beam. Consequently, the basic transport process to be described is the three dimensional diffusion of excess minority carriers. In a first approximation, their additional drift due to the depletion field can be neglected; the collection effect of the field obviously cannot, but can be represented by a condition of infinite surface recombination velocity. By neglecting drift one has the advantage of obtaining a pure diffusion problem, however at the cost of reduced accuracy in the description of the EBIC contrast due to defects that lie in the depletion region. Only the time-independent diffusion problem needs to be considered, since carrier motion is rapid enough for a stationary situation to be established at each beam position.

We consider first minority carrier diffusion in the semiconductor without defects. The carrier generation due to the electron beam can be represented by a function $g(\underline{r})[cm^{-3}s^{-1}]$, which gives the pair generation rate per unit volume. For analytical convenience, this function has been approximated, for instance, by a sphere of uniform generation

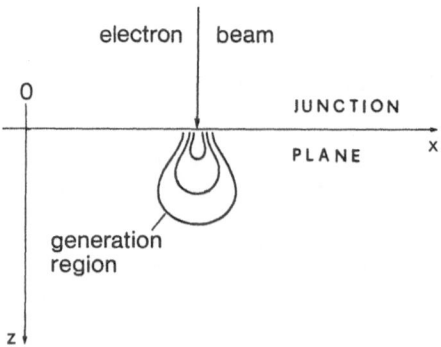

Fig. 2. Charge collection in a semi-infinite semiconductor.

(Bresse, 1972) or even by a simple point source. The beam injection produces a steady-state excess minority carrier concentration $p_o(\underline{r})$ which is related to $g(\underline{r})$ by the diffusion equation:

$$D\nabla^2 p_o(\underline{r}) - (1/\tau)p_o(\underline{r}) = -g(\underline{r}) \qquad (1)$$

where D and τ are the diffusion coefficient and lifetime of minority carriers, respectively. The boundary condition corresponding to infinite recombination velocity at the surface z=0 is:

$$p_o(\underline{r})=0 \quad ; \quad z=0 \qquad (2)$$

The collected (particle) current is the diffusion current to the surface:

$$I_o = D \int\limits_{-\infty}^{+\infty}\int\limits_{-\infty}^{+\infty} \frac{\partial p_o}{\partial z}\bigg|_{z=0} dx\,dy \qquad (3)$$

The solution of Eq.(1) with (2) can be expressed in terms of the Green's function (i.e. the solution for a point generation) that satisfies Eq.(2); the method of images (Morse and Feshbach, 1953) yields:

$$G(\underline{r},\underline{r}') = (4\pi D)^{-1}[(1/r_1)\exp(-r_1/L) - (1/r_2)\exp(-r_2/L)] \qquad (4)$$

where $L = (D\tau)^{\frac{1}{2}}$ is the minority carrier diffusion length and $r_1 = |\underline{r} - \underline{r}'|$, $r_2 = |\underline{r} - \underline{r}''|$, $\underline{r}'$ and $\underline{r}''$ being the position of the point source and its image in the plane z=0, respectively. The solution is:

$$p_o(\underline{r}) = \int\limits_{V} G(\underline{r},\underline{r}')\, g(\underline{r}')dV' \qquad (5)$$

where V is the half-space z $\geq$ 0. An example of the distribution of $p_o(\underline{r})$ for the simple scheme of the uniform generation sphere is shown in Fig.3. Substitution Eq.(5) into Eq.(3) yields the final result:

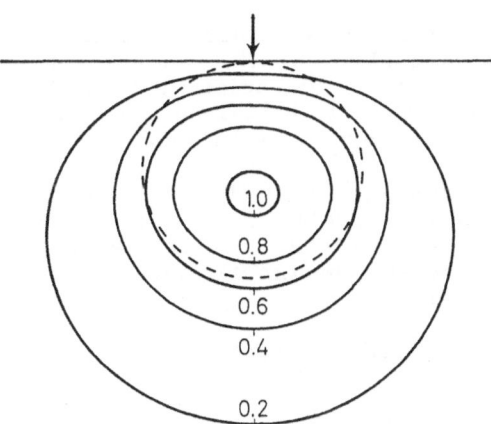

Fig. 3. Distribution of the density of excess minority carriers (continuous lines) for uniform generation over a sphere (dashed line) and L = ∞ .

$$I_0 = \int_V g(\underline{r})\exp(-z/L)dV \tag{6}$$

This equation expresses I_0 as the sum of elementary contribution in the generation volume. In a volume dV centered at $\underline{r}$, $g(\underline{r})dV$ carriers per second are generated: of them, only a fraction $\exp(-z/L)$ is able to reach the junction (here the surface) and thus contribute to the current. In view of this interpretation, the function:

$$\phi(z) = \exp(-z/L) \tag{7}$$

can be called charge-collection probability at the depth z in the (idealized) structure of Fig.2; obviously $\phi(z)$ also gives the current produced by a unit point source at a depth z. Since ϕ does not depend on x,y, as a consequence of the translational symmetry of the device along these directions, we can perform in Eq.(6) the corresponding integrations, to get:

$$I_0 = \int_0^\infty h(z)\phi(z)dz \tag{8}$$

where

$$h(z) = \int_{-\infty}^{+\infty}\int_{-\infty}^{+\infty} g(x,y,z)dxdy \tag{9}$$

is the function describing the depth distribution of the generation.

The expression (8) for I_0 involves one-variable functions only. Hence one might wonder whether it is possible to obtain Eq.(8) without going through the three-variable function $p_0(\underline{r})$. The answer is positive and relies upon a reciprocity principle that yields an equation directly for ϕ and is discussed in the Appendix. The calculation of I in the presence of a defect, however, requires in general the evaluation of $p_0(\underline{r})$, as will be shown in the next Section.

The EBIC Contrast of Defects

The approach to the EBIC contrast of defects given here is essentially that proposed earlier by the author (Donolato, 1978/79), but with some simplification of the original formalism. An analogous approach has been used recently by Wilson and McCabe (1987) for the related light beam induced current technique.

Semiconductor defects usually show up as dark features in an EBIC image, i.e. they reduce locally the collected current by increasing carrier recombination. Therefore, a defect can be represented as a region F where the minority carrier lifetime has a value $\tau'<\tau$ (Fig.4). Equation (1) then becomes:

$$D\nabla^2 p(\underline{r}) - [1/\tau(\underline{r})]p(\underline{r}) = -g(\underline{r}) \tag{10}$$

with

$$\tau(\underline{r}) = \begin{cases} \tau & \text{outside F} \\ \tau' & \text{inside F} \end{cases} \tag{11}$$

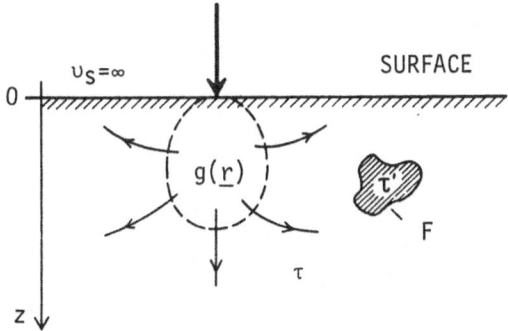

Fig. 4. Model for the formation of the EBIC contrast at a semiconductor defect.

By introducing the function $e(\underline{r})$ that has value 1 inside F and vanishes elsewhere, we can rewrite Eq.(10) as:

$$D\nabla^2 p(\underline{r}) - (1/\tau)p(\underline{r}) = - g(\underline{r}) + [1/\tau'-1/\tau]e(\underline{r})p(\underline{r}) \qquad (12)$$

In comparison to Eq.(1), Eq.(12) contains one more term that describes the additional recombination introduced by the presence of the defect. Since Eq.(12) is difficult to solve exactly, it is reasonable to look for an approximate solution, by treating the term involving $e(\underline{r})$ as a perturbation. Hence the first-order approximation to $p(\underline{r})$ in the presence of F obeys the equation:

$$D\nabla^2 p(\underline{r}) - (1/\tau)p(\underline{r}) = - g(\underline{r}) + \gamma\, e(\underline{r})p_0(\underline{r}) \qquad (13)$$

where the perturbation term has been approximated by replacing $p(\underline{r})$ with $p_0(\underline{r})$ inside F. The constant

$$\gamma = 1/\tau' - 1/\tau \qquad (14)$$

gives a measure of the additional carrier recombination due to the presence of F and can therefore be called the "recombination strength" of the defect.

Equation (13) shows that in the first-order approximation the influence of a defect on the distribution of minority carriers injected by the electron beam can be described by adding to the external source term $g(\underline{r})$ a sink term $\gamma\, e(\underline{r})p_0(\underline{r})$; this term is known, since the defect shape is given and p_0 can be calculated from Eq.(5). Thus Eq.(13) becomes formally similar to Eq.(1), and the collected current corresponding to $p(\underline{r})$ will be given, in analogy to Eq.(6), by

$$I = \int_V [g(\underline{r}) - \gamma\, e(\underline{r})p_0(\underline{r})]\exp(-z/L)dV = I_0 - I^* \qquad (15)$$

where I_0 is given by Eq.(6) or (8), and

$$I^* = \gamma \int_F p_0(\underline{r})\exp(-z/L)dV \qquad (16)$$

represents the induced current reduction due to the defect. Therefore, in Eq.(15), I_0 corresponds to the background current and I^* to the signal by which the defect is imaged (Fig.5). The EBIC contrast of the defect is defined by:

$$i^* = I^*/I_0 \tag{17}$$

Equations (16),(17) show that in the first-order approximation the defect contrast is proportional to γ; for this reason, this approximation has also been called the linear contrast model. The linear dependence of i^* on γ is clearly a good approximation only for sufficiently "weak" defects; for large γ the linear model can yield values of i^* larger than unity, a physically unacceptable result.

Equation (16) strictly applies to a volume defect, but can be easily adapted to defects mainly extended in one or two dimensions. For instance, for a surface defect like a grain boundary, we may write in Eq.(16) $dV = h \, d\sigma$, h being the (infinitesimal) boundary thickness, and replace the volume integration by an integration over the surface Σ of the grain boundary:

$$I^* = \gamma h \int_{\Sigma} p_0(\underline{r}) \exp(-z/L) d\sigma \tag{18}$$

It is convenient to characterize the grain boundary strength by the single parameter γ_s, which is the limiting value of the product γh for $h \to 0$ and $\gamma \to \infty$, and has the correct dimensions [cms^{-1}] of a surface recombination velocity. Table I summarizes the relation between the geometrical dimension of a defect and the resulting physical dimensions of its recombination strength.

We may see more clearly how Eqs.(16),(17) describe the contrast distribution in the EBIC image of a defect, by specifying in $p_0(\underline{r})$ of Eq.(16) its dependence on the position (x_0,y_0) of the electron beam on the sample surface:

$$i^*(x_0,y_0) = (\gamma/I_0) \iiint_F p_0(x-x_0,y-y_0,z) \exp(-z/L) dx \, dy \, dz \tag{19}$$

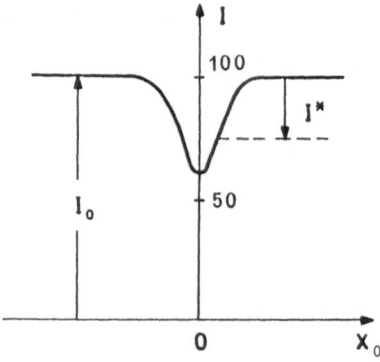

Fig. 5.   EBIC line scan across a defect.

Table I. Relation between defect dimensions and units of the recombination activity γ .

Defect dimensions	Units of γ	Example
▱ (3)	s^{-1}	inhomogeneity
▱ (2)	$cm\ s^{-1}$	grain boundary
| (1)	$cm^2\ s^{-1}$	dislocation
• (0)	$cm^3\ s^{-1}$	small precipitate

Since p_0 is also expressed as a volume integral, the practical use of Eq.(19) is not easy. We may obtain a closed form expression for p_0 by using for $g(\underline{r})$ the simplified scheme of the uniform generation sphere. Moreover, if the minority carrier diffusion length L is large in comparison to the diameter 2R of the generation sphere (this diameter is taken equal to the primary electron range Rp), we can use instead of the general expression for G of Eq.(4) its limit form for $L \rightarrow \infty$. With these simplifications

$$p_0(x-x_0,\ y-y_0,\ z) = \frac{g_0}{4\pi D} \cdot \begin{cases} 1/d - 1/f & d \geq R \\ [3-(d/R)^2]/(2R) - 1/f & d \leq R \end{cases} \qquad (20)$$

where g_0 [s^{-1}] is the total generation rate, and

$$d(-),\ f(+) = [(x-x_0)^2 + (y-y_0)^2 + (z\pm R)]^{\frac{1}{2}} \qquad (21)$$

Thus i^* can be calculated by integrating, in general numerically, the simple function of Eq.(20) over the region occupied by the defect.

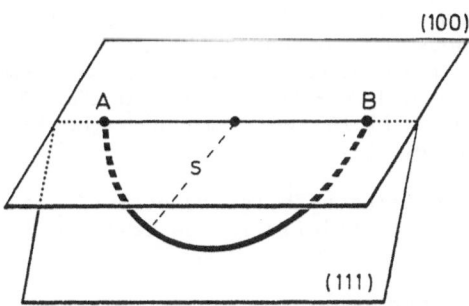

Fig. 6. Schematic illustration of an oxidation-induced stacking fault in (100) silicon.

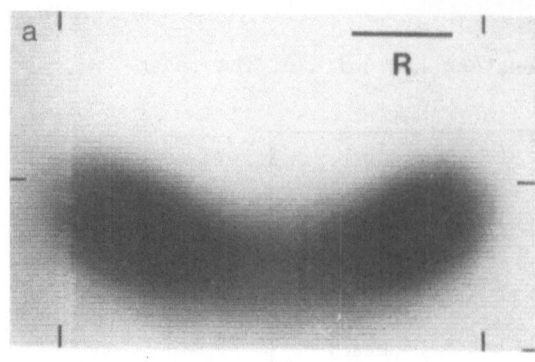

Fig.7a. Simulated EBIC image of the stacking fault of Fig.6. R is here the electron beam range.

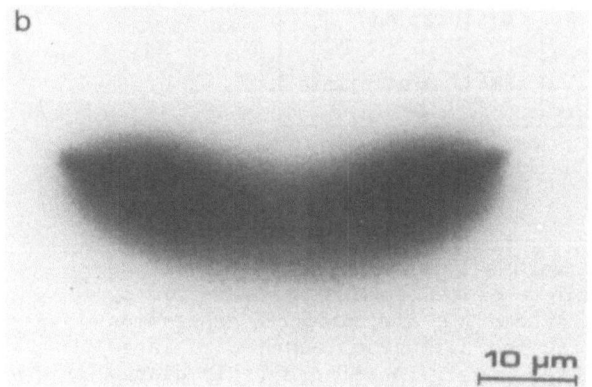

Fig.7b. Schottky barrier EBIC image of an oxidation-induced stacking fault in (100) Si (ϱ = 40-60 Ωcm).

To check whether the model, in spite its many simplifications, gives a satisfactory description of actual EBIC images, we can use Eqs.(16),(20),(21) to produce computer simulated images of simple defects and compare the results to experiment.

This has been done for instance for a stacking fault in silicon (Donolato and Klann, 1980; Donolato and Venturi, 1982). Since only the bounding partial dislocation (A B in Fig.6) has been observed to be electrically active, the fault can be modelled for EBIC as a line defect. Figure 7a shows the resulting computer simulated image and Fig.7b shows the corresponding EBIC image obtained with the configuration of Fig.1 at E=40 KeV. The comparison shows that the model accounts for the overall behaviour of the experimental EBIC contrast. However:
  a) the shape of the dark features in the simulations are less elongated than in the experimental images, probably because of the uniform generation sphere approximation: the actual beam generation is higher near the beam axis and close to the surface;
  b) the model does not reproduce the observed apexes near the fault boundary outcrops; these features are due to the combined effect of the actual form of the generation (which is pear-shaped rather than spherical) and the depletion field, which reduces the lateral spread of carriers near the surface.

Recent studies dealing with the removal of some the above approximations will be discussed briefly in the last Section.

HIGHER-ORDER / EXACT TREATMENTS

Pasemann (1981) first considered the problem of solving Eq.(13) to higher order and gave an iterative solution in the form of a perturbation series. Obtaining a manageable expression from this formal solution is in general difficult, although this may prove feasible for defects with simple configuration, and actually has been accomplished by Pasemann for the case of a straight dislocation parallel to the surface.

For illustration purposes, Pasemann's iterative procedure will be applied here to a plane grain boundary perpendicular to the surface (Fig.8). For this configuration it is possible to sum the perturbative series, and the sum reproduces the expression obtained by solving exactly the related diffusion problem.

Instead of building up the iterative solution for Eq.(13), it is simpler to do this for the corresponding equation for the charge collection probability (see Appendix), since in this way one obtains directly the EBIC profile for a point generation. The equation for ϕ corresponding to Eq.(13) is:

$$\nabla^2 \phi(\underline{r}) - (1/L^2)\phi(\underline{r}) = (\gamma/D)e(\underline{r})\phi(\underline{r}) \tag{22}$$

The boundary condition is:

$$\phi(\underline{r}) = 1 \; ; \; z = 0 \tag{23}$$

We assume for ϕ a perturbation expansion with the form:

$$\phi(\underline{r}) = \phi_0 + (\gamma/D)\phi_1 + (\gamma/D)^2\phi_2 + \ldots \tag{24}$$

Substituting Eq.(24) into Eq.(22) and equating the terms containing equal powers of /D yields:

$$\phi_0(\underline{r}) = \exp(-z/L) \tag{25}$$

and the recursion equations for ϕ_n with $n \geq 1$:

$$\nabla^2 \phi_n(\underline{r}) - (1/L^2)\phi_n(\underline{r}) = e(\underline{r})\phi_{n-1}(\underline{r})$$
$$\phi_n(\underline{r}) = 0; \quad z = 0 \tag{26}$$

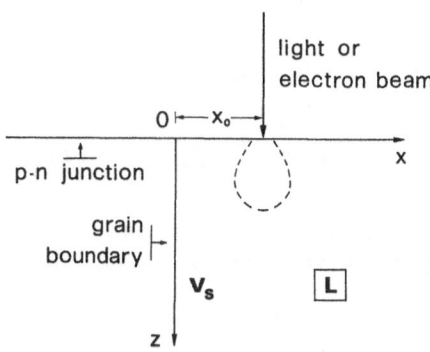

Fig. 8. Scheme of beam induced current characterization of a grain boundary; v_s is the same as γ_s.

By means of the Green's function (4) we obtain from these equations the recurrence relations:

$$\phi_n(\underline{r}) = - \int_F \phi_{n-1}(\underline{r}') \; G(\underline{r},\underline{r}')dV' \qquad n = 1,\ldots \qquad (27)$$

which specify all terms of the series (24).

For a grain boundary the volume integrals of Eq.(27) can be replaced by surface integrals as in Eq.(18), by simultaneously replacing γ with γ_s in the expansion (24). If the grain boundary is coincident with the half-plane x=0, z>0, Eq.(27) becomes:

$$\phi_n(x,z) = - \int_0^\infty \phi_{n-1}(0,z') \; G_2(x,z,0,z')dz' \qquad (28)$$

where the ϕ_n's (and hence ϕ) do not depend upon y since the configuration has translational symmetry along y. In Eq. (28) G_2 denotes the two-dimensional Green's function obtained by integrating Eq.(4) along the y axis, and is given by:

$$G_2(x,z,x',z') = \{K_0[\lambda \sqrt{(x-x')^2+(z-z')^2}\,] -$$
$$- K_0[\lambda \sqrt{(x-x')^2+(z+z')^2}\,]\}/(2\pi) \qquad (29)$$

where K_0 is the modified Bessel function of second kind of order zero, and $\lambda=1/L$. By using the integral representation for K_0 (Gradshteyn and Ryzhik,1980,p.498) we can write G_2 as:

$$G_2(x,z,x',z') = (1/\pi) \int_0^\infty (e^{-\mu|x-x'|}/\mu)\sin kz \, \sin kz' \, dk \qquad (30)$$

where $\mu = (k^2 + \lambda^2)^{1/2}$. Using this expression in Eq.(28) we obtain the simple relation:

$$\widetilde{\phi}_n(x,k) = - (1/2)(e^{-\mu|x|}/\mu) \; \widetilde{\phi}_{n-1}(0,k) \qquad (31)$$

where the tilde denotes the Fourier sine transform with respect to z:

$$\widetilde{\phi}_n(x,k) = \int_0^\infty \phi_n(x,z) \; \sin kz \, dz \qquad (32)$$

Since $\widetilde{\phi}_0(k) = k/\mu^2$, the series expansion for $\widetilde{\phi}$ is

$$\widetilde{\phi}(x,k) = \widetilde{\phi}_0(k) + (k/\mu^2)e^{-\mu|x|} \cdot \sum_{n=1}^\infty [-s/(2\mu)]^n \qquad (33)$$

where $s = \gamma_s/D$. The geometric series in Eq.(33) converges for any k if sL<2; performing the summation and inverting the Fourier sine transform yields finally:

$$\phi(x,z) = e^{-\lambda z} - (2s/\pi) \int_0^\infty (k/\mu^2)[e^{-\mu|x|}/(2\mu+s)]\sin kz \, dk$$
$$(34)$$

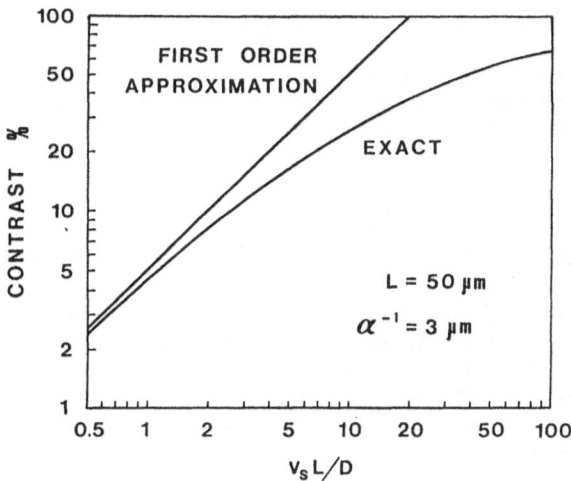

Fig. 9. Comparison between first order and exact calculation of the light beam induced current contrast of a grain boundary.

This is just the expression that is obtained by solving exactly the equation:

$$\nabla^2 \phi(\underline{r}) - (1/L^2)\phi(\underline{r}) = 0 \tag{35}$$

with the condition (23) at z=0, and describing the grain boundary not as a volume perturbation (Eq.22), but by an interface recombination velocity, i.e. by the additional boundary condition (Donolato, 1983):

$$\left.\frac{\partial \phi}{\partial x}\right|_{x=0^+} - \left.\frac{\partial \phi}{\partial x}\right|_{x=0^-} = s\ \phi(0,z) \tag{36}$$

Since the latter treatment does not introduce any restriction on the value of s, we see that Eq. (34) actually holds for any value of sL.

Starting from Eq.(34) we can compute the contrast profile (and hence the maximum contrast) for an arbitrary generation function. For instance, for a thin parallel beam of light we may write:

$$g(x,z) \propto \delta(x-x_0)\ \exp(-\alpha z) \tag{37}$$

where α is the light absorption coefficient and δ is the Dirac delta function; for this source function the contrast can be computed in closed form (Donolato, 1983). This makes it easier to compare the exact dependence of the contrast upon s with its linear approximation. The example of Fig.9 shows that the linear approximation is good for small s, as expected, but fails in the region of large s, where it yields contrast values larger than unity.

A comparison between exact and first-order theoretical contrast calculations have been given by Pasemann and Hergert (1987) for a straight dislocation perpendicular to the surface. In the next section an example will illustrate the results obtained by applying to the same contrast measurements theoretical expressions of different order of approximation.

In conclusion, higher-order/exact treatments are required to describe realistically defects of high strength, but are possible only for configurations of high symmetry. The first-order approximation can be applied quite easily to a defect of arbitrary shape, but becomes inaccurate for strongly recombining defects.

RECOVERY OF THE DEFECT STRENGTH FROM CONTRAST MEASUREMENTS

Having outlined how the EBIC image of a defect can be computed, let us consider the inverse problem, i.e. the recovery of the recombination strength of a defect from its EBIC images. Here 'image' means 'contrast distribution of the image', i.e. the function $i^*(x_0,y_0)$, which should be measured experimentally according to the definition of Eq.(17).

An EBIC image contains information both about the shape and strength of a defect, but recovering simultaneously geometrical and electrical properties would be generally rather difficult. However, for pointlike defects, which are characterized by a single geometrical parameter, i.e. their depth, depth and strength could be determined simultaneously from a sequence of EBIC images taken at different beam energies (Mil'dviskii et al, 1985). More frequently, the configuration of the defect is determined independently, and EBIC data are used only for the electrical characterization; the following discussion will be limited to this case.

If the linear approximation is applicable, Eq.(19) provides a direct relation between maximum image contrast and defect strength, with the form:

$$c = \gamma\, F(E,\ L,\ configuration) \tag{38}$$

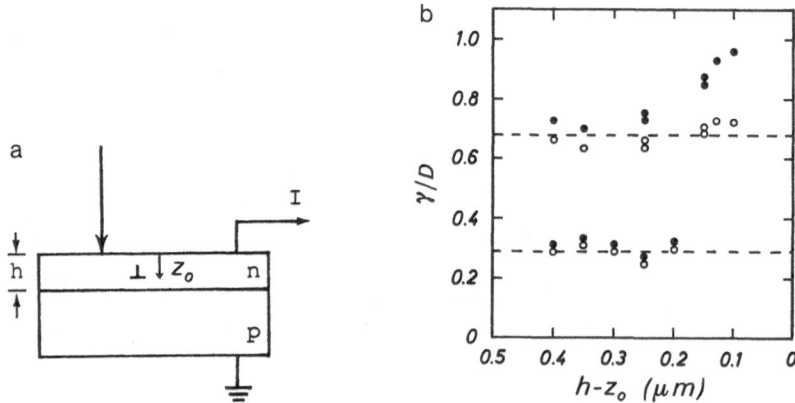

Fig. 10. Determination of the dislocation strength by Pasemann et al.(1982). a) EBIC configuration; b) Normalized recombination strength vs distance from the p-n junction, as calculated by: first-order (full symbols) and higher-order (open symbols) contrast theory.

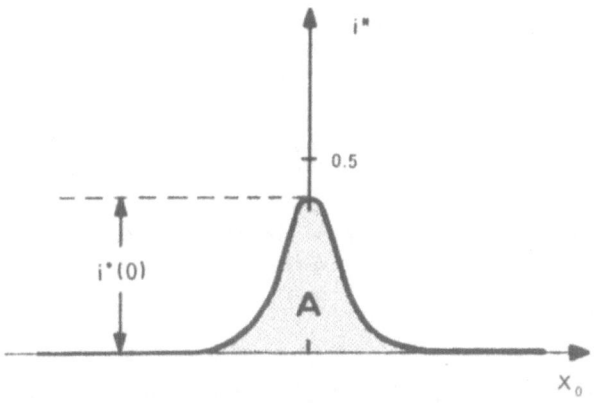

Fig.11. Contrast maximum $i^*(0)$ and contrast profile area A.

The function F involves both properties of the excitation, as the electron beam energy, device parameters, as the minority carrier diffusion length, and the defect-device configuration. As pointed out by Kittler and Seifert (1981), F plays the role of a correction factor by which the measured maximum contrast must be divided to obtain γ.

An example of the use of higher-order contrast expression to determine the recombination strength of dislocations is given by the work of Pasemann et al.(1982). The charge-collection configuration was that of Fig.10a: the dislocations were lying parallel to the surface in the emitter of silicon p-n diode, and their depth was determined by transmission electron microscopy. Figure 10b shows the values of the dislocation recombination strength γ [cm^{2}s^{-1}] normalized to D, versus their depth z_0. The plot shows for comparison both the values obtained by Pasemann's higher-order treatment and those obtained according to the linear approximation starting from the same EBIC contrast data. Both treatments allow the classification of these defects in two groups according to their strength; for the lower strength group (measured contrast 2-4 %) both approaches give essentially the same result; for the higher strength group (measured contrast 4-9 %) the linear model overestimates the strength and also retains some apparent dependence of γ on the depth, which is corrected by the higher order theory (+).

The maximum contrast is not the only characteristic of the EBIC image of a defect that can be used to determine γ. Another useful parameter is the integral of the contrast function over the image plane. Let us consider for simplicity the case of a defect whose image is completely described by a single line scan along x, e.g. a straight dislocation or a plane boundary perpendicular to the sample surface. It is then sufficient to consider the contrast profile area A [cm] (Fig.11):

$$A = \int_{-\infty}^{+\infty} i^*(x_0)dx_0 \qquad (39)$$

(+) There has been some controversy on the interpretation of the recombination strength obtained by the first-order approximation. For details the reader is referred to Donolato (1983b) and Pasemann (1984).

The contrast profile for an extended generation $g(x,z)$ (the projection of $g(\underline{r})$ onto the y plane) is obtained by weighting with this function the contrast profile due to a point source i^*_{ps}:

$$i^*(x_0) = \int\limits_0^\infty dz \int\limits_{-\infty}^{+\infty} g(x-x_0,z)i^*_{ps}(x,z)\ dx \qquad (40)$$

and the corresponding contrast profile area is given by:

$$A = \int\limits_0^\infty dz\ h(z) \int\limits_{-\infty}^{+\infty} i^*_{ps}(x,z)\ dx \qquad (41)$$

where $h(z)$ is the same as in Eq.(9). Hence we see that A, unlike the maximum contrast $i^*(0)$ (see Eq.(40)), does not depend on the lateral distribution of the generation: this gives both practical and computational advantages.

In the exact contrast expressions, i^*_{ps} and hence A do not have a simple dependence on γ (see, e.g. the dependence on s in Eq.(34) for the grain boundary); in this case, graphical methods can be useful to determine γ from A (Donolato, 1983). If the linear approximation applies, A becomes proportional to γ and the expression of the correction factor (Eq.38) also becomes simpler. For the grain boundary, it is easy to obtain from the first-order expansion of Eq.(34) that:

$$A = \int\limits_0^\infty h(z)(sLz/2)dz = sL(z^*/2) \qquad (42)$$

where z^* is the center of gravity of the known function $h(z)$; this equation yields a rapid method to estimate the value of sL (and hence of s if L is evaluated independently) from the measured contrast profile area.

More generally, it can be shown from Eq.(19) that in the linear approximation the integral of i^* over the image plane is only dependent on $p(z)$ and can therefore be calculated for any defect by solving a one dimensional diffusion problem only. An application of this method to contrast measurements on dislocations can be found in Donolato and Bianconi (1987).

RECENT DEVELOPMENTS

Recent theoretical studies on EBIC contrast of defects mainly consider the extension of the contrast model described in the previous sections, but a substantially different approach has also been proposed (Jakubowicz, 1985).The most intensively studied problem deals with the description of the contrast of defects that lie, at least in part, in the depletion layer.

Joy (1986) computed EBIC profiles of a straight dislocation parallel to the surface and lying in the depletion layer of a Schottky diode. The influence of the field on recombination was described by scaling the defect strength by an amount proportional to the field at the defect. With this generalization, the influence of an applied bias on the contrast can be computed: the main result is that the defect contrast reaches a maximum when the boundary of the depletion layer reaches the defect. The influence of the depletion field on the recombination activity of dislocations was described by Sieber (1987) by representing the dislocation as a cylinder with variable radius ε. Thus a dislocation perpendicular to the surface of a Schottky diode was attributed a smaller radius (and hence a smaller strength, since $\gamma \propto \varepsilon^2$) near the surface, where the electric field is higher. This model accounted for the experimentally observed maximum of the contrast vs beam energy of dislocations in CdTe diodes, a feature that cannot be explained if the presence of the depletion layer is neglected.

Another improvement concerns the electron beam generation, which has been represented more realistically by using Monte Carlo simulation (Joy, 1986). This technique associated with a description of field effects should allow a more accurate simulation, for instance, of the EBIC image of Fig.7b; such an approach, however, has not yet been attempted. A possible further extension should include the observed influence of the injection level on defect images (Kittler, 1980; Leamy, 1982; Wilshaw and Booker, 1985).

The formulation of a different, non-linear EBIC model by Jakubowicz (1985) originated from measurements of the contrast of dislocations as a function of the temperature (Ourmazd, Wilshaw and Booker, 1983), which were interpreted as being incompatible with a linear contrast model. The incompatibility was shown later to be only apparent (Donolato, 1986), and actually subsequent similar measurements were interpreted with the linear model (Wilshaw and Booker, 1985).

In Jakubowicz's scheme, the basic point-like defect is represented as a spherical surface with s = ∞ (s is the surface recombination velocity) and the generation by a point source. With this geometry, the related diffusion problem for minority carriers in a semi-infinite semiconductor can be treated by the method of images. A defect with finite s is represented by a sphere with a radius reduced by a factor γ, with $0 \leq \gamma \leq 1$: γ is defined as 'strength of the defect'. The contrast turns out to be a non-linear function of γ; this non-linearity can give a qualitative explanation of the results of Ourmazd et al. (1983). This scheme is open to some objections, the most serious one being that it produces a divergent contrast when the source is at the defect, which corresponds to the best observation conditions in actual EBIC experiments. Moreover, it cannot be extended easily to defect of different shapes, because non-linearity prevents the description of an extended defect as a superposition of point-like defects.

APPENDIX: RECIPROCITY

The Green's theorem (Morse and Feshbach, 1953) shows that the general formal solution to Eq.(1) in the half-space $z \geqslant 0$ with arbitrary boundary conditions is given by:

$$p(\underline{r}) = \int_V G(\underline{r},\underline{r}')g(\underline{r}')dV' + D \int_\Sigma \left(G \; \frac{\partial p}{\partial n} - p \; \frac{\partial G}{\partial n} \right) d\sigma \qquad (A1)$$

where V is the half space $z \geq 0$ and Σ is the plane z = 0. If we chose the Green's function G to be zero on Σ (i.e. that given by Eq.(4)), the first term in the surface integral vanishes.

Let us call $\phi(\underline{r})$ the function that satisfies the homogeneous version of Eq.(1)(i.e. with $g(\underline{r})=0$) and has unit value on Σ. Equation (A1) then gives:

$$\phi(\underline{r}) = - D \int_{\Sigma} \frac{\partial G}{\partial n} \, d\sigma \qquad (A2)$$

Since the external normal $\underline{n}$ points to $-z$, we see by comparison with Eq.(3) that this expression just gives the current collected by Σ in the presence of a unit point source at $\underline{r}$, i.e. by definition the charge collection probability at $\underline{r}$. The function $\phi(\underline{r})$ can therefore be found directly by solving the boundary value problem (Donolato,1985):

$$\nabla^2\phi(\underline{r}) - (1/L^2)\phi(\underline{r}) = 0$$

$$\phi(\underline{r}) = 1 \; ; \; z = 0 \qquad (A3)$$

that reduces to the one-dimensional equation:

$$\phi''(z) - (1/L^2)\phi(z) = 0$$

$$\phi(0) = 1 \qquad (A4)$$

The solution of this equation vanishing at infinity is $\exp(-z/L)$, in agreement with Eq.(7).

Equation (A3) expresses a reciprocity property analogous to that following from the Green's reciprocity theorem of electrostatics (Jackson, 1975): it states that the current collected by the junction plane Σ in the presence of a unit point source of carriers at $\underline{r}$ is the same (apart from a unity dimensional factor) as the excess minority carrier density at $\underline{r}$ due to a unit carrier density on Σ.

REFERENCES

Bresse, J. F., 1972, Scanning Electron Microsc. 1972:105.
Donolato, C., 1978/79, Optik, 52:19.
Donolato, C., and Klann, H., 1980, J.Appl.Phys'., 51:1624.
Donolato, C., and Venturi, P., 1982, Phys.Stat.Sol.(a), 73:377.
Donolato, C., 1983a, J.Appl.Phys., 54:1314.
Donolato, C., 1983b, J.Physique Coll., 44(C4):269.
Donolato, C., 1985, Appl.Phys.Lett., 46:270.
Donolato, C., 1986, J.Physique, 47:171.
Donolato, C., and Bianconi, M., 1987, Phys.Stat.Sol.(a), 102:k7.
Donolato, C., 1988, Scanning Microscopy, 2:801.
Gradshteyn, I. S., and Ryzhik, I. M., 1980, "Table of Integrals, Series and Products", Academic Press, New York.
Grove, A.S., 1967, "Physics and Technology of Semiconductor Devices", J.Wiley, New York.
Hanoka, J. I., and Bell, R. O., 1981, Ann.Rev.Mater.Sci., 11:353.
Holt, D. B., and Lesniak, M., 1985, Scanning Electron Microsc. 1985; I:67.
Jackson, J. D., 1975, "Classical Electrodynamics", J.Wiley, New York.
Jakubowicz, A., 1987, Scanning Microscopy, 1:515.
Joy, D. C., 1986, J.Microsc., 143:233.
Kittler, M., 1980, Kristall und Technik, 15:575.
Kittler, M., and Seifert, W., 1981, Phys.Stat.Sol.(a), 66:573.

Leamy, H. J., 1982, J.Appl.Phys., 53:R51.

Mil'vidskii, M. G., Osvenskii, V. B., Reznik, V. Ya., and Shershakov, A. N., 1985, Sov.Phys.Semiconductors, 19:22.

Morse, P. M., and Feshbach, H,, 1953, "Methods of Theoretical Physics", McGraw-Hill, New York.

Ourmazd, A., Wilshaw, P. R., and Booker, G. R, 1983, J.Physique Coll., 44(C4):289.

Pasemann, L., 1981, Ultramicroscopy, 6:237.

Pasemann, L., Blumtritt, H., and Gleichmann, R., 1982, Phys. Stat. Sol. (a), 70:197.

Pasemann, L., 1984, J.Physique, 45:L-133.

Pasemann L, and Hergert, W., 1987, Izvestiya Akademii Nauk SSSR, Fiz.Ser., 51:1528.

Sieber, B., 1987, Phil.Mag., 55:585.

Wilshaw, P. R., and Booker, G. R., 1985, in "Microscopy of Semi-conducting Materials 1985", Conf.ser.No.76, A. G. Cullis and D. B. Holt, eds., Institute of Physics, Bristol and Boston, p.329.

Wilson, T., and McCabe, E. M., 1987, J.Appl.Phys., 61:191.

RECOMBINATION AT DISLOCATIONS IN SILICON AND GALLIUM ARSENIDE

P.R. Wilshaw, T.S. Fell, and G.R. Booker

Department of Metallurgy and Science of Materials
University of Oxford, Parks Road, Oxford OX1 3PH

INTRODUCTION

For many years it has been known that dislocations in semiconductors are associated with energy levels within the band gap. Such levels have been detected experimentally in many ways. For example charge trapped at the dislocation level both alters the free carrier concentration which can be measured using the Hall effect and also introduces unpaired electrons whose spin can be detected using EPR. The thermal capture and re-emission of carriers at the defect energy levels can be measured using DLTS. The emission of photons when carriers make a transition to the dislocation level can be observed in photoluminescence experiments. In each case the experimental results obtained can be directly related to the position or concentration of the energy levels present, or the relevant theory describing the experiment is sufficiently complete that the data may be interpreted in such terms with a good degree of confidence[1]. In respect to deformation induced dislocations in silicon, all of the above techniques have been widely used to characterise the parameters describing energy levels at dislocations. It is thus surprising that the cause of the energy levels is still not understood. They have been variously attributed to the dislocation strain field, dangling bonds at the dislocation core, kinks, jogs, faults in the reconstruction process of dangling bonds, and impurities or point defect centres present at the dislocation. It seems likely that this list does contain the actual causes of activity and also that the major cause of the activity will be different for dislocations in different environments and which have undergone different thermal treatments. In some cases, particularly when the dislocations are known to be decorated, it seems likely that the activity can be attributed to impurities[2]. However in most other cases the situation is unclear and in particular the question of whether or not a perfectly clean, straight dislocation would be electrically active remains unanswered.

If the cause of the electrical activity of dislocations is to be understood it is necessary to use different techniques in addition to those mentioned above and in particular new information must be gained about the energy levels at dislocations. For this reason increasing effort has been directed towards the study of dislocations using techniques which provide a degree of spatial resolution. In this way individual segments of dislocations can be measured rather than obtaining averages over large numbers. The main method used in this respect is the EBIC technique[3] whereby the process of electron hole recombination at dislocations is measured with a spatial resolution of ~ 1μm. However in this case, unlike those mentioned above, the theory relating experimental results to the fundamental parameters of the defect energy levels has not been well developed. As a result, the intepretation of EBIC data has largely been made only on a phenomological basis and the full potential of the technique has not yet been exploited. It is the purpose of the present work to describe a new theory for recombination at dislocations which, it is believed, allows the experimental EBIC data obtained from dislocations in silicon to be interpreted in terms of the concentration of energy levels along the dislocation.

Experimental results are also presented for many different dislocations in silicon and also the first measurements of the temperature and minority carrier concentration dependence of recombination at dislocations in GaAs.

RECOMBINATION AT DISLOCATIONS

Several authors have considered theoretically the process of carrier recombination at dislocations using different approaches. The most straightforward is to consider the action of a dislocation to be the sum of independent, point defect like, recombination centres which lie along the length of the dislocation. However, this approach, used by Kittler[4] and others, neglects the effect of the band bending which is produced around a line of charged defect states (see next section for further details). Kimerling[5] first considered directly the effect that the charge on a dislocation and hence the band bending would have on its efficiency as a recombination centre. He deduced that the recombination efficiency of dislocations in silicon would be roughly constant or decrease with increasing temperature. This is because the Fermi level would move towards the centre of the band gap, decrease the charge on the dislocation and hence reduce the probability of minority carrier capture. Some experiments have indeed produced results which demonstrate this behaviour but others have found that over some range of temperature the recombination efficiency actually increases with increasing temperature before reaching a plateau and then decreasing again[6] [7]. Figielski[8] considers the case where the charge on the dislocation changes with both temperature and also the excitation level, i.e. the concentration of excess minority carriers. He derives equations which predict the steady-state excess free carrier concentration for a given rate of carrier generation and then uses these to describe photoconductivity at both high and low injection levels. However, Figielski's work, which concentrates on majority carrier lifetime, assumes the net capture efficiency of minority carriers to be independent of the excitation level and constant except at very high temperatures. Ourmazd[6] uses a two-stage model for deep level capture to describe recombination at dislocations in order to interpret EBIC results from edge and Frank partial dislocations. He considers dislocations in heavily doped material for which he assumes the band bending is constant. Using this model Ourmazd finds no dependence of the recombination efficiency on the excitation level and finds that it is either constant or varies exponentially as a function of temperature.

The theory now presented initially follows the approach of the Figielski theory, but thereafter differs significantly in the assumptions made, the form in which it is developed and the parameters which are calculated. This model is the first so far developed which predicts the

observed dependence of recombination at dislocations on the excess
minority carrier concentration[15].

In the following a dislocation in n-type Si which introduces
acceptor states into the band gap normally below the position of the
Fermi level is assumed, however the model is equally applicable to dislocations in p-type material where the states introduced show donor type
behaviour and lie above the Fermi level. Figure 1 shows the band
structure of a dislocation where some of the dislocation states are
occupied and so the dislocation becomes negatively charged. This
negative charge, Q per unit length, increases the energy of electrons in
the vicinity of the dislocation and is represented on a band diagram by
the conduction and valence bands bending up. It is normally assumed that
the addition of electrons to the dislocation energy level does not alter
its position relative to the band edges. In addition it is assumed that
the spatial extent of the wave functions associated with the dislocation
states is small compared to the long range electrostatic potential which
normally extends in the region of 0·1μm or more. Thus the entire dislocation level can be considered to be rigidly shifted by the electrostatic
potential of its charge.

Recombination of an electron hole pair via such a dislocation level
can be considered to take place in 4 steps.

(1) Diffusion of a hole to the space charge region.
(2) Capture of the hole into the bound hold states at the top of the
    valence band which are produced by the electrostatic potential well
    of the space charge region.
(3) Transition of the hole to the primary dislocation level.
(4) Capture of an electron to the primary dislocation level via the
    repulsive potential barrier of the space charge region.

Step 1 is not a "bottleneck" for recombination at dislocations in
Si[9]. The degree of EBIC contrast normally measured (< 12%) at
dislocations in silicon implies that they can be regarded as "weak"
defects[10] in the sense that the diffusion of minority carriers to the
dislocation is sufficiently fast that recombination there only weakly
perturbs the surrounding carrier distribution.

It is assumed that step 3 is not rate limiting. This seems likely
since the holes in the bound hole band are physically trapped by the
electrostatic field at the same position in the crystal as the primary
dislocation states, thus the probability of this transition is likely to
be high. If step 3 is not rate limiting then thermal emission of trapped
holes back into the valence band will not be important provided the
excess minority carrier concentration is much higher than its thermal
equilibrium value. This requirement is normally satisfied experimentally
and so thermal emission of holes is not considered further.

Thus in this model it is step 2, i.e. the capture of holes into the
bound hole states and step 4, the excitation of electrons over the
electrostatic potential barrier which control the recombination of
carriers at dislocations. At steady state the rate of capture of
electrons and all subsequent transitions must equal the rate of capture
of holes by the bound hole state. Thus a detailed analysis of steps 2
and 4 will provide a complete description of the kinetics of the overall
process of recombination at dislocations. Thus in this model the number
of dislocation levels in the band gap and the nature of the transitions
to and between them, whether multiphonon or radiative, is irrelevant to
the overall recombination process. Thus this particular model allows a

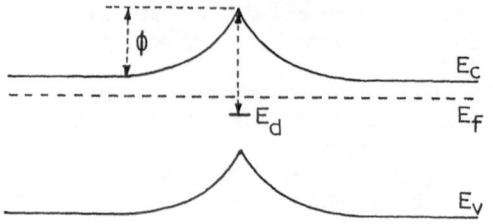

Fig.1. Band structure of a charged dislocation in an n-type semiconductor.

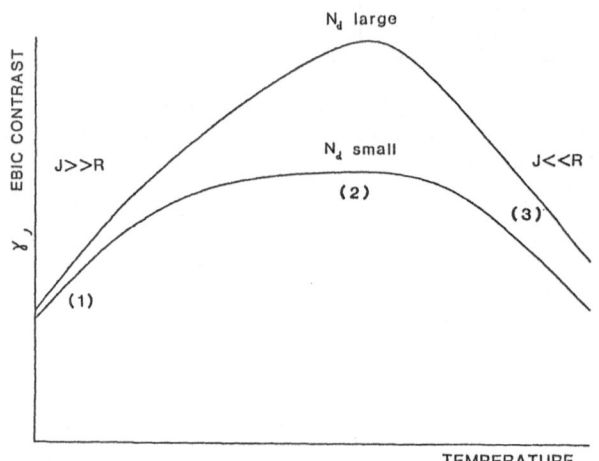

Fig.2. Theoretical curves for dislocation recombination strength versus temperature. The parameters E_0, N_d and C_e have been chosen to illustrate the different regimes of behaviour on one set of curves. (1) Q mainly determined by the recombination process. $f < 1$. (2) Small N_d. All dislocation states occupied without the excess charge raising the states to E_f. Q determined mainly by N_d. $f \simeq 1$. (3) Dislocation states pinned to the Fermi level. Q determined mainly by E_f. $f < 1$.

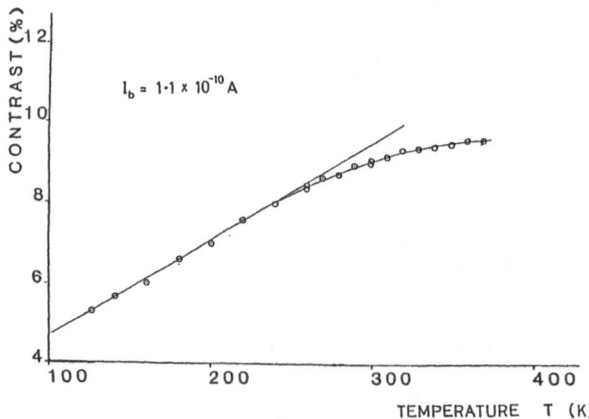

Fig.3. EBIC contrast versus temperature for a screw dislocation in n-type, 10^{15} cm^{-3}, silicon.

great simplification of the overall recombination process which can now be described by considering only hole capture and electron capture.

Hole Capture   Once holes have diffused to the space charge region of radius r_d which surrounds the dislocation they are captured by the attractive potential into the bound hole states at the top of the bent valence band.   This capture of holes results in a reduced minority carrier lifetime τ' within the space charge region.   The hole capture rate per unit dislocation length J_h is related to τ' by:

$$J_h = \pi r_d{}^2 \Delta P / \tau' \tag{1}$$

where ΔP is the excess minority carier concentration.   It is convenient at this stage to describe the recombination at a dislocation by the dislocation recombination strength γ [10] which is the recombination rate normalised by the excess minority carrier concentration and diffusivity, hence:

$$\gamma = J_h/\Delta P D_h = \pi r_d{}^2/D_h \tau' \tag{2}$$

where D_h is the hole diffusivity.

The recombination strength of the dislocation thus depends on r_d, the radius of the space charge region.   An expression for r_d is obtained by using the condition of charge neutrality at the dislocation for which it is assumed that the screening is due to the expulsion of majority carriers from the space charge region:

$$r_d = (Q/\pi n_0 q)^{\frac{1}{2}} \tag{3}$$

where n_0 is the majority carrier concentration, Q is the net excess negative charge per unit dislocation length and it is assumed that the electrostatic barrier height is large compared with kT.

Electron Capture   The electron capture rate per unit dislocation length J_e for a nondegenerate electron gas is determined by detailed balance and is given by:

$$J_e = C_e N_d \left[(1-f) n_0 \exp\left(-\frac{E}{kT}\right) - f N_c \exp\left(-\frac{E_0}{kT}\right) \right] \tag{4}$$

where C_e is the probabilty of the transition of an electron from the conduction band to an empty dislocation level, N_d is the number of states per unit dislocation length, N_c is the effective density of states in the conduction band and f is the occupancy factor of the dislocation states. E is the Gibbs free enthalpy required to excite electrons from the conduction band over the electrostatic barrier to the dislocation level and E_0 is the Gibbs free enthalpy for the reverse process[11].

The expression for J_e consists of two terms.   The first represents the rate of capture by the dislocation level of electrons thermally excited over the potential barrier.   In the following it is assumed that the free energy, E, required to accomplish this is equal to $q\phi$ where ϕ is the potential barrier height of the dislocation in volts.   In this way the change in entropy associated with this process and the effect of electron tunnelling through the top part of the barrier[12] are assumed to be small.   The second term of equation (4) describes the thermal re-emission of electrons from the dislocation level back into the conduction band.

Electron Hole Recombination

For the steady state recombination conditions, the overall recombination rate J, and the hole and electron capture rates described above must be equal:

$$J = J_h = J_e \tag{5}$$

In the above equations hole capture was found to be dependent on the dislocation charge Q, whilst electron capture depended on the electrostatic barrier height ϕ. In order to solve equation (5) an expression for ϕ in terms of Q must be used and to ensure that a simple algebraic expression for γ is obtained the following approximation is made:

$$\phi \simeq AQ, \quad A = \text{constant} \tag{6}$$

Equations (1) to (6) may then be solved together as follows:

$$\phi = -\frac{kT}{q} \ln \left(\frac{\Delta p \gamma D_h + C_e N_d f N_c \exp(-E_0/kT)}{(1-f) n_0 C_e N_d} \right)$$

$$\gamma = \frac{Q}{n_0 q D_h \tau'} \simeq \frac{\phi}{A n_0 q D_h \tau'}$$

$$\gamma \simeq \frac{-T}{A n_0 q D_h \tau'} \ln \left(\frac{\Delta P \gamma D_h + C_e N_d f N_c \exp(-E_0/kT)}{(1-f) n_0 C_e N_d} \right) \tag{7}$$

This algebraic expression for the recombination strength γ, is useful in that it allows the general trends of recombination behaviour to be followed both as a function of the experimental variables $(T, \Delta P, n_0)$ and the fundamental dislocation parameters (E_0, N_d, C_e, τ'). However if a quantitative analysis of dislocation behaviour is required then a numerical solution for γ has to be found. In this case the more accurate expression for ϕ and Q obtained by Masut et al[13] can easily be employed:

$$\phi = \frac{Q}{2\pi\epsilon\epsilon_0} \left(\ln \frac{\lambda_D Q}{q} - 0.5 \right) \tag{8}$$

where λ_D is the Debye length.

A numerical solution for γ can be found as follows. The hole capture rate can be expressed in terms of the dislocation line charge Q:

$$J_h = \frac{\Delta PQ}{\tau' n_0 q} \tag{9}$$

In equation (4) the dislocation occupancy factor f, can be expressed in terms of Q using $f = Q/(N_d q)$ and ϕ in terms of Q using equation (8). Thus both hole and electron capture rates are described in terms of one undefined variable, Q. Iteration can then be used to obtain numerically a value for Q which allows the experimentally realised condition at steady state, $J_h = J_e$, to be obtained within the numerical model. The value of Q so obtained can then be used to calculate values for γ, f, ϕ and r_d as required. The value of γ produced by this numerical solution follows the same trends as described by equation (7) but is more accurate due to the use of the accurate relation between Q and ϕ. The theoretical plots of γ shown in figures 2 and 4 were produced using this numerical method of solution.

Variation of Dislocation Recombination Strength With Experimental Parameters

Having gained a theoretical description of the recombination process it is interesting to follow what happens as various experimental parameters are changed. To do this, equation (7) is simplified as follows. It is assumed that holes which are captured into the bound hole states are not re-emitted and so the reduced minority carrier lifetime, τ' depends on the cascade capture of holes into the potential well surrounding the dislocation. This has been treated theoretically by Sokolova[14] who found that τ' varies as $T^{1.5}$. Experiment has shown that D_h varies as $T^{-1.4}$ and consequently the product $\tau'D_h$ is closely independent of temperature. Thus the recombination strength can be written as:

$$\gamma \sim \frac{AT}{n_0} \ln\left(\frac{(1-f)n_0 C_e N_d}{J+R}\right) \qquad (10)$$

where $J = \Delta P \gamma D_h$, the recombination rate and $R = C_e N_d f N_c \exp(-E_0/kT)$, the rate of re-emission of electrons into the conduction band. It is noted that since $\tau'D_h$ is nearly independent of temperature equations (2) and (3) show that γ is almost directly proportional to Q.

Figure 2 shows plots of the recombination efficiency of a dislocation for two different values of N_d. Three regimes of behaviour can be defined.

In the first regime at low temperatures or large excess minority carrier concentrations the dislocation recombination strength is approximately proportional to temperature for a constant excess minority carrier concentration ΔP, and varies as $-\ln(\Delta P)$ at constant temperature. This behaviour may be understood by considering a dislocation initially at equilibrium with line charge Q_0 and a zero net rate of electron and hole capture. If an excess minority carrier concentration ΔP is introduced the rate of capture of holes is increased, being proportional to the product of ΔP and Q. However, the rate of capture of electrons is initially unchanged because $\Delta n \ll n_0$ for the conditions considered here. Thus $J_h > J_e$ which will result in the dislocation line charge decreasing from its equilibrium value Q. As Q decreases the electrostatic barrier to electron capture also decreases and thus J_e increases according to equation (4). Reducing Q also has the effect of decreasing J_h. This process will continue until $J_e = J_h$ and then steady state is reached. Thus for increasing hole concentrations the dislocation charge has to be progressively decreased such that the extra rate of electron capture matches the increased hole capture. Since the recombination strength γ is proportional to the dislocation line charge this gives rise to a reduction in recombination strength with increasing minority carrier concentration according to $\gamma \propto -\ln(\Delta P)$. This feature is not predicted by any other existing theory but is observed experimentally[15]. By a similar argument if the temperature is increased, thermal activation of the majority carriers over the potential barrier is also increased and so a given electron capture rate is achieved for a higher value of Q. This leads to γ increasing approximately linearly with T.

As the temperature is increased further or the minority carrier concentration is reduced the charge on a dislocation initially in regime 1 will increase and will tend to approach the equilibrium charge it would have in the absence of recombination. If the concentration of states at the dislocation N_d, and the position of the Fermi level are such that at equilibrium all the dislocation states are occupied and the dislocation energy level is below the Fermi level, then the recombination behaviour for sufficiently small values of ΔP lies in regime 2. In this regime, in which the dislocation charge is close to equilibrium, f is nearly unity and so Q is nearly constant and thus the recombination strength is approximately independent of both temperature and the excess minority carrier concentration.

If however, the concentration of states at the dislocation is such that the band bending is sufficient for the dislocation level to intersect the Fermi level at values of f < 1, then at equilibrium the dislocation level will be "pinned" to the Fermi level.

Thus, regime 3 occurs when the temperature is high enough and the minority carrier concentration low enough that the dislocation level is pinned to the Fermi level. In this regime the recombination strength decreases with increasing temperature as the Fermi level moves towards the centre of the band gap. The recombination strength is however independent of minority carrier concentration. For these conditions f < 1 and R >> J in equation 10.

Discussion

In the theory developed here for dislocations in n-type silicon the main features of recombination are not dependent on the detailed quantum mechanical processes of the transitions taking place, nor indeed on the detailed mechanism of minority carrier capture. Instead the important features are the exponential dependence of electron capture on the electrostatic barrier surrounding the dislocation and the dependence of hole capture on the dislocation line charge. This approach results in three regimes of dislocation recombination behaviour which should be applicable, in some form, not only to dislocations in n-type Si but also to p-type Si and also to dislocations in other semiconductors. This is the only model of dislocation recombination yet developed which predicts the dependence of recombination on the excess minority carrier concentration at dislocations.

The behaviour predicted in this work shows that the recombination activity of a dislocation is not related in any simple way to the fundamental dislocation parameters, rather the recombination strength of a dislocation is a monitor of the dislocation charge which may be close to, or far from equilibrium depending on experimental conditions. Thus to gain information concerning such parameters as N_d and E_0 it is essential to measure the dislocation recombination strength γ over a large range of both T and ΔP. To interpret the results it is also necessary to know the doping concentration as this will determine the position of the Fermi level at a given temperature.

Unless such a procedure is carried out it is not possible to compare experimental data obtained from dislocations in different specimens and under different conditions. For example, changing the electron beam current in an EBIC experiment has the effect of changing ΔP and thus may easily be sufficient to halve the recombination efficiency of a dislocation without altering its fundamental parameters. Similarly, dislocations in differently doped specimens, which are otherwise identical, are expected to show greatly different behaviour. Clearly great caution must be exercised when different sets of recombination data are compared.

RECOMBINATION AT DISLOCATIONS IN GaAs

An analysis similar to that used for dislocations in silicon may also be applied to dislocations in GaAs. It is thus expected that recombination may exhibit three regimes of behaviour. However, unlike silicon the details of hole capture are not known. In particular it is not understood whether dislocations behave as strong or weak defects nor is the temperature dependence of hole capture known. Thus although in the first regime it is expected that the dislocation charge will be far from equilbrium and that γ will increase with increasing temperature and decrease with increasing minority carrier excitation, it is not possible to predict the exact form of this dependence. Similarly for regimes 2 and 3 it is not possible, as yet, to predict theoretically the temperature dependence of γ, although as with dislocations in silicon γ will be independent of Δp.

EXPERIMENTAL

In the present work the recombination of carriers at dislocations has been studied using the EBIC technique. Donolato[10] has shown that the measured EBIC contrast, C, of a dislocation is directly proportional to its recombination strength γ. Thus, the EBIC contrast of dislocation is given by:

$$C = \frac{\beta T}{n_0} \ln\left(\frac{(1-f)n_0 C_e N_d}{J+R}\right),$$

where β is a constant.

The recombination flux J, for an EBIC experiment can be written as $J = \eta I_b C / q l$ where I_b is the incident beam current of the SEM, η is the number of electron hole pairs generated for each incident electron and l is the length of the dislocation over which recombination takes place. Thus in an EBIC experiment ΔP is changed by changing the electron beam current. In this work contrast measurements were made using an EBIC system similar to the one described previously[16] which is now based around a Philips 505 SEM with a LaB$_6$ gun.

Dislocations in Silicon

High purity, float zone, n-type, 10^{15} cm^{-3} Si that has been deformed under clean conditions by two stage compression at 850°C and 420°C has been studied. The specimens contain hexagonal loops and straight segments of dislocation up to several hundred microns in size. Contrast measurements were made on individual straight, dissociated, screw and 60° dislocations using an accelerating voltage of 15kV. Figures 3 and 4 show some results obtained from the measurement of dislocation contrast as a function of specimen temperature and incident beam current. All the data are taken from the same 2μm segment of one screw dislocation. Similar results have been obtained for a 60° dislocation. (The specimen used for this work was provided by Professor H. Alexander and colleagues, Cologne University, FRG.) Since $\Delta P \propto I_b$ these results are in agreement with the theory which predicts that C varies as $\ln(\Delta P)$ and varies linearly with T at low temperatures or at high beam currents. This is regime 1 of the recombination behaviour described above. At high temperatures and small beam currents the contrast tends to become independent of both temperature and beam current, (ΔP), as described by regime 2 of the recombination theory. The best fit of the theory to the experimental data is shown by the solid lines of Figure (4). These were obtained with a value for N_d of ~ 2×10^6 cm^{-1} for this particular screw dislocation segment. The curves taken from the 60° dislocation were best fitted using a value of N_d ~ 10% higher than for the screw. The values of N_d obtained using the EBIC method are probably not accurate to more than 20 or 30% due to considerations such as the non-uniformity of the minority carrier generation volume and the difficulty in estimating the length of the dislocation segment along which recombination takes place.

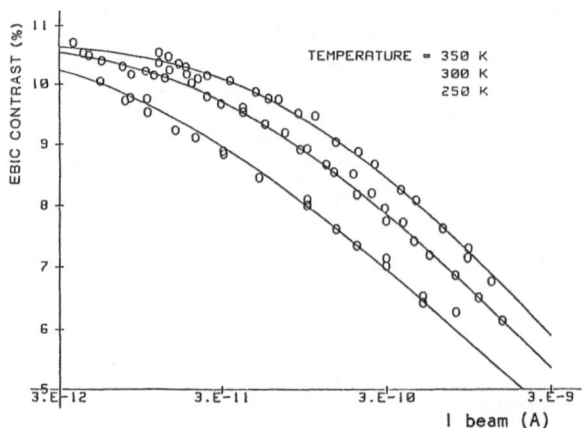

Fig.4. EBIC contrast versus $\ln(I_b)$ for the screw dislocation of figure 3. Points are experimental, lines are theoretical.

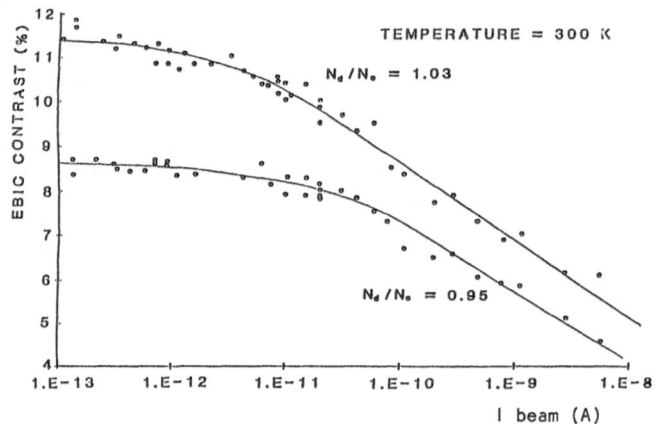

Fig.5. EBIC contrast versus $\ln(I_b)$ for two screw dislocation in n-type 10^{15} cm^{-3}, silicon.

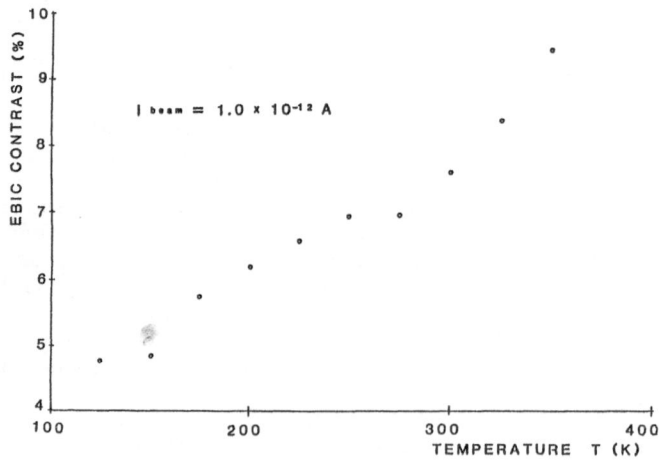

Fig.6. EBIC contrast versus temperature for a screw dislocation in n-type, 5×10^{15} cm^{-3}, GaAs.

However, the technique should be sensitive to small changes in N_d along the length of a dislocation and between different dislocations. Thus in the following all values of N_d are quoted relative to the value of the screw dislocation described in figure 4 which will henceforth be denoted as N_0.

Apart from the specimen used for figures 3 and 4 all others were deformed and prepared in Oxford. Figure 5 shows typical contrast versus beam current curves for two screw dislocations produced in Oxford. The data is in good agreement with the theory for more than four orders of magnitude change in beam current. These curves demonstrate how the C versus $\ln(I_b)$ curves change as N_d changes. In total 22 different screw and 27 different 60° dislocation segments have so far been measured, including 7 segments lying in the depletion region. The general shape of all the C vs I_b curves obtained for screw and 60° dislocations lying in the bulk was the same as those shown in figure 4 and so it is assumed that the differences in recombination strength observed between different dislocations is due only to differences in the concentration of active recombination centres N_d, and not due to differences in their type. In this way the value of C_e obtained from fitting the theory to the data in figure 4 can also be used for other dislocations. Thus provided other experimental parameters stay the same, measurements of C versus I_b are required only at one temperature to obtain a value of N_d for a particular dislocation. This procedure has been used on the following.

Experimental Results

(i) Measurements at different positions on a dislocation

The value of N_d was approximately constant along the length of individual dislocations. In particular measurements of N_d were made at 10 equally spaced intervals along a 300μm long, 60° dislocation. The maximum deviation from the mean value was +8% and -6%. Ten equally spaced measurements were made on a 90μm segment of screw dislocation with the maximum deviation from the mean value of N_d being +4% and -3%. Four equally spaced measurements along an 80μm segment of 60° dislocation gave ± 1% deviation.

(ii) Measurement of screw and 60° dislocation pairs

When a dislocation changes its line direction from a <110> direction parallel to b to another <110> direction, its character changes from a screw dislocation to a 60° dislocation. Measurements made either side of such a change in direction are shown in the following table. In each case the 60° dislocation showed a higher value of N_d than its corresponding screw portion.

Table 1 Measurement of N_d for screw, 60° pairs

N_d screw/N_0	N_d sixty/N_0	N_d sixty/N_d screw
1·04	1·11	1·07
0·99	1·09	1·11
0·96	1·01	1·06
1·00	1·12	1·12

(iii) Measurements of different length dislocation segments

The deformation induced dislocations studied in this work are nucleated as loops during the first stage of deformation at high temperature and then these loops expand during the second stage of deformation at low temperature. Thus, in general terms, the longer the length of the straight segment of dislocation between the points at which its line direction changes, the further through the crystal it will have moved. Table 2 shows measurements of N_d from various different length dislocation segments. In some cases one end of the dislocation extended beyond the edge of the Schottky barrier and so it was only possible to put a lower limit on the dislocation length.

Table 2   Measurement of N_d for different dislocation lengths

Screw dislocations		60° dislocations	
Segment length	N_d/N_0	Segment length	N_d/N_0
> 130μm	0·97	300μm	0·96
> 110	0·95	> 80	0·98
90	0·91	53	1·16
40	0·98	50	1·11
40	1·04	37	1·09
34	1·05	> 30	1·14
24	1·09	30	1·14

From these results it can be seen that there is little difference in the concentration of recombination centres with the length of the dislocation segment. However there is some evidence of increased activity at small loops rather than large loops.

Discussion

The EBIC results produced show excellent agreement with the theory and demonstrate that the EBIC technique can be used to measure the concentration of recombination centres along a dislocation with a spatial resolution of ~ 1μm. It is interesting that dislocations present in FZ silicon specimens produced from different starting materials and deformed in different apparatus are indistinguishable in terms of their recombination behaviour. This seems to indicate that the electrical behaviour is not too sensitive to the small differences in oxygen, nitrogen and transition metal impurity concentrations present in the different specimens. Indeed the variation in N_d from the most active to the least active dislocation measured is only ~ 30%. This variation is much less than the relative variation of the dislocation contrast measurements and demonstrates that measurements of the absolute level of dislocation contrast are not an accurate indication of N_d. It is instead necessary to measure the contrast for a range of experimental conditions for N_d to be deduced.

In order to gain information about the fundamental cause of the recombination activity at dislocations it is useful to exploit the spatial resolution of the EBIC technique to look at the behaviour of different dislocation segments. The variation in N_d observed between different straight dislocations indicates that the activity is not intrinsic to a straight, perfect dislocation in the sense that it would be if, for example, the activity was due to the strain field surrounding the dislocation. In such a case the activity observed would be closely the same for each dislocation. Instead it appears that the activity is associated with a small concentration ~ 2×10^6 cm^{-1}, of "special sites" distributed along the dislocation. Such "special sites" may be associated with the dislocation itself, e.g., kinks or jogs or the constriction of the two partials to form a segment of perfect dislocation. Or they may be due to impurities segregated to the dislocation. It is clear however that the activity is not due to impurities "swept up" by the dislocation as it moves through the crystal because this would imply that the dislocations which have moved the furthest should be most active, whereas the experimental data indicates that, generally, dislocations which have moved least far are more active. To explain this behaviour it is possible to postulate that a larger concentration of kinks or jogs or constrictions is present on the smaller loops than on larger loops, though why this should be so and why similar differences are not observed along the length of a dislocation is not clear. A more probable explanation is that the smaller loops, which have moved more slowly during the deformation period (~ 2 hours), are more likely to have had impurities diffuse to them. Slow moving dislocations may also be able to carry with them higher concentrations of impurities than fast moving dislocations. If this is the case then the electrical activity observed is due to impurities and should change according to the dislocation velocity during specimen deformation. Experiments are at present underway to see whether this is so. The differences between screw dislocations which comprise two 30° partials, and 60° dislocations comprising a 30° and 90° partial, are then explained by the 90° partial being more effective at gettering impurities than the 30° partial.

In conclusion the EBIC technique has been successfully used to measure N_d at individual dislocation segments in silicon. The results obtained so far allow some possibilities for the electrical activity of dislocations to be ruled out. At present the experimental data indicate that the most likely cause of dislocation activity is the diffusion of impurities to the dislocation during deformation. However, there is as yet no conclusive evidence to this effect and the question of whether a "clean" dislocation in silicon would be electrically inactive remains unanswered.

Dislocations in GaAs

Dislocations were introduced into 5×10^{15} cm^{-3}, n-type GaAs using a microhardness indentor. The indents were made on a (111) face and the dislocations lay along <110> directions, the screw dislocation segments running almost parallel to the surface at a depth of a few microns. Experiments have so far only been performed on screw dislocation segments which in this experiment extended for distances of over 100μm from the indent.

EBIC measurements were performed using a surface Schottky barrier and an accelerating voltage of 15kV. Figure 6 shows preliminary results of the variation of the dislocation contrast as a function of temperature. The measurements were made with a beam current of 10^{-12}A. As expected from the theoretical considerations discussed earlier, the contrast increases with increasing temperature. However, more experimental work is required to improve the accuracy of these measurements so that the precise form of the dependence on temperature can be ascertained.

Figure 7 shows the dependence of the EBIC dislocation contrast in GaAs on the incident beam current at 300K. The recombination strength of the dislocation γ, decreases with increasing excitation but unlike the behaviour in regime 1 for dislocations in silicon, γ is not proportional

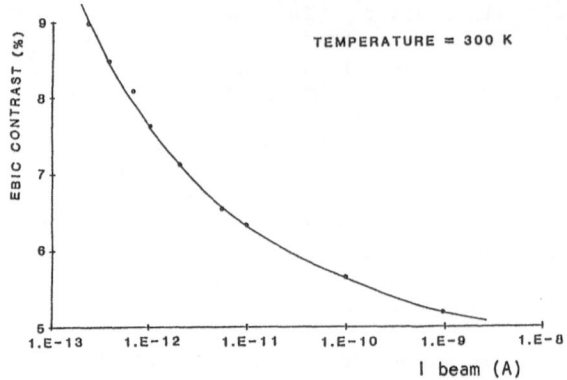

Fig.7. EBIC contrast versus lin(I_b) for the screw dislocations of fig. 6.

to -ln(ΔP). Instead the dependence on beam current, or equivalently ΔP, becomes stronger as the beam current is reduced. Even at a beam current as low as 2×10^{-13}A, γ is still dependent on ΔP. This indicates that the dislocation charge is far from equilibrium which is contrary to the behaviour of screw and 60° dislocations in silicon. Work is currently being undertaken to explain these results in terms of possible processes for hole capture and also in terms of the higher concentration of dislocation levels expected for dislocations in GaAs than are associated with dislocations in silicon.

Acknowledgements

The authors would like to thank Dr P.D. Warren for providing the indented GaAs specimen and SERC for financial support.

References

1. See for example P. Omling, E. R. Weber, L. Montelius, H. Alexander and J. Michel, Phys. Rev. B 32, 6571 (1985).
2. J. Heydenreich, H. Blumtritt, R. Gleichmann and H. Johansen, Cryst. Res. and Tech. 16, 133 (1981).
3. A. Ourmazd, P. R. Wilshaw and G. R. Booker, Physica 116B 600 (1983).
4. M. Kittler and W. Seifert, Phys. Stat. Sol. (a) 66, 573 (1981).
5. L. C. Kimerling, H. J. Leamy and J. R. Patel, Appl. Phys Lett. 30, 217 (1977).
6. A. Ourmazd, Cryst. Res. Tech. 16, 137 (1981).
7. A. Jakubowicz, H. U. Harbermeier, A. Eisenbeiss and D. Käss, Phys. Stat. Sol. (a) 104, 635 (1987).
8. T. Figielski, Solid State Electron. 21, 1403 (1978).
9. D. R. Wight, I. D. Blenkinsop, W. Harding and B. Hamilton, Phys. Rev. B 23, 5495 (1981).
10. C. Donolato, Optik 52, 19 (1978).
11. O. Engström and A. Anders, Solid State. Electron. 21, 1571 (1978).
12. R. Labusch, J. Physique 40, C6-81 (1979).
13. R. Masut, C. M. Penchina and J. L. Farvacque, J. Appl. Phys. 53 (7), 4964 (1982).
14. E. B. Sokolova, Sov. Phys. Semiconductors 3, 1266 (1970).
15. P. R. Wilshaw and G. R. Booker, Inst. Phys. Conf. Ser. 76, 329 (1985).
16. P. R. Wilshaw, A. Ourmazd and G. R. Booker J. Physique 44, C4-445 (1983).
17. R. Jones S. Öberg and S. Marklund, Phil. Mag. B. 43, 839 (1981).

IMAGING OF EXTENDED DEFECTS BY QUENCHED INFRA-RED BEAM INDUCED CURRENTS (Q-IRBIC)

A. Castaldini and A. Cavallini

University of Bologna, Department of Physics, CISM-GNSM
Via Irnerio 46, I-40126 Bologna, Italy

INTRODUCTION

The study of the electrical properties by photoabsorption and photoconductivity kinetics measurements is already a classical method in the analysis of semiconducting materials. Likewise, the Beam Induced Current mode of Scanning Microscopy is a well established technique in experimental physics, in spite of its recent birth, when dealing both with an electron beam (EBIC) and a light beam (LBIC).

A substantial difference exists between the two above mentioned methods, concerning both the operating way and the materials parameters investigated.

As is well known, the former implies the examination of the samples as a whole, i.e. it is a bulk technique. Its field of study is the behaviour of non-equilibrium excess carriers, investigated by carrier generation, motion and recombination processes. The presence of trapping centers and recombination centers in the bulk, as well as their position in the forbidden gap, may be determined by the photoelectric effect kinetics. However, no information about the spatial distribution of these centers in the sample may be achieved.

In contrast, the beam induced current methods, EBIC and LBIC, are a very sensitive tool for local analysis, and for this reason they are now widely used in semiconducting materials assessment. Point and extended defects are imaged by the diffusion and collection of the beam generated excess carriers. Determination of the minority carrier lifetime (or diffusion length) can be inferred from the analytical treatment of the defect image formation. In this case, however, there is no information on the nature of the recombination activity. Indeed, it is possible, in principle, to deduce the position of the energy level associated with an individual defect by monitoring the variation of the contrast as a function of sample conditions (temperature, doping concentration, etc.)[1]. However, in practice what happens now is that usually the excess minority lifetime only is measured by beam induced current methods.

The method described in this paper (which is a development of the IRBIC one[2]) consists in combining the two above said techniques, so as to obtain the information specific of both of them. The aim of the work performed is, therefore, to determine the defect energy levels lying within the forbidden energy gap and, at the same time, their spatial position. For this purpose, the photoelectric effect used in association with the induction of charges by a light beam is the so called "optical quenching by long wavelength secondary illumination". Optical excitation from the valence band to imperfection energy levels or from these levels to the conduction band can lead, under the proper circumstances[3], to the quenching of the intrinsic photoconductivity induced by band-gap light. The location of the imperfection levels in the forbidden gap can be determined from the wavelength dependence of this process.

PRINCIPLE OF OPERATION

A scanning band-gap light probe impinges the front surface of the sample (Fig.1), generating charge carrier pairs inside it. Electrons and holes so produced diffuse into the specimen, where they are divided by the built-in electric field due to the depletion zone of a Schottky barrier. Hence they give rise to an induced current, modifying the junction current (Fig.2). If the specimen is homogeneous, the TV grey-scale image obtained by converting the current values to grey levels is uniformly bright. Instead, if recombination or trap centers are present, a local current reduction due to the enhanced recombination occurs as the beam scans across them, and the defects appear as dark spots if they are point-like, or dark lines if they are extended.

At the same time, the opposite surface of the sample is irradiated by light, whose wavelength can be selected by a monochromator in such a way that the photon energy sweeps the whole forbidden energy gap.

When the photon energy of the back surface irradiation coincides with the distance E_T of a level within the forbidden gap from one of the allowed energy bands, a level-band transition occurs. So the concentration of the efficient recombination centers at the defect is modified and, therefore, its occupancy factor f, too. In turn, this change affects the probe induced current reduction at the defect as well as its grey-scale image .

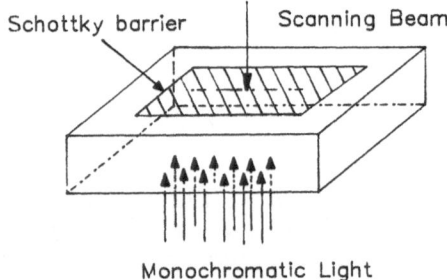

Fig. 1. Principle of the quenched infra-red beam induced current method.

Thus the concurrent application of scanning optical microscopy and photoconductivity quenching makes it possible to perform localized investigations on energy levels lying in the forbidden gap as well as to map them, thus overcoming the lack of spatial resolution inherent to photoconductivity measurements.

Since both the beam and quenching lights used have wavelengths in the range of the infra-red irradiation and the beam induced photocurrent is quenched by the

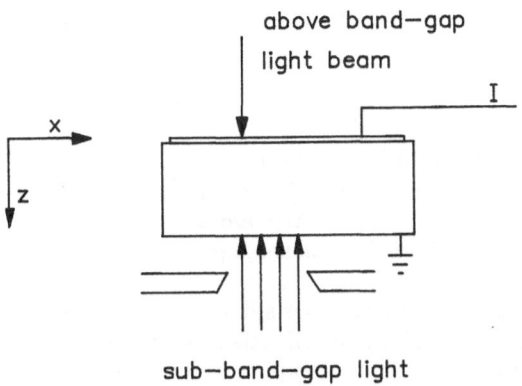

Fig. 2. Configuration used in EBIC and Q-IRBIC observations of extended defects.

secondary illumination, from now on we shall name the method described here Q-IRBIC (Quenched Infra-Red Beam Induced Current).

EXPERIMENTAL PROCEDURE

Silicon single crystals, floating zone grown and doped with phosphorous $(5 \cdot 10^{14} cm^{-3})$, were plastically deformed by creep along the $\langle 112 \rangle$ direction in an argon atmosphere at $650^{\circ}C$ ($\approx 0.54 T_m$). The sample crystallografic orientation was (111) and their thickness was about $200 \mu m$.

For sample examination, the light from a quartz-halogen lamp is chopped and focused by a conventional optical microscope onto the sample surface (Fig.2). The beam diameter at the sample surface is reduced by means of a pin-hole ($\Theta = 20 \mu m$) placed at the field aperture plane. Interferential filters may intercept the light path for selecting the beam wavelength. The spot size, experimentally determined, ranges from $1.2 \mu m$ up to $1.5 \mu m$ at the sample surface, according to the wavelength used. A gold metallization 150Å thick is evaporated on the sample surface to provide a Schottky barrier.

Since the basic formulation of the method implies low injection level[3] for the illumination of both the opposite sample sides, as well as the knowledge of the light intensity at the semiconductor surface, the determination of the beam energy loss by the gold film is required. For this aim reflectance and transmittance of the metal deposition were calculated in the wavelength range used by the matrix method[4,5]. The resulting reflectance R and transmittance T diagrams[6] corresponding to a 150Å Schottky barrier are shown in Fig.3.

As this practical arrangement involves the use of a suitable high numerical aperture microscope objective, the specimen is illuminated with an Airy disk spot which is attenuated within the specimen[7]. By assuming that the attenuation is sufficiently strong that we can regard the function as simply an Airy spot multiplied by an $exp(-\alpha z)$ attenuation term (z denotes the distance from the semiconductor surface), the carrier generation function is given by:

$$g(r) = \left[\frac{2J_1(v)}{v}\right]^2 \cdot exp(-\alpha z) \qquad (1)$$

where r is a radial variable, $(x^2 + y^2)^{1/2}$, α is the intensity attenuation coefficient,

J_1 is the Bessel function of first kind and order one, and v is an optical coordinate given by:

$$v = \frac{2\pi}{\lambda} r \sin\theta \qquad (2)$$

where λ is the wavelength and $\sin\theta$ the numerical aperture.

The raster scanning of the sample is achieved by mounting it on an x-y holder moved by computer driven step-by-step motors . Scanned sample area and spatial resolution are computer controlled, since the step number and size can be varied by software.

The light from a glow-bar crosses a line grating monochromator and is conveyed onto the sample back-surface by an off-axis mirror. To control its intensity, the cone of light is cut down by an adjustable aperture in front of the glow-bar mirror.

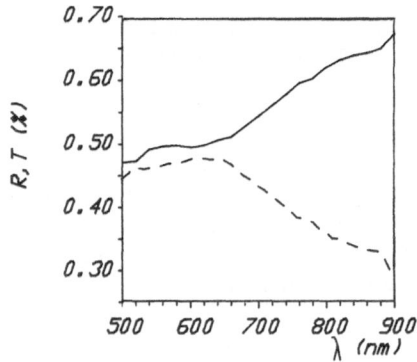

Fig. 3. Reflectance R (——) and transmittance T (- - - -) of a gold film 150 Å thick on a Si substrate.

On the sample back surface an ohmic contact, that collects the charge carriers, is connected to a transimpedance preamplifier and, through this one, to an SR530 lock-in amplifier. The procedure to build-up the defect image is as follows: when the sample scans across the light beam, the signal derived from the photo-induced current $I(x,y)$ is computer stored as a rectangular array, so as to build-up a map representing the Q-IRBIC topography of the sample. When the current sampling is very noisy, a data smoothing is performed to bring into evidence the meaningful variation patterns. The filtering algorithm adopted is the Sheppard's five-term equation:

$$\hat{Y}_i = \frac{1}{35}[17Y_i + 12(Y_{i+1} + Y_{i-1}) - 3(Y_{i+2} + Y_{i-2})] \qquad (3)$$

($i = 3$ to $n - 2$, if n are the experimental observations Y) when a current profile has to be examined, whilst the digital filtering is performed by the median filter based on a ranking procedure[8] when a sample area has to be imaged.

Before imaging the defects in the Q-IRBIC mode, a surface sample picture is made in EBIC mode, usually with beam voltages equal to 10 or 15kV, corresponding to maximum electron ranges of 1 and $2\mu m$[9], respectively. Thus the Q-IRBIC sample observations are performed by using a beam wavelength equal to 505 or 633nm, so as to have light beam penetration depths[4] equal to 0.66 and 2.62μm, respectively, comparable to the EBIC electron range.

Bearing in mind that in a sample which is illuminated with monochromatic light of wavelength λ, the intensity of light decreases according to the law:

$$I(\lambda, z) = I_0(\lambda) exp[-\alpha(\lambda)z] \qquad (4)$$

where I_0 is the intensity at the sample surface, the actual light intensity of the back-surface illumination at the defect under examination has to be accurately determined, since the number of photons that hit the defect is required for a meaningful measurement of the quenched beam induced current.

SAMPLE ABSORPTION AND SECONDARY ILLUMINATION INTENSITY

Light absorption characterized by the electron excitation from the valence band to the conduction band (either by direct transitions or by indirect transitions and characterized by an absorption edge) is typical of semiconductors. Here we refer to the main absorption band. In practice only this kind of absorption is in evidence from the absorption edge outwards the higher photon energies. However beyond the absorption edge, towards greater wavelengths, further kinds of absorption, listed below, can be observed[10,11]:

1) where the conduction or valence bands are made up of several bands, as in silicon, the electrons may cross from one band into the other meeting the selection rule of momentum conservation.
2) Impurities provide a further source of light absorption. A photon may cause an electron to cross from a neutral donor into the conduction band, or from the valence band to a neutral acceptor, resulting in the creation of a hole.
3) Light absorption by carriers in a single band also occurs, even though free electrons cannot at the same time absorb light and remain in the same band. If interaction with photons is to take place, defects in the perfect periodicity of the lattice are essential to satisfy the requirement of momentum conservation.
4) Finally, the direct interaction of photons with the semiconductor lattice may result in light absorption. This kind of absorption can be accounted for by the lattice vibrations corresponding to the optical modes, producing a deformation in the charge density surrounding the atoms. Silicon exhibits an absorption band in the region from 8 to 25μm, due precisely to lattice vibrations.

In the context of the problem dealt with in this paper it is important to bring into evidence whether or not current carriers are formed by light absorption. The formation of free carriers during irradiation is referred to as the internal photoelectric effect. The absorption is photoelectrically active if it leads to electron-transition from the valence to the conduction band. Light absorption by impurities or lattice imperfections, too, is photoelectrically active since it can produce either an electron in the conduction band, or a hole in the valence band. Instead, the other kinds of absorption reported above ((1) and (4)) do not produce free charge carriers.

Because of the above mentioned complexity of the absorption mechanisms, the absorption coefficient as a function of the photon energy does not depend only on the semiconductor material itself, but also on the sample history.

Therefore, to know the secondary illumination intensity at the defect (eq.4), it is necessary to measure the absorption coefficient α vs. λ for each sample. For this purpose measurements of the transparence $T(\lambda)$ were carried out, since the following relationship holds:

$$T(\lambda) = \frac{I(\lambda, z)}{I_0(\lambda)} = \frac{\{exp[-\alpha(\lambda)z](1 - R)^2\}}{\{1 - R^2 exp[-2\alpha(\lambda)z]\}} \qquad (5)$$

261

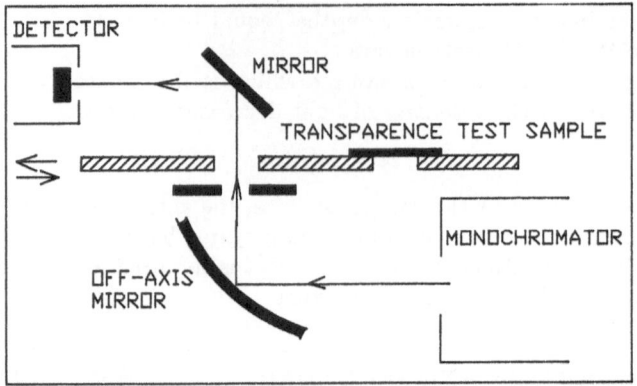

Fig. 4. Layout of the transparence measurement system.

Here $\mathcal{R}$ denotes the reflection coefficient given by:

$$\mathcal{R} = \frac{[(1-n)^2 + n^2k^2]}{[(1+n)^2 + n^2k^2]} \qquad (6)$$

where n and k, depending, in turn, on the light wavelength, denote the real and imaginary part of the refractive index, respectively.

The transparence curves were obtained by the arrangement shown in Figure 4: the light from the monochromator impinges directly the detector for measuring $I_0(\lambda)$, then it crosses the sample so as to determine $I(\lambda)$. Absorption diagrams obtained by the expression (5) by probing as-grown, heated at 650^oC and deformed at 650^oC samples are reported in Fig.5.

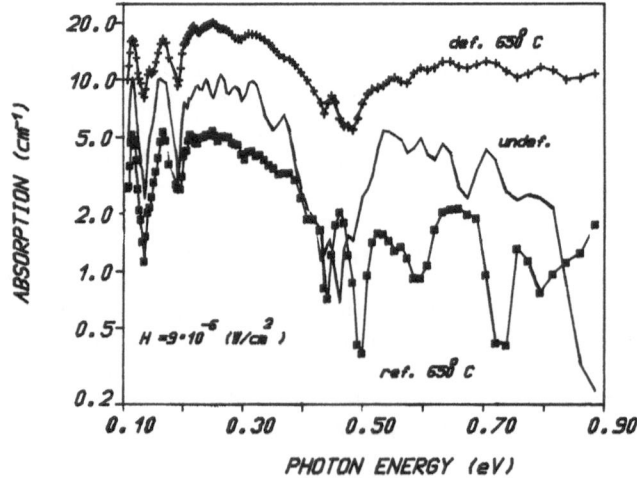

Fig. 5. Absorption coefficient vs. quenching light photon energy for as-grown(——), heated at 650^oC (——) and deformed at 650^oC (-+-+-) samples.

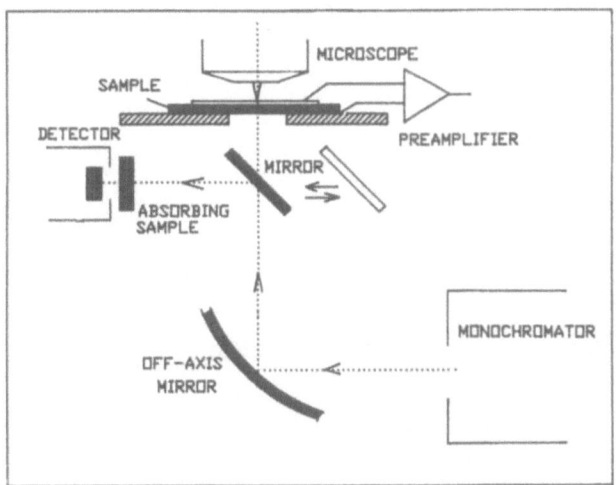

Fig. 6. Optical arrangement of the Q-IRBIC system.

The sample back-surface illumination intensity is then adjusted according to the absorption curves so as to keep the photon flux constant at the defect examined. The connection between the Q_i photons/(cm²s) incident upon the surface z and the incident light intensity $I(\mu W/cm^2)$ at the same surface is given by:

$$Q_i = \frac{\lambda I}{hc} = 5.035 \cdot 10^{12} \lambda I \tag{7}$$

for λ expressed in microns.

Q-IRBIC LAYOUT

The Q-IRBIC layout for local level probing has recently been modified to determine as accurately as possible the charge carrier generation rate at the defect under examination and, at the same time, to improve the method sensitivity, overcoming the experimental complexity of the procedure described above. Nevertheless, the absorption curves are still required for further numerical evaluations of analytical modelling of the method.

A direct evaluation of the density of the secondary illumination photons impinging the defect per unit time can be made by considering that the maximum depth of the defects observed is 2μ m below the front surface, since the beam penetration ranges from $\approx$ 0.6 to 2μm with the light wavelengths used. Thus, to measure the incident photon rate after the photons have crossed the specimen implies a deviation from the true value "at the defect" below 0.1% (bearing in mind that the sample is 200μm thick). Therefore the intensity of the secondary illumination light, deflected by a mirror (Fig.6), is measured after crossing a test sample, and from its value the incident photon rate is evaluated. The general layout of the Q-IRBIC system is shown in Figure 7, together with the instrumentation for the data acquisition.

RECOMBINATION AT DISLOCATIONS

As pointed out above, light of a suitable wavelength lying on the long wavelength side of the absorption edge may produce transitions between levels in the forbidden gap and the allowed bands.

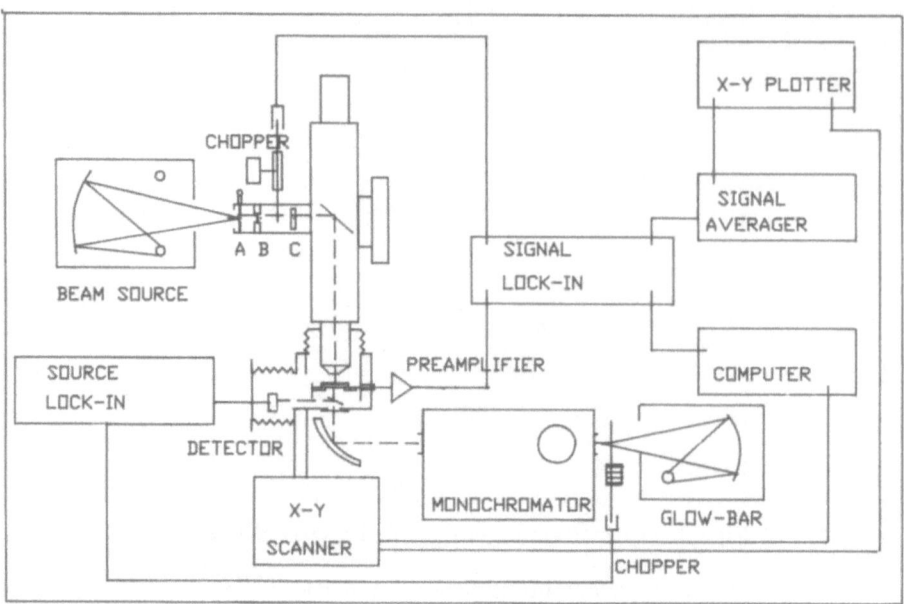

Fig.7. Block diagram of the Q-IRBIC system.

This means that the illumination not only does it change the density of free electrons and/or holes in the bands but also alters the occupancy factor of local imperfection levels in the forbidden band, that is the fraction of dislocation states which are occupied by electrons. Therefore, for the Q-IRBIC measurements the occupancy statistics, as well as the continuity equation about the transport behavior of excess carriers, has to be considered under steady-state illumination conditions.

We are dealing with extrinsic n-type material and in our observations we adopted low injection level conditions, that is $n_0 \gg p_0$ and $n_0 \gg \delta p$, where n_0 denotes the majority carrier concentration, p_0 the minority carrier concentration and δp the excess hole density. Due to the configuration used (Fig. 2), in a large thin sample the excess carrier concentration varies essentially only along the z-direction, allowing one to use the one-dimensional form of the continuity equation. Therefore in the steady-state condition the continuity equation takes the form:

$$\frac{d^2(\delta p)}{dz^2} - \frac{\delta p}{L_p^2} = -\frac{g_L}{D_p} \tag{8}$$

Here L_p and D_p are the hole carrier diffusion length and diffusion coefficient, respectively, and g_L is the generation rate of excess carriers per unit volume due to the long wavelength illumination.

The general expression for g_L is

$$g_L = \frac{\alpha \mu I}{\hbar \omega} \tag{9}$$

where μ is the dimensionless quantum yield and $\hbar \omega$ is the photon energy.

If the secondary illumination photon energy $\hbar \omega$ corresponds to an electronic transition between a level lying in the forbidden gap, for example an acceptor level, and an allowed band, then the product $\alpha\mu$ becomes[12]

$$\alpha\mu = \sigma_t N_t (1 - f) \tag{10}$$

and the following relationship holds:

$$g_L = \sigma_t N_t (1 - f) Q_i \tag{11}$$

where N_t is the acceptor state concentration, σ_t their capture cross-section and f the electron occupancy probability of the level.

The solution of equation (8) may be expressed as the general solution of the corresponding homogeneous equation plus a particular solution of (8). It is clear that $\delta p = g_L \tau_p$ (with τ_p hole lifetime, equal to L_p^2/D_p) is just such a particular solution, and thus the general solution of (8) can be written as

$$\delta p = A \, \cosh\left(\frac{z}{L_p}\right) + B \, \sinh\left(\frac{z}{L_p}\right) + g_L \tau_p \tag{12}$$

where A and B are constants determined by the working conditions.

Let us suppose, for example, that the spot and quenching light intensities are equal, and the sample thickness z_o is very small so as at the first order of approximation the light intensity may be considered uniform throughout the crystal. In that case, by the symmetry of the sample geometry, inside the volume situated below the beam δp must be an even function of z, whereby $B = 0$ and (12) becomes

$$\delta p = A \cosh\left(\frac{z}{L_p}\right) + g_L \tau_p \tag{13}$$

Applying at either surfaces $z = \pm z_o/2$ the surface boundary condition

$$-D_p \left(\frac{\partial(\delta p)}{\partial z}\right)_s = s \cdot (\delta p)_s \tag{14}$$

where the s subscripts indicate that the quantities concerned are evaluated at the surfaces and s is the surface recombination velocity, one may easily obtain

$$A = \frac{- s g_L \tau_p}{s \cosh\left(\dfrac{z_o}{2L_p}\right) + \dfrac{D_p}{L_p} \sinh\left(\dfrac{z_o}{2L_p}\right)} \tag{15}$$

and

$$\delta p = g_L \tau_p \left[1 - \frac{s \cosh\left(\dfrac{z}{L_p}\right)}{s \cosh\left(\dfrac{z_o}{2L_p}\right) + \dfrac{D_p}{L_p} \sinh\left(\dfrac{z_o}{2L_p}\right)} \right] \tag{16}$$

The photocurrent of reference and dislocated samples as a function of the photon energy are shown in Figure 8.

Let us now consider the imaging process of the sample defective state by the beam induced current. Because of the recombination mechanisms at the lattice imperfections, a current reduction occurs when the beam crosses a defect and the defect

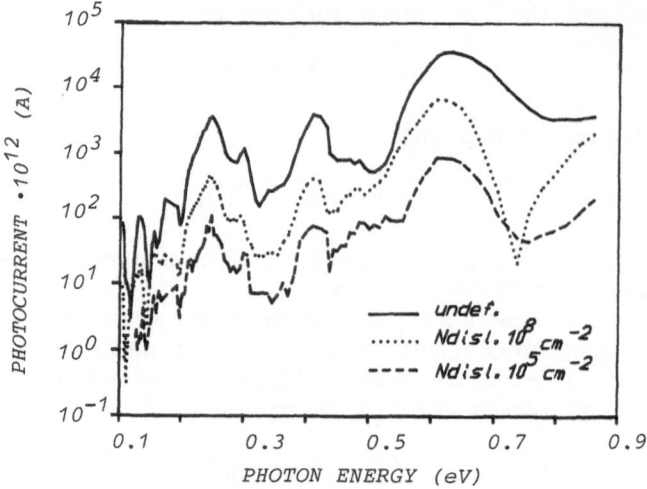

Fig.8. Photocurrent of reference and deformed samples at 650 ^{o}C vs. photon energy (photon flux=$1.2 \cdot 10^{15}$photons $cm^{-2}s^{-1}$).

contrast $c = (I_0 - I_d)/I_0$ may be defined, I_0 and I_d being the signal levels corresponding to the background and to the defect, respectively. When applied to dislocations, for the related EBIC contrast Donolato's model yields the following expression[13]:

$$c = \gamma\Big[\frac{1}{I_0} \int_\Gamma \delta p(r_l)\phi(z_l)dl\Big] \tag{17}$$

where ϕ is the carrier collection probability and $\gamma \, [cm^2s^{-1}]$ the recombination strength (or line recombination velocity) along the dislocation axis Γ.

The line recombination velocity $\gamma = \pi\epsilon^2/\tau'$ (where ϵ is the defect radius and τ' is the minority carrier lifetime along the defect) can be related to the line density of

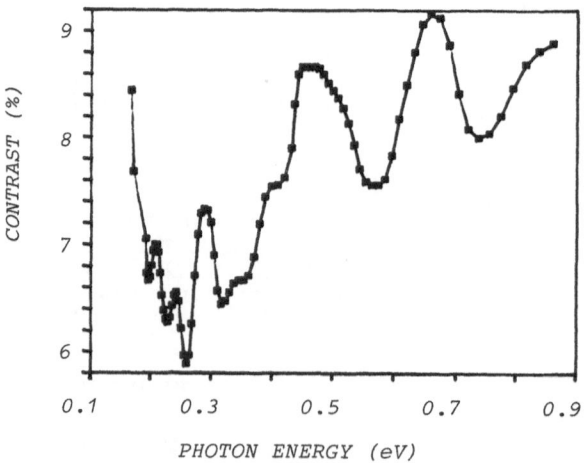

Fig. 9. Q-IRBIC contrast vs. photon energy.

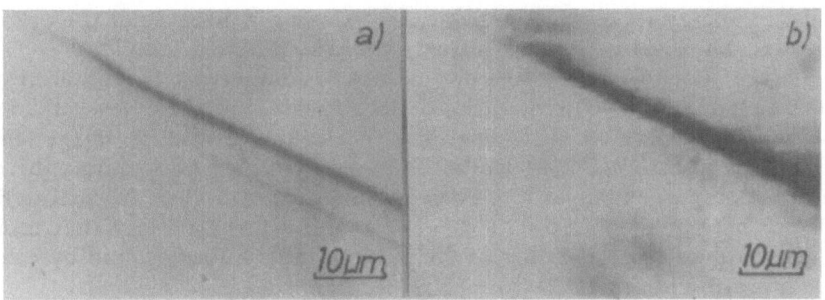

Fig. 10. Dislocation micrograph in EBIC mode (a) and IRBIC mode (b).

active recombination centers N' by the following expression[14]:

$$\gamma = N'\sigma v_{th} \qquad (18)$$

where v_{th} denotes the carrier thermal velocity. Since v_{th} is known, the experimental value of γ yields the value of the product $N'\sigma$.

The density of efficient recombination centers N' along the dislocation axis depends on the level occupancy factor f. Therefore, if f experiences a variation as in the quenching process, also the recombination strength changes, as well as, in turn, the contrast at the defect.

Thus, significant changes in the contrast trend vs. the secondary illumination photon energy can be interpreted as occurring level-band transitions. In this way

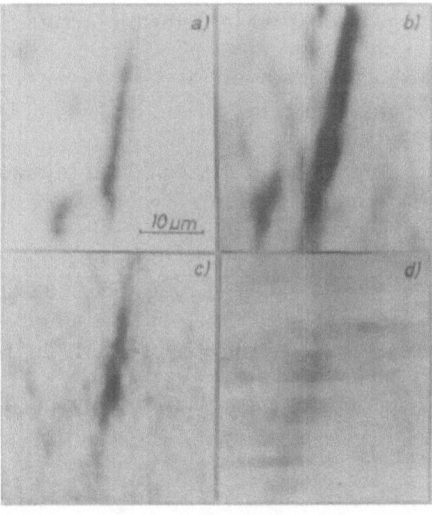

Fig. 11. EBIC (a) and Q-IRBIC (b, c, d) dislocation micrograph. The Q-IRBIC images were obtained with a beam wavelength=550nm, without secondary illumination (b) and with a secondary illumination photon energy=0.64eV (c, d). The light intensity of the quenching light is $8\mu W/cm^2$ and $12\mu W/cm^2$, in (c) and (d), respectively.

the defect energy levels may be identified. Moreover, because of the above reported expression, the level occupancy factor f could be derived, too.

Even though quantitative evaluations are underway, it is interesting just the same to examine the contrast variation as a function of the energy of the quenching photons (Fig.9). The contrast diagram refers to the dislocation imaged in Figure 10 in EBIC (a) and IRBIC[2] (b) mode. This was obtained by scanning the dislocation in a direction perpendicular to its axis, with a beam wavelength equal to $550nm$ and an irradiance equal to $1.16 * 10^{-3} W/cm^2$. It has to be noticed that minima occur at energy values corresponding to dislocation levels already found by other authors (wide literature in reference 15).

The effect of the quenching photon flux on the recombination strength, and then on the Q-IRBIC contrast, is evident in Figure 11. Here a dislocation is imaged by EBIC (a) and Q-IRBIC (b, c and d) methods. The picture in (b) was obtained by the beam induced current with a probe light wavelength equal to $550nm$, without secondary illumination, whereas the defect images (c) and (d) have been obtained in the same scanning probe conditions as (b) plus the quenching irradiation. The photon energy is the same (E=0. 64eV) for both, whereas the light intensity I is $8\mu W/cm^2$ in (c) and $12\mu W/cm^2$ in (d). The reduction of the recombination activity of the defect is evident.

REFERENCES

1. P. R. Wilshaw and G. R. Booker, The Theory of Recombination at Dislocations in Silicon and an Interpretation of EBIC Results in Terms of Fundamental Dislocation Parameters, in : "Structure and Properties of Dislocations in Semiconductors," Isz. Akad Nauk, USSR, (1987)
2. A. Castaldini, A. Cavallini, P. Gondi and E. Bonetti, Generation and Recombination Images of Dislocations in Si by Scanning Microscopy, Phys. Stat. Sol.(a) 73:617 (1982).
3. A. G. Milnes, "Deep Impurities in Semiconductors," J. Wiley & Sons, New York (1973).
4. M. Born and E. Wolf, "Principles of Optics," Pergamon Press, Oxford, (1987).
5. "Handbook of Optical Constants of Solids," E. D. Palik ed., Academic Press, Orlando (1985).
6. A. Castaldini and A. Cavallini, Infrared Beam Induced Contrast with Double Illumination, in:"Scanning Microscopy Technologies and Applications," E. C. Teague, ed., SPIE, Los Angeles (1988). (1988).
7. T. Wilson and E. M. McCabe, Theory of Optical Beam Induced Current Images of Defects in Semiconductors, J.Appl.Phys. 58: 2638 (1986).
8. R. Gonzales and P. Wint, "Digital Image Processing," Addison-Wesley, (1977).
9. C. Donolato, Contrast and Resolution of SEM Charge-Collection Images of Dislocations, Appl. Phys. Lett. 34:80 (1979).
10. R. H. Bube, "Photoconductivity of Solids," J. Wiley & Sons, New York (1960).
11. W. Jones and N. H. March, "Theoretical Solid State Physics," J. Wiley & Sons, New York (1973).
12. K. Seeger, "Semiconductor Physics," Springer-Verlag, Berlin (1985).
13. C. Donolato, Quantitative Evaluation of EBIC Contrast of Dislocations, J.de Physique, C4:269 (1983).
14. M. Kittler and W. Seifert, On the Sensitivity of the EBIC Technique as Applied to Defect Investigations in Silicon, Phys.Stat.Sol.(a) 66:573 (1981).
15. H. Alexander, Dislocations in Covalent Crystals, in: "Dislocations in Solids," F. R. N. Nabarro, ed., North-Holland, Amsterdam (1986).

CONTRIBUTORS

PARTICIPANTS

H. ALEXANDER   II Phys. Institute University
                                      Zuelpicher SN.77
                                      D-5000 Koln, FRG

F. ALLEGRETTI   Dip. Chimica Fisica & Elettrochimica
                                      Via Golgi 19
                                      20133 Milano, Italy

J. AULEYTNER   Institute of Physics of the
                                      Polish Academy of Sciences
                                      Al Istnikow 32/46 Pav.9
                                      Warszawa, Poland

M. BALKANSKI   Laboratoire de Physique des Solides
                                      Université P. & M. Curie
                                      4, Place Jussieu, Tour 13
                                      F-75252 Paris, Cedex 05, France

G. BENEDEK   Dipartimento di Fisica
                                      Via Celoria 16
                                      I-20133 Milano, Italy

F. BORSANI   Dip. Chimica Fisica & Elettrochimica
                                      Via Golgi 19
                                      I-20133 Milano, Italy

F. CATANIA   Dip. di Fisica
                                      Corso Italia 57
                                      I-95129 Catania, Italy

A. CASTALDINI   Dip. di Fisica
                                      Via Irnerio 46
                                      I-40126 Bologna, Italy

D. CAVALCOLI   Dip. di Fisica
                                      Via Irnerio 46
                                      I-40126 Bologna, Italy

G.F. CEROFOLINI

ENICHEM
Via dei Medici del Vascello 26
I-20138 Milano, Italy

A. CAVALLINI

Dip. di Fisica
Via Irnerio 46
I-40126 Bologna, Italy

S. DEL SORDO

Via Ignazio Silvestri 29
I-90135 Palermo, Italy

C. DONOLATO

Istituto LAMEL del CNR
Via Castagnoli 1
I-40126 Bologna, Italy

G. FEUILLET

Centre d'Etudes Nucleaires de
Grenoble - D.R.F./S.PH/PSc.
F-38041 Grenoble, France

A. GAUZZI

Ecole Politechnique Fédérale
Dép. de Matériaux
Lab. de Metallurgie Chimique
Ch. de Bellerive, 36
CH-1007 Lausanne

P. HAASEN

Institut für Metalphysik
Hospital str. 3/5
D-3400 Göttingen, FRG

J. HARTUNG

MPI für Festkörperforschung
Heisenbergstrasse 1
D-7000 Stuttgart 80, FRG

T. HEISER

C.N.R.S.
Laboratoire Phase
F-67037 Strasbourg Cedex, France

T. HEYDENREICH

Academy of Sciences of the GDR
Inst. State Phys. EM
PF 250
DDR-4020 Halle/S., G.D.R.

D.B. HOLT

Dept. of Materials
Imperial College
London, SW7 2BP, Great Britain

I.A. HÜMMELGEN

IV Physikalisches Inst.
Bunsenstrasse 11-15
D-3400 Göttingen, FRG

L.C. KIMERLING

AT & T Bell Laboratories
600 Mountain Avenue
Murray Hill, N.J. 07974, USA

M. KITTLER

Inst. für Halbleiterphysik der AdW
W. Korsing strasse 2
DDR-Frankfurt/Oder, GDR 1200

J. KUMAR

MASPEC-CNR
Via Chiavari 18/A
43100 Parma, Italy

R. LABUSCH

Inst. für Angewandte Physik der TK
Arnold Sommerfeld - SN 6
D-3392 Clausthal Zellerfeld, FRG

N.Q. LIEM

Institute of Physics of VN
Nghia de- Tu Liem
Hanoi, Vietnam

Z. LILIENTAL-WEBER

Lawrence Berkeley Lab. 62/203
Berkeley, CA. 94705, USA

S. ONURLU

Sumer Koop. Evleri
Kasimpati Sok. 31
Tarabya-Istanbul, Turkey

A. OURMAZD

Rm 4E-414 AT & T Bell Labs.
Holmdel, N.J. 07733, USA

S. OZEN

Aydin SOK. 25/3
Suadiye
8110 Istanbul, Turkey

N. OZER

Istanbul Technical University
Physics Dept. (Fen-Edebiyat)
Maslak-Istanbul 80626, Turkey

S. PIZZINI

Dip. Chimica Fisica & Elettrochimica
Via Golgi 19
I-20133 Milano, Italy

A. POGGI

LAMEL-CNR
Via de' Castagnoli 1
40126 Bologna, Italy

D. POHL

IBM Research Division
Zurich Research Laboratory
CH-8803 Rüschlikon, Switzerland

M.L. POLIGNANO

SGS-Thomson
Via Olivetti 2
20041 Segrate, Milano, Italy

M. PORTO   Dipartimento di Fisica
                                Corso Italia 57
                                I-95129 Catania, Italy

T. PRESCHA   Max-Planck-Institut für FKF
                                Heisenbergstrasse 1
                                D-7000 Stuttgart 80, FRG

J.L. PUTAUX   Centre d'Etudes Nucléaires
                                DRF-SPh-S - 85X
                                F-38041 Grenoble, France

L. RIVA DI SANSEVERINO   Dipartimento di Scienze Mineralogiche
                                P.za Porta San Donato, 1
                                I-40126 Bologna, Italy

G. ROOS   Institut für Angewandte Physik
                                Gluckstrasse 9
                                D-8520 Erlangen, FRG

H. SALEMINK   IBM Research Laboratory
                                Säumerstrasse 4
                                CH-8803 Rüschlikon, Switzerland

C. SCHRÖDER   PSI/RCA
                                Badenerstrasse 569
                                CH-Zurich, Switzerland

W. SCHRÖTER   IV Physikalisches Institut
                                Bunsenstrasse 11-1p5
                                D-34 Göttingen, FRG

R.C. SPINELLA   Dipartimento di Fisica
                                Corso Italia 57
                                I-95125 Catania, Italy

K. SUMINO   Institute for Materials Research
                                Tohoku University
                                Sendai 980, Japan

J.M. STETTIN   Delfin System
                                134 Moffet Park DR.
                                Sunnyvale, CA. 94089, USA

E. SUSI   LAMEL-CNR
                                via de' Castagnoli 1
                                I-40126 Bologna, Italy

L. TARRICONE   Dip. di Fisica dell'Università
                                Via Celoria 16
                                I-20133 Milano, Italy

J. THIBAULT

Dép. de Recherche Fondamentale
Service de Physique
Centre d'Etudes Nucléaires
85X - F-38041 Grenoble, France

M. TINCANI

ENICHEM, Corporate R/D
Via Medici del Vascello 26
I-20138 Milano, Italy

A. VITTADINI

ICTR-CNR
Corso Stati Uniti 4
35100 Padova, Italy

E. WEBER

Dept. of Materials Science
University of California
Berkeley, CA 94720, USA

P. WILSHAW

Dept. of Metallurgy
Parks Road
Oxford OX1 3PH, U.K.

ACRONYMS

AFM	atomic force microscope
APB	antiphase boundary
APD	antiphase domains
ARM	atomic resolution microscopy
BIC	beam induce current
C-V	capacitance-voltage
CB	conduction band
CCM	charge collection microscopy
CDW	charge density wave
CL	cathodoluminescence
CTF	contrast transfer function
CVD	chemical vapor deposition
CZ	Czochralski (method)
DAP	donor-acceptor pair
D-C	depletion-confinement (interface)
DLTS	deep level transient spectroscopy
DSC	discrete shift complete
DX	donor-unknown partner pair
EBIC	electron beam induced current
EBIV	electron beam induced voltage
EDX	energy dispersion X-ray spectroscopy
EG	electronic grade
EL2	defect absorption band in GaAs at 1.2 eV
ENDOR	electron-nuclear double resonance
EPMA	electron probe microanalysis
EPR	electron paramagnetic resonance
ESR	electron-spin resonance
FET	field-effect transistor
FTIR	Fourier-transform infrared spectroscopy
FZ	float-zone (method)
GB	grain boundary
GBD	grain boundary dislocation
GL	gettering layer
HPI	hexagonally-packed intermediate
HREM	high resolution electron microscopy
HRTEM	high resolution TEM
I-V	current-voltage
IR	infrared (spectroscopy)
IRBIC	infrared beam induced current
KM	kink migration
KPF	kink pair formation
LBIC	light (laser) beam induced current
LC	Lomer-Cottrell

LCAO	linear combination of atomic orbitals
LEC	liquid encapsulated Czrochalski
LED	light emitting diode
LO	longitudinal optical
LT/HS	low-temperature high-stress
MBE	molecular beam epitaxy
MBE	molecular beam epitaxy
MIM	metal-insulator-metal
MODFET	fast modulation doped FET
MOS	metal-oxide-semiconductor
MW	microwave
MWC	microwave conductivity
NAA	neutron activation analysis
NBE	near band edge
NO	near-field optics
OBIC	optical beam induced current
OD-ENDOR	optically-detected ENDOR
ODMR	optically detected magnetic resonance
P-F	Poole-Frenkel interaction
PDG	phosphorous diffusion gettering
PEP	polarized excitation photocapacitance
PL	photoluminescence
PSG	(amorphous) phosphosilicate glass
Q-IRBIC	quenched IRBIC
RBS	Rutherford backscattering
RTA	rapid thermal annealing
SCM	scanning capacity microscopy
SEAM	scanning electroacoustic microscopy
SEM	scanning electron microscopy
SF	stacking fault
SFM	scanning force microscope
SGB	subgrain boundary
s.i.	semi-insulating
SIMS	secondary ion mass spectroscopy
SIS	semiconductor-insulator-semiconductor
SLM	scanning laser microscope
SMM	micropipette microscope
SNOM	scanning near-field optical microscopy
SPV	surface photovoltage method
SRPL	spatially resolved photoluminescence
STEM	scanning transmission electron microscopy
STHM	scanning thermal microscopy
STM	scanning tunneling microscopy
SXM	any scanning microscopy
TB	twin boundary
TD	thermal donor
TEM	transmission electron microscopy
TEMSCAN	transmission electron microscopy scanner
TSCAP	thermally stimulated capacitance
TV	television
UHV	ultra-high vacuum
VLSI	very large scale integration

INDEX

mobility, 88
n AC conductivity, 25, 26, 35
networks, 26, 213
one-dimensional bands, 52, 58
optical inhomogeneities, 83
pairs, 253
partials, 53, 155, 156, 159, 162, 232, 255
Peierls gap, 33
Peierls potentials, 65, 71-72
photoluminescence of., 61, 243
pinning, 73
potential barrier, 31, 246
quasi-metallic conductivity, 52
recombination rate, 248
recombination strength, 237, 239, 244, 247-252, 255
scanning microscopy, 61
screw, 53, 58, 61, 154, 157, 162, 207, 216, 251, 253
segments, 254-255
segregated impurities, 25
Shockley partials, 160, 162
straight, 233, 235, 238, 243, 251, 254
strain field, 243
unlocking (release) stress, 86-87
velocity, 65, 73, 255
α-type, 90-92, 209, 252, 255-256
β-type, 90-92, 209, 255
Divacancies, 9
Donor-acceptor pair, 218
Double acceptor, 97
Double illumination IR induced contrast, 259
DX centers, 39, 41, 45
in AlGaAs:Se(Te), 39
in AlGaAs:Si, 39, 42, 44

in CdTe, 02
in GaAs:S, 40
in GaAsP:S(Se,Te), 39
in GaSb:S(Se), 39, 40
in Te, 40
in ZnS:Ga(In,Tl), 39

E-centers, 98
EL2 defects, 44-46, 65, 74, 84
in GaAs, 44-46, 48
metastability, 46
OD-ENDOR, 45, 46
optical absorption, 44, 45
photo-EPR, 45
photocapacitance, 45
pseudopotential calculations, 45
Electron beam induced current (EBIC), 12, 61, 83, 207, 212, 214, 225-240, 257-260
contrast, 56, 207, 215-217, 226-232, 245, 251, 266, 268
higher-order, 232, 237, 239
linear model, 225, 230, 236
of defects, 228, 238
image, 228, 236
of dislocations, 61, 216-217, 266
of grain boundaries, 216, 233
of point-like defects, 236, 239
of stacking faults, 232
phenomenological models, 225
theory, 208, 215, 225-240
time-resolved, 215
Electron beam induced voltage (EBIV), 212
Electron injection energy, 202
Electron paramagnetic resonance, 52, 55-57, 207

Electron probe microanalysis, 212 of defects in Si, 52-54
Emission barrier, 43
Epitaxial films, 211
    GaAs/Ge(100), 211
    GaAs/Si(100), 211
    InGaAsP/InP(100), 212
Epitaxial multilayers, 213
Etching, 211
Extrinsic self-trapping, 41, 47

FeSi2, 98
FR1 defects, 46
Frank loop, 157, 160, 244
Frankel pair, 124

GaAs(110), 188, 201, 204
    ic surface states, 201
    oxygen adsorption, 203
Generation sphere, 231
Gettering, 81, 85, 95, 208
    by silicide formation, 101
    in silicon, 82, 95
    kinetics, 81
    layer, 96, 100-102
    mechanisms, 95-103
    of metallic impurities, 95
    strength, 102
Glide set structure, 53, 207-209
Grain boundary, 15, 17, 105, 153-159, 208, 231
    -impurity interaction, 109
    activation energy, 22
    activation, 109
    Auger recombination, 109-110, 118
    band bending, 216
    conductivity, 15, 107, 108
    deactivation, 109
    double Schottky junction, 106
    EBIC studies, 216, 230, 233-234
    electrical activity, 116-118

ic states, 17, 20, 21, 106
    equivalent strength, 118
    in GaP, 216-217
    in Ge, 218
    in polycrystalline Si, 208, 218
    in ZnS, 216
    in ZnSe, 216
    LBIC contrast, 235
    local elastic deformation, 114, 118
    parallel conductance, 16, 25, 26
Grain boundary (continued)
    perpendicular conductance, 16, 20, 23
    potential barrier, 17, 19-22, 24, 106-108
    recombination at., 108-109, 216, 235
    space charge region, 105-107
    strength, 230
    strongly recombining., 117-118
    subgrain boundary, 159
    theory of conductivity, 16
    tracking tunnel microscopy, 194
    $\sigma 11$ structure, 17, 19
    σ 25, 25
Graphite, 202
Green's function, 227, 234, 237
Green's reciprocity theorem, 240
Gundlach states, 194

Heterodyne interferometry, 201-202
Heterojunctions, 203
    AlGaAs/GaAs, 203
Heterostructures, 202-205
    depletion region, 202
    p-n junction in., 202
    confinement region, 202

ing, 95, 98-102
channeling spectroscopy,
    98
DLTS of., 98
neutron activation
    analysis, 98
Rutherford backscattering,
    98, 209
Phosphorsilicate glass, 95
Photo-EPR, 54-55
Plastic bending, 209-210
    in InSb, 210-213
Plastic deformation, 51, 65,
    207, 209
    electrical properties,
        51-61
    in Ge, 65, 207
    in InSb, 207
    in silicon, 51-61, 207
Plastic deformation
    (continued)
    LT/HS, 53-54, 58
Point defects, 1-13,
    123-133
    at dislocations, 51
    carrier mobility, 128-129,
        131
    clusters of., 51-52, 55
    extended nature, 123-133
    in GaAs, 39-49
        AsGaAsi complex, 45
        BE1, 46-48
        EL2, 44-48
        FR1, 46
        VAsVGaAsGa complex, 45
    in silicon, 4
        A center, 5
        deep states, 8
        divacancy, 10, 11
        iron-boron pair, 12-13
        oxygen donor, 4-6
        shallow states, 7
    local Debye temperature,
        126-127
    off-site, 129
    relaxation field (displa-
        cement cloud), 123-
        130, 132
    substitutional, 126, 128
    superlattice of., 129-130

Point-contact spectroscopy,
    196
Polar bending, 210-211
Polarized excitation photoca-
    pacitance (PEP), 1,
    9-10
Polaronic conduction, 108
Polycrystalline silicon, 208,
    216
    electrical conductivity,
        108
    grain boundaries, 208
    iron-doped, 109-110
Poole-Frenkel interaction, 6
Profilometers, 201
    thermal., 202, 204

Quadrupole doublet,
    99, 103
Quenched IRBIC, 257-260, 263-
    260
    at dislocations, 266-267
    contrast, 266-267
    defect ic states
        from., 268
    in silicon, 257, 262, 266
    layout, 263-264
    photocurrent, 265-266

Recombination
    at defects, 52, 207,
        215, 231, 236, 266-
        267
    at dislocations, 217, 236,
        239, 243-256, 263,
        266
        in GaAs, 243-256
    in silicon, 243-256
    line recombination
        velocity, 248, 266-
        267
    theory, 244-251
    at grain boundaries, 108-
        109, 216, 235
    at surfaces, 214, 216,
        226, 242